DYNAMICAL SYSTEMS AND IRREVERSIBILITY

A SPECIAL VOLUME OF ADVANCES IN CHEMICAL PHYSICS

VOLUME 122

EDITORIAL BOARD

BRUCE J. BERNE, Department of Chemistry, Columbia University, New York, New York, U.S.A.
KURT BINDER, Institut für Physik, Johannes Gutenberg-Universität Mainz, Mainz, Germany
A. WELFORD CASTLEMAN, JR., Department of Chemistry, The Pennsylvania State University, University Park, Pennsylvania, U.S.A.
DAVID CHANDLER, Department of Chemistry, University of California, Berkeley, California, U.S.A.
M. S. CHILD, Department of Theoretical Chemistry, University of Oxford, Oxford, U.K.
WILLIAM T. COFFEY, Department of Microelectronics and Electrical Engineering, Trinity College, University of Dublin, Dublin, Ireland
F. FLEMING CRIM, Department of Chemistry, University of Wisconsin, Madison, Wisconsin, U.S.A.
ERNEST R. DAVIDSON, Department of Chemistry, Indiana University, Bloomington, Indiana, U.S.A.
GRAHAM R. FLEMING, Department of Chemistry, The University of California, Berkeley, California, U.S.A.
KARL F. FREED, The James Franck Institute, The University of Chicago, Chicago, Illinois, U.S.A.
PIERRE GASPARD, Center for Nonlinear Phenomena and Complex Systems, Université Libre de Bruxelles, Brussels, Belgium
ERIC J. HELLER, Department of Chemistry, Harvard-Smithsonian Center for Astrophysics, Cambridge, Massachusetts, U.S.A.
ROBIN M. HOCHSTRASSER, Department of Chemistry, The University of Pennsylvania, Philadelphia, Pennsylvania, U.S.A.
R. KOSLOFF, The Fritz Haber Research Center for Molecular Dynamics and Department of Physical Chemistry, The Hebrew University of Jerusalem, Jerusalem, Israel
RUDOLPH A. MARCUS, Department of Chemistry, California Institute of Technology, Pasadena, California, U.S.A.
G. NICOLIS, Center for Nonlinear Phenomena and Complex Systems, Université Libre de Bruxelles, Brussels, Belgium
THOMAS P. RUSSELL, Department of Polymer Science, University of Massachusetts, Amherst, Massachusetts
DONALD G. TRUHLAR, Department of Chemistry, University of Minnesota, Minneapolis, Minnesota, U.S.A.
JOHN D. WEEKS, Institute for Physical Science and Technology and Department of Chemistry, University of Maryland, College Park, Maryland, U.S.A.
PETER G. WOLYNES, Department of Chemistry, University of California, San Diego, California, U.S.A.

DYNAMICAL SYSTEMS AND IRREVERSIBILITY: PROCEEDINGS OF THE XXI SOLVAY CONFERENCE ON PHYSICS

ADVANCES IN CHEMICAL PHYSICS
VOLUME 122

Edited by

IOANNIS ANTONIOU

International Solvay Institutes for Physics and Chemistry, Brussels, Belgium

Series Editors

I. PRIGOGINE

Center for Studies in Statistical Mechanics
and Complex Systems
The University of Texas
Austin, Texas
and
International Solvay Institutes
Université Libre de Bruxelles
Brussels, Belgium

STUART A. RICE

Department of Chemistry
and
The James Franck Institute
The University of Chicago
Chicago, Illinois

AN INTERSCIENCE PUBLICATION
JOHN WILEY & SONS, INC.

This book is printed on acid-free paper. ∞

Copyright © 2002 by John Wiley & Sons, Inc., New York. All rights reserved.

Published simultaneously in Canada.

No part of this publication may be reproduced, stored in a retrieval system or transmitted in any form or by any means, electronic, mechanical, photocopying, recording, scanning or otherwise, except as permitted under Section 107 or 108 of the 1976 United States Copyright Act, without either the prior written permission of the Publisher, or authorization through payment of the appropriate per-copy fee to the Copyright Clearance Center, 222 Rosewood Drive, Danvers, MA 01923, (978) 750-8400, fax (978) 750-4744. Requests to the Publisher for permission should be addressed to the Permissions Department, John Wiley & Sons, Inc., 605 Third Avenue, New York, NY 10158-0012, (212) 850-6011, fax (212) 850-6008, E-Mail: PERMREQ@WILEY.COM.

For ordering and customer service, call 1-800-CALL-WILEY

Library of Congress Catalog Number: 58-9935

ISBN 0-471-22291-7

Printed in the United States of America.

10 9 8 7 6 5 4 3 2 1

CONTRIBUTORS TO VOLUME 122

L. ACCARDI, Centro Vito Volterra, Polymathematics, Facolta di Economia, Universita degli Studi di Roma Tor Vergata, Rome, Italy

Y. AIZAWA, Department of Applied Physics, Faculty of Science and Engineering, Waseda University, Tokyo, Japan

I. ANTONIOU, International Solvay Institutes for Physics and Chemistry, Free University of Brussels, Brussels, Belgium; and Theoretische Natuurkunde, Free University of Brussels, Brussels, Belgium

F. T. ARECCHI, Department of Physics, University of Florence, Florence, Italy; and National Institute of Applied Optics (INOA), Florence, Italy

R. BALESCU, Department of Physical Statistics–Plasma, Free University of Brussels, Brussels, Belgium

A. BOHM, Department of Physics, University of Texas, Austin, Texas, U.S.A.

LUIS J. BOYA, Center for Particle Physics, Department of Physics, The University of Texas, Austin, Texas, U.S.A. *Permanent address*: Department of Theoretical Physics, Faculty of Science, University of Zaragoza, Zaragoza, Spain

PIERRE GASPARD, Center for Nonlinear Phenomena and Complex Systems, Free University of Brussels, Brussels, Belgium

KARL GUSTAFSON, Department of Mathematics, University of Colorado, Boulder, Colorado, U.S.A.; and International Solvay Institutes for Physics and Chemistry, University of Brussels, Brussels, Belgium

HIROSHI H. HASEGAWA, Department of Mathematical Sciences, Ibaraki University, Mito, Japan; and Center for Statistical Mechanics, University of Texas, Austin, Texas, U.S.A.

KUNIHIKO KANEKO, Department of Pure and Applied Sciences, College of Arts and Sciences, University of Tokyo, Tokyo, Japan

E. KARPOV, Center for Studies in Statistical Mechanics and Complex Systems, University of Texas, Austin, Texas, U.S.A.; and International Solvay Institutes for Physics and Chemistry, Free University of Brussels, Brussels, Belgium

S. V. KOZYREV, Centro Vito Volterra, Polymathematics, Facolta di Economia, Universita degli Studi di Roma Tor Vergata, Rome, Italy

MIKIO NAMIKI, Department of Physics, Waseda University, Tokyo, Japan

G. ORDONEZ, Center for Studies in Statistical Mechanics and Complex Systems, University of Texas, Austin, Texas, U.S.A.; and International Solvay Institutes for Physics and Chemistry, Free University of Brussels, Brussels, Belgium

T. PETROSKY, Center for Studies in Statistical Mechanics and Complex Systems, University of Texas, Austin, Texas, U.S.A.; International Solvay Institutes for Physics and Chemistry, Free University of Brussels, Brussels, Belgium; and Theoretical Physics Department, University of Vrije, Brussels, Belgium

I. PRIGOGINE, Center for Studies in Statistical Mechanics and Complex Systems, The University of Texas, Austin, Texas, U.S.A.; and International Solvay Institutes for Physics and Chemistry, Free University of Brussels, Brussels, Belgium

Z. SUCHANECKI, International Solvay Institutes for Physics and Chemistry, Free University of Brussels, Brussels, Belgium; Theoretische Natuurkunde, Free University of Brussels, Brussels, Belgium; and Institute of Mathematics, University of Opole, Opole, Poland

E. C. G. SUDARSHAN, Center for Particle Physics, Department of Physics, The University of Texas, Austin, Texas, U.S.A.

P. SZÉPFALUSY, Department of Physics of Complex Systems, Eötvös University, Budapest, Hungary; and Research Institute for Solid State Physics and Optics, Budapest, Hungary

S. TASAKI, Department of Physics, Nara Women's University, Nara, Japan; and Institute for Fundamental Chemistry, Kyoto, Japan. *Present address*: Department of Applied Physics and Advanced Institute for Complex Systems, Waseda University, Tokyo, Japan

H. WALTHER, Sektion Physik der Universität München and Max Planck Institut für Quantenoptik, Garching, Federal Republic of Germany

ADMINISTRATIVE BOARD OF THE INTERNATIONAL SOLVAY INSTITUTES FOR PHYSICS AND CHEMISTRY

J. Solvay	President of the Administrative Board
F. Bingen	Vice-President of the Administrative Board
I. Prigogine	Director of the Solvay Institutes
I. Antoniou	Deputy Director of the Solvay Institutes
F. Lambert	Secretary of the Administrative Board
A. Bellemans	Secretary of the Scientific Committee of Chemistry
M. Henneaux	Secretary of the Scientific Committee of Physics
D. Janssen	
A. Jaumotte	
G. Nicolis	
J. M. Piret	
J. Reisse	
R. Lefever	

SCIENTIFIC COMMITTEE FOR PHYSICS OF THE INTERNATIONAL SOLVAY INSTITUTES FOR PHYSICS AND CHEMISTRY

A. ABRAGAM	Professeur Honoraire au Collège de France, Paris, France
P. W. ANDERSON	Department of Physics, Princeton University, Princeton, NJ, U.S.A.
F. T. ARECCHI	Directeur, Istituto Nazionale de Ottica Applicata, Florence, Italy
M. HENNEAUX	Département de Physique, Université Libre de Bruxelles, Bruxelles, Belgium
I. M. KHALATNIKOFF	Russian Academy of Sciences, Landau Institute of Theoretical Physics, Moscow, Russia
Y. NE'EMAN	Sackler Institute for Advanced Study, Tel-Aviv University, Tel Aviv, Israel
D. PHILLIPS	35, Addisland Court, Holland Villas Road, London, England
R. Z. SAGDEEV	East West Place Science Center, University of Maryland, College Park, MD, U.S.A.
G. SETTI	European Southern Laboratory, Munich, Germany
G. t'HOOFT	Institute for Theoretical Physics, Utrecht University, Utrecht, Netherlands
S. C. TONWAR	Tata Institute for Fundamental Research, Bombay, India
H. WALTHER	Max-Planck-Institute für Quantenoptik, Münich, Germany
V. F. WEISSKOPF	Department of Physics, Massachusetts Institute of Technology, Cambridge, MA, U.S.A.

THE SOLVAY CONFERENCES ON PHYSICS

The Solvay conferences started in 1911. The first conference on radiation theory and the quanta was held in Brussels. This was a new type of conference and it became the tradition of the Solvay conference; the participants are informed experts in a given field and meet to discuss one or a few mutually related problems of fundamental importance and seek to define the steps for the solution.

The Solvay conferences in physics have made substantial contributions to the development of modern physics in the twentieth century.

1. (1911) "Radiation theory and the quanta"
2. (1913) "The structure of matter"
3. (1921) "Atoms and electrons"
4. (1924) "Electric conductivity of metals"
5. (1927) "Electrons and photons"
6. (1930) "Magnetism"
7. (1933) "Structure and properties of the atomic nuclei"
8. (1948) "Elementary particles"
9. (1951) "Solid state"
10. (1954) "Electrons in metals"
11. (1958) "The structure and evolution of the universe"
12. (1961) "The quantum theory of fields"
13. (1964) "The structure and evolution of galaxies"
14. (1967) "Fundamental problems in elementary particle physics"
15. (1970) "Symmetry properties of nuclei"
16. (1973) "Astrophysics and gravitation"
17. (1978) "Order and fluctuations in equilibrium and nonequilibrium statistical mechanics"
18. (1982) "High-energy physics. What are the possibilities for extending our understanding of elementary particles and their interactions to much greater energies?"
19. (1987) "Surface science"
20. (1991) "Quantum optics"
21. (1998) "Dynamical systems and irreversibility"

For more information, visit the website of the Solvay Institutes
http://solvayins.ulb.ac.be

XXIst INTERNATIONAL SOLVAY CONFERENCE IN PHYSICS, KEIHANNA PLAZA, NOVEMBER 1–5, 1998
DYNAMICAL SYSTEMS AND IRREVERSIBILITY

(Top row) M. Miyamoto, H. Takahashi, H. Nakazato, G. Ordonez, H. Fujisaka, S. Sasa, H. Hasegawa, Y. Ootaki, A. Oono

(Fourth row from bottom) M. Gadella, A. Bohm, R. Willox, K. Sekimoto, T. Arimitsu, K. Kaneko, D. Driebe, S. Tasaki, Y. Ichikawa

(Third row from bottom) F. Lambert, K. Gustafson, J. R. Dorfman, M.Ernst, S. Pascazio, T. Hida, B. Pavlov, Y. Aizawa, Yu. Melnikov, T. Petrosky, A. Awazu

(Second row from bottom) K. Kitahara, Ya. Sinai, I. Antoniou, L. Accardi, H. Hegerfeldt, O'Dae Kwon, P. Szepfalusy, M. Namiki, L. Boya, K. Kawasaki, H. Posch, P. Gaspa

(Bottom row) R. Balescu, Hao Bai-lin, H. Mori, H. Walther, J. Kondo, I. Prigogine, J. Solvay, L. Reichl, N. G. van Kampen, T. Arecchi, S. C. Tonwar

CONTENTS

ADMINISTRATIVE BOARD OF THE INTERNATIONAL SOLVAY INSTITUTES FOR PHYSICS AND CHEMISTRY	vii
SCIENTIFIC COMMITTEE FOR PHYSICS OF THE INTERNATIONAL SOLVAY INSTITUTES FOR PHYSICS AND CHEMISTRY	ix
THE SOLVAY CONFERENCES ON PHYSICS	xi
PREFACE	xv
OPENING SPEECH BY J. SOLVAY	xvii
INTRODUCTORY REMARKS BY ILYA PRIGOGINE	xxi

PART ONE
DISCRETE MAPS

NON-MARKOVIAN EFFECTS IN THE STANDARD MAP By R. Balescu	3
THERMODYNAMICS OF A SIMPLE HAMILTONIAN CHAOTIC SYSTEM By Hiroshi H. Hasegawa	21
HARMONIC ANALYSIS OF UNSTABLE SYSTEMS By I. Antoniou and Z. Suchanecki	33
PROPERTIES OF PERMANENT AND TRANSIENT CHAOS IN CRITICAL STATES By P. Szépfalusy	49
FROM COUPLED DYNAMICAL SYSTEMS TO BIOLOGICAL IRREVERSIBILITY By Kunihiko Kaneko	53

PART TWO
TRANSPORT AND DIFFUSION

IRREVERSIBILITY IN REVERSIBLE MULTIBAKER MAPS — TRANSPORT AND FRACTAL DISTRIBUTIONS By S. Tasaki	77
DIFFUSION AND THE POINCARÉ–BIRKHOFF MAPPING OF CHAOTIC SYSTEMS By Pierre Gaspard	109

TRANSPORT THEORY FOR COLLECTIVE MODES AND GREEN–KUBO
FORMALISM FOR MODERATELY DENSE GASES   129
   By T. Petrosky

NEW KINETIC LAWS OF CLUSTER FORMATION IN N-BODY
HAMILTONIAN SYSTEMS   161
   By Y. Aizawa

PART THREE
QUANTUM THEORY, MEASUREMENT, AND DECOHERENCE

QUANTUM PHENOMENA OF SINGLE ATOMS   167
   By H. Walther

QUANTUM SUPERPOSITIONS AND DECOHERENCE: HOW TO DETECT
INTERFERENCE OF MACROSCOPICALLY DISTINCT OPTICAL STATES   199
   By F. T. Arecchi and A. Montina

QUANTUM DECOHERENCE AND THE GLAUBER DYNAMICS FROM THE
STOCHASTIC LIMIT   215
   By L. Accardi and S. V. Kozyrev

CP VIOLATION AS ANTIEIGENVECTOR-BREAKING   239
   By K. Gustafson

PART FOUR
EXTENSION OF QUANTUM THEORY AND FIELD THEORY

DYNAMICS OF CORRELATIONS. A FORMALISM FOR BOTH INTEGRABLE
AND NONINTEGRABLE DYNAMICAL SYSTEMS   261
   By I. Prigogine

GENERALIZED QUANTUM FIELD THEORY   277
   By E. C. G. Sudarshan and Luis J. Boya

AGE AND AGE FLUCTUATIONS IN AN UNSTABLE QUANTUM SYSTEM   287
   By G. Ordonez, T. Petrosky, and E. Karpov

MICROPHYSICAL IRREVERSIBILITY AND TIME ASYMMETRIC QUANTUM
MECHANICS   301
   By A. Bohm

POSSIBLE ORIGINS OF QUANTUM FLUCTUATION GIVEN BY
ALTERNATIVE QUANTIZATION RULES   321
   By Mikio Namiki

AUTHOR INDEX   331

SUBJECT INDEX   339

PREFACE

This volume contains the contributions to the XXIst Solvay Conference on Physics, which took place at the Keihanna Interaction Plaza in the Kansaï Science City. The topic was Dynamical Systems and Irreversibility.

The conference has been made possible thanks to the support of the Keihanna Foundation, the Honda Foundation, and the International Solvay Institutes for Physics and Chemistry, founded by E. Solvay.

IOANNIS ANTONIOU

OPENING SPEECH BY J. SOLVAY

Ladies and Gentlemen,

It is a great pleasure and honor to open here the XXIst Solvay Conference on Physics. Generally, the Conferences are held in Brussels. There were also a few organized in the United States. This is the first Solvay meeting organized in Japan. I would like to interpret this conference as a sign of admiration for the creativity of Japanese scientists. May I first tell you an anecdote? Ernest Solvay, my great-grandfather, was a man of multiple interests. He was equally attracted by physics, chemistry, physiology, and sociology. He was in regular correspondence with outstanding people of his time, such as Nernst and Ostwald. This was a period where the first difficulty had appeared in the interpretation of the specific heat by classical physics. Ernest Solvay was bold enough to have his own opinion on this subject. He thought there were surface tension effects, and he expressed his view in a meeting with Nernst in 1910. Nernst was a practical man. He immediately suggested that Ernest Solvay should organize an international meeting to present his point of view. This was the starting point for the Solvay Conferences, the first of which took place in 1911. The Chairman was the famous physicist H. A. Lorentz. At the end of the conference, Lorentz thanked Ernest Solvay not only for his hospitality but also for his scientific contribution. However, in fact his contribution was not even discussed during the meeting. Ernest Solvay was not too disappointed. He thought he had just to continue to work and appreciated greatly the first conference dealing with radiation theory and quanta. He therefore decided to organize the "Solvay Institute for Physics," which was founded in May 1912. He called it the "Institut International de Physique" with the goal "to encourage research which would extend and deepen the knowledge of natural phenomena." The new foundation was intended to concentrate on the "progress of physics." Article 10 of the statutes required that "at times determined by the Scientific Committee a 'Conseil de Physique,' analogous to the one convened by Mr. Solvay in October 1911, will gather, having for its goal the examination of significant problems of physics." A little later, Ernest Solvay established another foundation "Institut International de Chimie." The foundations were ultimately united into "Les Instituts Internationaux de Physique et de Chimie," each having its own Scientific Committee.

The first Solvay Conference on Physics had set the style for a new type of scientific meetings, in which a select group of the most well informed experts in a given field would meet to discuss the problems at the frontiers and would seek to identify the steps for their solution. Except for the interruptions caused by the two World Wars, these international conferences on physics have taken place almost regularly since 1911, mostly in Brussels. They have been unique occasions for physicists to discuss the fundamental problems that were at the center of interest at different periods and have stimulated the development of physical science in many ways. This was a time where international meetings were very exceptional. The Solvay Conferences were unexpectedly successful. In his foreword to the book by Jadgish Mehra, *"The Solvay Conferences on Physics,"* Heisenberg wrote:

> *I have taken up these reminiscences in this foreword in order to emphasise that the historical influence of the Solvay Conferences on the development of physics was connected with the special style introduced by their founder. The Solvay Meetings have stood as an example of how much well-planned and well-organised conferences can contribute to the progress of science.*

It was often said that the people who met at Solvay Conferences went subsequently to Stockholm to receive the Nobel Prize. This is perhaps a little exaggerated, but there is some truth. It is also at the Solvay Conference in 1930 that one of the most famous discussions in the history of science took place. This was the discussion between Einstein and Bohr on the foundations of Quantum theory. Nearly 70 years later it is remarkable to notice that physicists seem not to agree on who won in this discussion.

There is another, more personal aspect that influences the development of the Solvay Conferences. When my friend Ilya Prigogine some 40 years ago in 1958 was nominated Director of the Institutes, he extended their activities from organizing conferences to doing research in a direction that encompasses today's theme, "Irreversible Processes and Dynamical Systems."

The Institutes evolved into a mini Institute for Advanced Study centered around complex systems, nonlinear dynamics, and thermodynamics. In that role, they were an impressive success. Work done within the Institutes shows that far from equilibrium, matter acquires new properties that form the basis of a new coherence. These results introduced the concept of auto-organization, which is echoed into economic and social sciences. These innovations were the reason for Professor Prigogine's 1977 Nobel Prize.

We all know Professor Prigogine's passion for the understanding of time. The flow of time is present on various levels of observations, be it cosmology, thermodynamics, biology, or economics. Moreover, time is the basic existential dimension of man, and nobody can remain indifferent to the problem of time. We all care for the future, especially in the transition period in which we live today. Curiously, the place of time in physics is still a controversial subject. I hope that this conference will make a significant contribution to this vast subject.

My gratitude goes to the local committee that has organized this conference.

Finally, I want to thank Keihanna Plaza for the magnificent hospitality we have received there. I would also like to acknowledge the Honda Foundation, Unoue Foundation, L'Oreal Foundation, the Consul of Belgium and the European Commission for financial contributions that have made this conference possible.

INTRODUCTORY REMARKS
BY ILYA PRIGOGINE

I am happy to open the XXIst Solvay Conference on Physics, especially as it takes place in Japan, in this beautiful setting.

The organization of the Solvay Conference in Kansaï Science City is a fitting tribute to Japanese science. I want also to thank the staff of the Keihanna Interaction Plaza and especially Mr. Yasuki Takeshima for the hospitality and the local organization committee chaired by Professor Kitahara, who was many times our honored guest in Brussels. I am very grateful to Professor Ioannis Antoniou for his help in the organization of this Conference.

Over the years I had many Japanese students. The first was Professor Toda and the most recent were Professors Tasaki and Hasegawa, who are here. My Japanese co-workers had a decisive influence on the evolution of the work of the Brussels–Austin group.

The subject of the XXIst Conference, "Dynamical Systems and the Arrow of Time," is closest to the XVIIth conference, "Order and Fluctuations in Equilibrium and Nonequilibrium Statistical Mechanics," held in 1978. It is a pleasure to mention that a number of people who participated in the 1978 conference are here. Let me mention Professors Arecchi, Balescu, Hao Bai Lin, Kitahara, Reichl, Sinai,...; I hope I have not omitted anyone.

In the XVIIth Conference, much time was devoted to equilibrium critical phenomena and to macroscopic nonequilibrium dissipative structures. A high point was the discussion around the statement by Professor Philip Anderson that dissipative structures have no intrinsic character as they would depend on the boundary conditions. This led to hot discussions that have gone on for years. I believe that this question is now resolved by the experimental discovery of Turing structures with intrinsic wave lengths. At no previous Solvay Conference was the relation between irreversibility and dynamics systematically discussed. However, this is a fascinating subject as we discover irreversible processes at all levels of observations, from cosmology to chemistry or biology.

This is a kind of paradox. It is well known that classical or quantum dynamics lead to a time-reversible, deterministic description. In contrast, both kinetic theory and thermodynamics describe probabilistic processes with broken time symmetry. Kinetic theory and thermodynamics have been quite successful. It is therefore quite unlikely that they can be attributed to approximations introduced in dynamics. Many attempts have been now developed to give a deeper formulation to the problem.

From this point of view, there is some similarity between the goal of the first Solvay Conference held in 1911 and the present conference at the end of the century. In 1911, the question was how to formulate the laws of nature to include quantum effects. Now we ask if irreversibility is the outcome of approximations or if we can formulate microscopic basic laws that include time symmetry breaking.

In my long experience, I always found that the problem of time leads to much passion. So I look forward with great expectations to this conference.

PART ONE

DISCRETE MAPS

NON-MARKOVIAN EFFECTS IN THE STANDARD MAP

R. BALESCU

Department of Physical Statistics—Plasma, Free University of Brussels, Brussels, Belgium

CONTENTS

I. Introduction
II. Non-Markovian and Markovian Evolution Equations
III. Master Equation for the Standard Map
IV. Solution of the General Master Equation
V. Solution of the Standard Map Master Equation
VI. Conclusions
References

I. INTRODUCTION

Iterative maps have been extensively used for the study of evolution problems, as a substitute for differential equations. Of particular importance for the modeling of classical mechanical systems are *Hamiltonian (or area-preserving) maps*. A special case to which a great deal of attention has been devoted is the *Chirikov–Taylor standard map* [1–3] (we only quote here a few among the numerous works devoted to this subject). It is, indeed, the simplest two-dimensional Hamiltonian map, many properties of which can be derived analytically:

$$x_{\tau+1} = x_\tau - \frac{K}{2\pi} \sin 2\pi \theta_\tau$$
$$\theta_{\tau+1} = \theta_\tau + x_{\tau+1} \quad (\mod 1) \tag{1}$$

Dynamical Systems and Irreversibility: A Special Volume of Advances in Chemical Physics, Volume 122, Edited by Ioannis Antoniou. Series Editors I. Prigogine and Stuart A. Rice.
ISBN 0-471-22291-7. © 2002 John Wiley & Sons, Inc.

Here x_τ is a continuous variable ranging from $-\infty$ to $+\infty$, θ_τ is an angle divided by 2π, and τ is a "discrete time," taking integer values from 0 to ∞; K is a nonnegative real number, called the "*stochasticity parameter*." Iterative (in particular, Hamiltonian) maps prove to be useful tools for the study of transport processes. In order to treat such problems, one adopts a statistical description. The consideration of individual trajectories defined by Eq. (1) is then replaced by the study of a statistical ensemble defined by a distribution function in the phase space spanned by the variables x and θ: $f(x, \theta; \tau)$; this is a 1-periodic function of θ and is defined only for nonnegative integer values of τ. Of special physical interest is the phase-averaged distribution function, which will be called the *density profile* $n(x; \tau)$:

$$n(x; \tau) = \int_0^1 d\theta\, f(x, \theta; \tau) \tag{2}$$

It has been known for a long time [1–7] that in the limit of large K, the evolution described by the standard map has a *diffusive* character. This statement has to be made more precise, because it may address various aspects of the evolution.

In the pioneering work of Rechester and White [4], a Liouville equation for the distribution function is modified by adding (arbitrarily) an external noise. A calculation of the mean square displacement of x then yields, for large K, a diffusion coefficient. This derivation is unsatisfactory for two reasons: (a) the assumption of a *continuous-time* Liouville equation for the description of a discrete-time process and (b) the presence of noise, which introduces from the very beginning an artificial dissipation.

Abarbanel [5] gave a more transparent derivation, in which these two assumptions are no longer introduced. He used a projection operator formalism for the derivation of a kinetic equation. His formalism is close to ours, but uses a continuous time formalism and is used for a different purpose.

Hasegawa and Saphir [6] gave the first truly fundamental treatment of the standard map, showing that in the limit of large K, and simultaneously of large spatial scales, there exists an *intrinsic diffusive mode* of evolution of the standard map dynamics (this result was further developed by the present author [7]). No additional probabilistic assumption is necessary for obtaining this result. More specifically, these authors proved the existence (in this limit) of a pole of the resolvent (in Fourier representation) of the form $[-(2\pi q)^2 D]$, where q is the wave vector and D is identified with the diffusion coefficient.

Given this result, it appears desirable to study more globally the behavior of a system. In particular, we should like to determine how the system, starting from an arbitrary initial condition, and evolving by the exact standard map dynamics, reaches a regime in which the evolution is determined by a diffusion equation. This goal requires the study of the density profile, Eq. (2).

In "classic" statistical mechanics, such a study involves the solution of a kinetic equation—that is, a *closed* equation for a reduced distribution function. A corresponding equation for systems described by discrete time iterative maps was obtained in a recent paper by Bandtlow and Coveney [8]. They derived an *exact* closed equation for the density profile, analogous to the master equation obtained by Prigogine and Résibois [9] in continuous-time statistical mechanics. The most important characteristic of both equations is their *non-Markovian nature*: The evolution of the system at time τ is determined not only by its instantaneous state, but rather by its past history. It is well known in continuous-time kinetic theory that, whenever there exist two characteristic time scales that are widely separated (e.g., the duration of a collision, and the inverse collision frequency in a gas), the master equation reduces, for times much longer than the short time scale, to a *Markovian kinetic equation*.

The Bandtlow–Coveney equation is quite general; it appears that the standard map provides us with an ideal testing bench for studying its properties. It is interesting to investigate whether there exist here also two such characteristic time scales, and under which conditions a markovianization is justified. This will be the object of the present work.

II. NON-MARKOVIAN AND MARKOVIAN EVOLUTION EQUATIONS

The evolution of the distribution function of a system governed by the standard map in discrete time τ is determined by the *Perron–Frobenius operator U* [7]:

$$f(x, \theta; \tau + 1) = U f(x, \theta; \tau) \tag{3}$$

Alternatively, one may introduce the *propagator*, which relates the instantaneous distribution function to its initial value:

$$f(x, \theta; \tau) = U^\tau f(x, \theta; 0) \tag{4}$$

In continuous-time dynamics this propagator is related to the Liouville operator $\mathscr{L}$:

$$U(t) = \exp(\mathscr{L} t) \tag{5}$$

Here and below, Roman letters $t, s, \ldots$ denote real, continuous-time variables, whereas Greek letters $\tau, \sigma, \ldots$ denote discrete-time variables, taking only integer values.

The Fourier transform of the distribution function with respect to both phase space variables will be extensively used below:

$$f(x, \theta; \tau) = \sum_{m=-\infty}^{\infty} \int_{-\infty}^{\infty} dq \, e^{2\pi i (qx+m\theta)} \tilde{f}_m(q; \tau) \tag{6}$$

As explained in Section I, we are interested in deriving an equation of evolution for the reduced distribution function, or density profile $n(x; \tau)$ [Eq. (2)] or, equivalently, for its Fourier transform, which is simply the $m = 0$ Fourier component of the distribution function; it will be denoted by the notation $\varphi(q; \tau)$:

$$n(x; \tau) = \int_{-\infty}^{\infty} dq \, e^{2\pi i q x} \varphi(q; \tau)$$
$$\varphi(q; \tau) = \tilde{f}_0(q; \tau) \tag{7}$$

The density profile can also be obtained by acting on the full distribution function with a projection operator P whose effect is the average over the angle θ:

$$Pf(\tau) = \tilde{f}_0(\tau) \equiv \varphi(\tau) \tag{8}$$

(In forthcoming equations, the argument q of the distribution functions will not be written down explicitly whenever it is clearly understood.) Obviously, $P^2 = P$. Let Q be the complement of the projector P; thus $P + Q = I$, where I is the identity operator.

In order to derive a closed equation for the density profile, Bandtlow and Coveney [8] start from the trivial identity expressing the group property of the Perron–Frobenius operator:

$$U^{\tau+1} = UU^{\tau} \tag{9}$$

which is projected on the P and Q subspaces and rewritten in the form

$$PU^{\tau+1} = PUPU^{\tau} + PUQU^{\tau} \tag{10}$$

with a similar equation for $QU^{\tau+1}$. A Z-transformation (the analog of the Laplace transformation for discrete time) is performed on these equations, and some simple transformations (similar to those of Chapter 15 of Ref. 7) lead, without any approximations, to the following equation:

$$Pf(\tau + 1) = \sum_{\sigma=0}^{\tau} P\psi(\sigma) Pf(\tau - \sigma) + PD(\tau + 1)Qf(0) \tag{11}$$

The diagonal $(P - P)$ operator $\psi(\tau)$ is defined as follows (for simplicity we no longer write the P-projectors explicitly):

$$\psi(\tau) \equiv P\psi(\tau)P = \begin{cases} PUP, & \tau = 0 \\ 0, & \tau = 1 \\ PUQ(QUQ)^{\tau-1}QUP, & \tau \geq 2 \end{cases} \quad (12)$$

The nondiagonal "destruction operator" $PD(\tau)Q$ has a similar form, which is not written here, because it will not be needed in the forthcoming work. Note that Eq. (11) is not limited to the standard map: It is easily adapted to a general iterative map, in arbitrary dimensionality, subject only to some mathematical regularity conditions, discussed in Ref. 8.

Equation (11) is called the *Master equation in discrete time*. It is the closest analog to the Prigogine–Résibois master equation in continuous time [7,9] for the reduced velocity distribution function in a gas:

$$\partial_t \varphi(t) = \int_0^t ds\, \psi(s)\, \varphi(t-s) + D(t)\, Cf(0) \quad (13)$$

The most conspicuous characteristic of both equations is their *non-Markovian nature*, expressed by the convolution appearing in the first term of the right-hand side. Thus, the instantaneous change at time τ, leading to $\varphi(\tau + 1)$, is determined, in principle, by the whole past history. For obvious physical reasons, $\psi(\tau)$ must be a decreasing function of the time τ. The effective width of this function determines the range of the memory of the process; we therefore call $\psi(\tau)$ the *memory kernel*.

The second term in the right-hand side of Eq. (11) is a source term, describing the effect of the initial angle-dependent part of the distribution on the evolution at time τ of the density profile; it corresponds to the so-called *destruction term* acting on the initial correlations in the continuous-time master equation. Normally, it decreases in time over the same time scale as the memory kernel.

For simplicity, we shall assume here that the initial distribution function is independent of the angle θ, hence the destruction term is zero. Using also the simpler notation, Eq. (8), we rewrite Eq. (11) under this condition in the simpler form:

$$\varphi(\tau + 1) = \sum_{\sigma=0}^{\tau} \psi(\sigma)\, \varphi(\tau - \sigma) \quad (14)$$

As explained in Section I, whenever there exist two intrinsic, widely separated time scales, the master equation can be *markovianized*; that is, the retardation effects can be neglected for long times. We must now understand what this operation means in discrete-time dynamics. Mathematically, it implies that the right-hand side of Eq. (14) could be approximated by an expression containing only the distribution function at time τ.

(i) A straightforward, rather brutal way of achieving this goal consists of neglecting all terms corresponding to $\sigma \neq 0$. Equation (14) then reduces to

$$\varphi_0(\tau + 1) = \psi(0)\, \varphi_0(\tau) \tag{15}$$

This will be called the *zero-Markovian approximation*. It implies that the evolution of the density profile occurs without any memory: The memory kernel $\psi(\sigma)$ has strictly zero width. From the definition (12) it follows that this amounts to supposing that the "complementary" states Qf (which only appear in the terms with $\sigma \geq 2$) are completely excluded as intermediate states in the construction of the memory kernel. The latter reduces to

$$\psi(0) = PUP \tag{16}$$

that is, the diagonal $P - P$ element of the Perron–Frobenius operator. In continuous-time dynamics, this approximation corresponds to

$$\partial_t Pf(t) = P\mathscr{L}Pf(t) \tag{17}$$

which is the celebrated *Vlasov equation* [7,10].

(ii) A more subtle markovianization (called the *full Markovian approximation*) is performed when the following conditions are satisfied.

(a) *The memory kernel is a rapidly decaying function of time.* More precisely, there exists a characteristic time τ_M, called the *memory time*, such that

$$|\psi(\tau)| \approx 0 \quad \text{for} \quad \tau \gg \tau_M \tag{18}$$

This characteristic time is analogous to the duration of a collision in ordinary kinetic theory.

(b) *The density profile is slowly varying in time.* This implies the existence of a second time scale τ_R, the *relaxation time*, much longer than the memory time: $\tau_R \gg \tau_M$, such that

$$\varphi(\tau) \approx \varphi_{as}(\tau) \quad \text{for} \quad \tau \gg \tau_R \tag{19}$$

where $\varphi_{as}(\tau)$ is the asymptotic form of the distribution function, which is independent of the initial condition.

When these conditions are satisfied, then, for long times compared to τ_M, the following approximations are justified in Eq. (14):

- The retardation in the density profile is neglected on the right-hand side: $\varphi(\tau - \sigma) \approx \varphi(\tau)$
- The upper limit in the summation is pushed up to infinity.

The resulting equation is then

$$\varphi_M(\tau + 1) = \Psi \, \varphi_M(\tau) \tag{20}$$

The time-independent evolution operator appearing here is

$$\Psi = \sum_{\sigma=0}^{\infty} \psi(\sigma) \tag{21}$$

Equation (20) is a Markovian equation of evolution, which will be called the *kinetic equation* of the map. The name is suggested by the analogous kinetic equation of continuous-time statistical mechanics; the operator corresponding to Ψ is there the sum of the Vlasov operator and of the collision operator. The form of the kinetic equation is similar to the starting equation (3), and the kinetic operator Ψ plays a role similar to the Perron–Frobenius operator U. It must not be forgotten, however, that unlike Eq. (3), Eq. (20) is a *closed equation for the density profile*, that is, the P-component of the distribution function.

All the considerations of the present section are valid for arbitrary two-dimensional Hamiltonian maps (and can be easily generalized to higher dimensionality). We now illustrate the results of this section in the case of the standard map.

III. MASTER EQUATION FOR THE STANDARD MAP

The advantage of the standard map is that many quantities can be calculated analytically. Thus, the Fourier representation of Eq. (3) is [7]

$$\tilde{f}_m(q; \tau + 1) = \sum_{m'=-\infty}^{\infty} \int_{-\infty}^{\infty} dq' \, \langle q, m|U|q', m' \rangle \tilde{f}_{m'}(q'; \tau) \tag{22}$$

with the following expression for the matrix elements of the Perron–Frobenius operator:

$$\langle q, m|U|q', m' \rangle = \delta(q' - q - m) \, J_{m-m'}(q'K) \tag{23}$$

where $J_l(x)$ is the Bessel function of order l. It is clearly seen that this operator is nondiagonal in both q and m. Using now the definition (8) of the P-projector and the definition (12) of the memory kernel, a straightforward, but rather lengthy calculation similar to those of Refs. 6 and 7 leads to the following result (details will be published elsewhere [11]):

$$\psi(0) = J_0(qK)$$
$$\psi(1) = 0$$
$$\psi(\tau) = \sum_{m_1 \neq 0} \cdots \sum_{m_\tau \neq 0} \delta(m_1 + m_2 + \cdots + m_\tau) \times J_{-m_1}(qK) \quad (24)$$
$$\left\{ \prod_{j=1}^{\tau-1} J_{m_j - m_{j+1}} \left[\left(q - \sum_{l=1}^{j} m_l \right) K \right] \right\} J_{m_\tau}(qK), \quad \tau = 2, 3, \ldots$$

This expression is exact, but rather untransparent; in particular, the dependence on τ is not easily grasped. We now restrict the study of the evolution to a special domain of parameter space, which defines the *diffusive regime*:

$$\sqrt{K} \gg 1, \quad qK \ll 1 \quad (25)$$

Let us stress the fact that the mere condition of a large K is not sufficient for characterizing a diffusive regime. The second condition puts a limit on the wave vector q; it implies that the larger the stochasticity parameter, the larger the length scales ($\sim q^{-1}$) for which diffusive behaviour will (eventually) be observed. It follows from the well-known properties of the Bessel functions that in the diffusive regime the following orders of magnitude prevail:

$$J_m(qK) = O[(qK)^m], \quad m = 0, 1, 2, \ldots$$
$$J_m[(q-n)K] = O(K^{-1/2}), \quad n = \pm 1, \pm 2, \ldots, \quad m = 0, 1, 2, \ldots \quad (26)$$

Under these conditions, the expressions (24) can be approximated by retaining only a small number of terms in the summations. In the present work we approximate the memory kernel by retaining terms through order $(qK)^4$. We do not write down the explicit expressions, which will be published elsewhere [11].

We first check that the memory kernel $\psi(q, K; \tau)$ [$\equiv \psi(\tau)$] is a decreasing function of time. Choosing a rather extreme value for $q = 0.01$, we plot $\psi(2)$, $\psi(3)$, $\psi(4)$ against K in the range $15 \leq K \leq 50$ (Fig. 1).

Over this whole range, $\psi(0)$ varies very slowly from 1 to 0.98; thus it strongly dominates the remaining three components. The latter have a

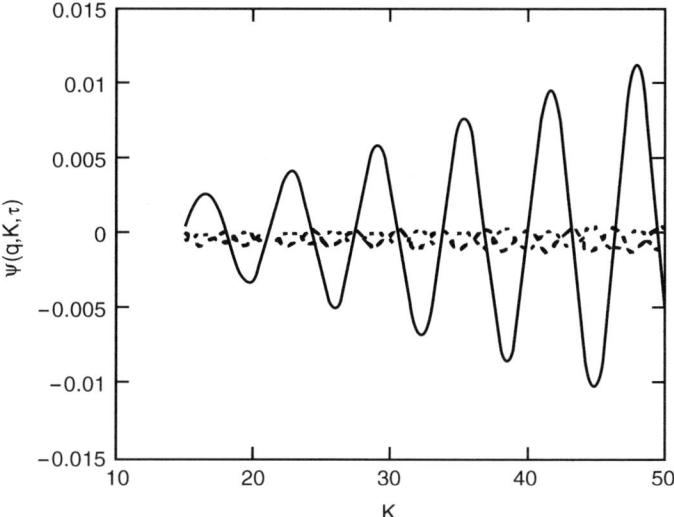

Figure 1. Dependence of the memory kernel $\psi(q, K; \tau)$ on the stochasticity parameter K ($q = 0.01$). Solid: $\tau = 2$; dot: $\tau = 3$; dash: $\tau = 4$.

characteristic oscillating behavior due to the Bessel functions. Their maximum amplitude (which increases with K) remains everywhere much smaller than $\psi(0)$; thus

$$|\psi(0)|_{Max} \gg |\psi(2)|_{Max} \gg |\psi(4)|_{Max} \tag{27}$$

[$\psi(3)$ is out of phase with $\psi(2)$ and $\psi(4)$]. The relative size of these functions is, however, a sensitive function of K [for instance, when $\psi(2)$ vanishes, the leading non-Markovian correction would be $\psi(3)$ or $\psi(4)$]. In spite of these details, the very rapid decay of $|\psi(\tau)|$ is obvious. The *memory time* defined in Eq. (18) is of the order $\tau_M \approx 4$ (this quantity is only defined, as usual, in order of magnitude: its value depends on the precision accepted in the calculations). Thus, for the present choice of parameters, the kernel $|\psi(\tau)|$ decreases by three orders of magnitude after $\tau = 4$.

IV. SOLUTION OF THE GENERAL MASTER EQUATION

We now take advantage of the rapid decrease of the memory kernel, expressed by Eq. (27), in order to obtain an approximate solution of the non-Markovian master equation (14). The latter is written in terms of a propagator:

$$\varphi(\tau) = W(\tau)\varphi(0) \tag{28}$$

We decide to truncate the convolution in the master equation at the level of $\psi(4)$. The propagator then obeys the following approximate equation:

$$W(\tau + 1) = \psi(0)W(\tau) + \psi(2)W(\tau - 2) + \psi(3)W(\tau - 3) + \psi(4)W(\tau - 4) \tag{29}$$

This equation is easily solved through order $\psi(4)$ (a detailed proof will be published separately):

$$W(\tau) = \psi(0)^\tau + (\tau - 2)\,\psi(0)^{\tau-3}\,\psi(2) + (\tau - 3)\,\psi(0)^{\tau-4}\,\psi(3)$$
$$+ (\tau - 4)\,\psi(0)^{\tau-5}\,\psi(4) + \frac{1}{2}(\tau - 5)(\tau - 6)\,\psi(0)^{\tau-6}\,\psi(2)^2 \tag{30}$$
$$\tau \geq 6$$

It is understood that in the first members ($\tau < 6$) of the sequence (30), the coefficients of the terms containing a negative power of $\psi(0)$ are set equal to zero.

This solution will be compared to the two Markovian approximations discussed in Section 2. The *zero-Markovian approximation* yields a trivially simple solution:

$$\begin{aligned}\varphi_0(\tau) &= W_0(\tau)\,\varphi(0) \\ W_0(\tau) &= [\psi(0)]^\tau\end{aligned} \tag{31}$$

that is, simply the first term in Eq. (30).

The *full Markovian approximation* is also easily obtained from Eqs. (20) and (21), truncated to order $\psi(4)$:

$$\begin{aligned}\varphi_M(\tau) &= W_M(\tau)\,\varphi(0) \\ W_M(\tau) &= \left[\sum_{\sigma=0}^{4}\psi(\sigma)\right]^\tau\end{aligned} \tag{32}$$

Let it be stressed at this point that all the results obtained in the present section are valid for an arbitrary Hamiltonian map dynamics, provided that the ordering (27) is valid and the truncation at the level $\psi(4)$ is justified. The truncation level can easily be extended to higher orders if necessary.

V. SOLUTION OF THE STANDARD MAP MASTER EQUATION

The general results obtained in Section IV are now applied to the standard map. The general expressions of the memory kernel, Eq. (24), are truncated at the appropriate level, considering the orders of magnitude (26) pertaining to the diffusive regime (25). The explicit calculations are somewhat tedious, but are facilitated by the use of a symbolic computer program, such as Maple. The result is then inserted into Eqs. (28)–(32), thus yielding the expressions of the non-Markovian solution, as well as of its Markovian approximations.

We choose on purpose a relatively small value of $K = 22.5$. The initial condition of the density profile (in Fourier representation) will be chosen as the following rectangular function:

$$\varphi(q;0) = \begin{cases} 1, & q \leq 0.01 \\ 0, & q > 0.01 \end{cases} \quad (33)$$

This choice ensures the validity of the second condition (25) over the whole range of wave vectors.

Figure 2 shows the solution $\varphi(q;\tau)$ of the non-Markovian master equation [Eq. (29)] for times $\tau = 10, 100, 1000, 5000$. The evolution from the initial rectangular distribution toward an asymptotic "Gaussian-like" distribution characteristic of the diffusive regime is quite evident. A more quantitative conclusion can be drawn from this figure. It is clearly seen that it takes a time $\tau \approx 1000$ for the initial shape of the density profile to be forgotten. This

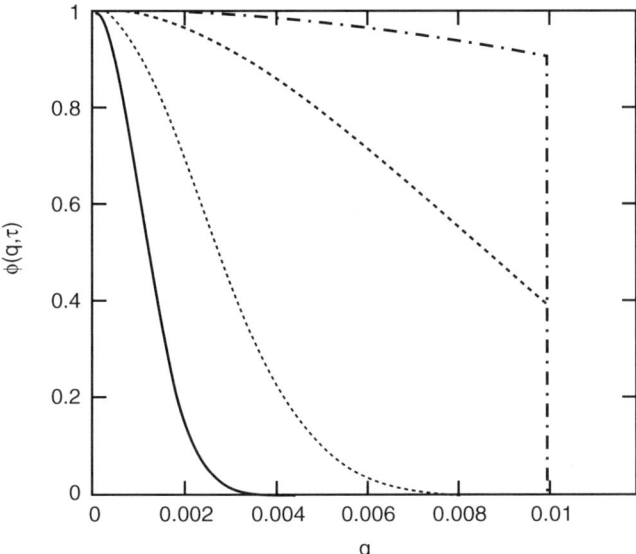

Figure 2. Non-Markovian solution $\varphi(q, K; \tau)$ at different times. $K = 22.5$. Dash-dot: $\tau = 10$; dot: $\tau = 100$; dash: $\tau = 1000$; solid: $\tau = 5000$.

value can be taken as the definition of the relaxation time τ_R, Eq. (19). Combining this result with the value of the memory time obtained in Section III, we see that the ratio of the characteristic times for the standard map in the diffusive regime is, in order of magnitude,

$$\frac{\tau_M}{\tau_R} \approx 10^{-2}-10^{-3} \qquad (34)$$

This very small value justifies the use of the markovianized kinetic equation (20) for $\tau \gg \tau_M$.

We now wish to examine in more detail the transition from the non-Markovian to the Markovian regime. We thus compare the non-Markovian solution $\varphi(q;\tau)$ (30) with the zero-Markovian solution $\varphi_0(q;\tau)$ (31) and with the full Markovian solution $\varphi_M(q;\tau)$ (32). The former has a simple expression for the standard map in the diffusive regime; using Eq. (24) we find

$$W_0(\tau) = [J_0(qK)]^\tau \approx \left[1 - \frac{1}{4}(qK)^2\right]^\tau \approx \exp\left(-\frac{1}{4}K^2 q^2 \tau\right) \qquad (35)$$

This is precisely of the form of the propagator associated with the diffusion equation (in Fourier representation), with a diffusion coefficient D:

$$W_0(\tau) \approx \exp[-D(2\pi q)^2 \tau] \qquad (36)$$

The diffusion coefficient appearing in the zero-Markovian approximation (35) is thus

$$D = D_{QL} = \frac{1}{4(2\pi)^2} K^2 \qquad (37)$$

The latter value is the well-known *quasilinear diffusion coefficient* [1–7]. Thus, the "Vlasov" approximation for the standard map in the diffusive regime is equivalent to the quasilinear approximation.

In Fig. 3 the three solutions are shown for a short time $\tau = 6$ (of the order of the memory time). As expected, the Markovian approximations deviate significantly from the non-Markovian one. The deviation is strongest for large q; the zero-Markovian is definitely not good, even at such short times.

In Fig. 4 the same three solutions are plotted for $\tau = 1000$ (of the order of the relaxation time). On the scale of this figure, the full Markovian solution is now very close to the "exact" non-Markovian one. On the other hand, the zero-Markovian (quasilinear) solution is significantly wrong. This is a quite interesting result. Recalling Eqs. (20) and (21), it is seen that the memory

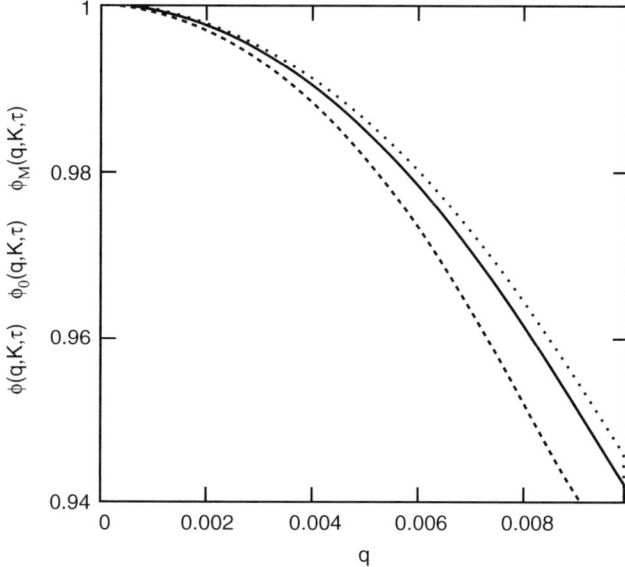

Figure 3. Non-Markovian and Markovian solutions for short time, $\tau = 6$. $K = 22.5$. Solid: Non-Markovian $\varphi(q, K; \tau)$; dash: Zero-Markovian $\varphi_0(q, K; \tau)$; dots: fully Markovian $\varphi_M(q, K; \tau)$.

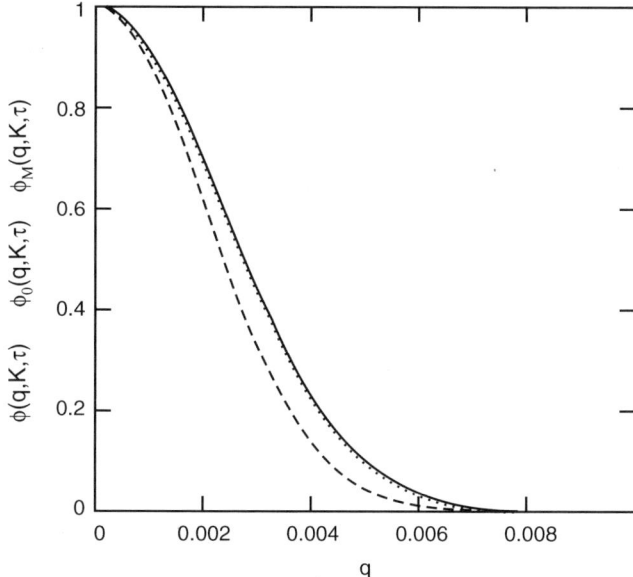

Figure 4. Non-Markovian and Markovian solutions for long time, $\tau = 1000$. $K = 22.5$. Solid: Non-Markovian $\varphi(q, K; \tau)$; dash: Zero-Markovian $\varphi_0(q, K; \tau)$; dots: fully Markovian $\varphi_M(q, K; \tau)$.

effect [i.e., $\psi(\sigma)$ for $\sigma > 0$] cannot be ignored in the markovianization of the evolution equation—that is, in the construction of the fully Markovian operator $W_M(\tau)$ [Eq. (32)] or Ψ [Eq. (21)]. Thus, the full Markovian approximation should not be understood as a "memoryless" evolution. The evolution operator Ψ is built up by the cumulative action of the exact operator over a finite time span of the order of the (short) memory time.

It is instructive to look more closely to the way in which the non-Markovian solution approaches the asymptotic Markovian solution as a function of time. In Fig. 5 we plot the difference $\Delta\varphi(q; \tau) = \varphi(q; \tau) - \varphi_M(q; \tau)$ for a fixed value of $q = 0.004$ (in the region of large deviation). The deviation is, of course, strongest for short time; it approaches zero asymptotically for times longer than the relaxation time $\tau_R \approx 1000$.

The final asymptotic density profile is expected to be a diffusive Gaussian of the form (36). The "true" diffusion coefficient is obtained from the "exact" non-Markovian density profile by the well-known relation

$$D = -\frac{1}{2(2\pi)^2} \frac{d}{d\tau} \frac{\partial^2 \varphi(q;\tau)}{\partial q^2}\bigg|_{q=0} \quad (38)$$

Let $\varphi_G(q; \tau)$ be the Gaussian (36) combined with the diffusion coefficient (38). In Fig. 6a it is seen that, for $\tau = 1000$, the deviation of the non-Markovian

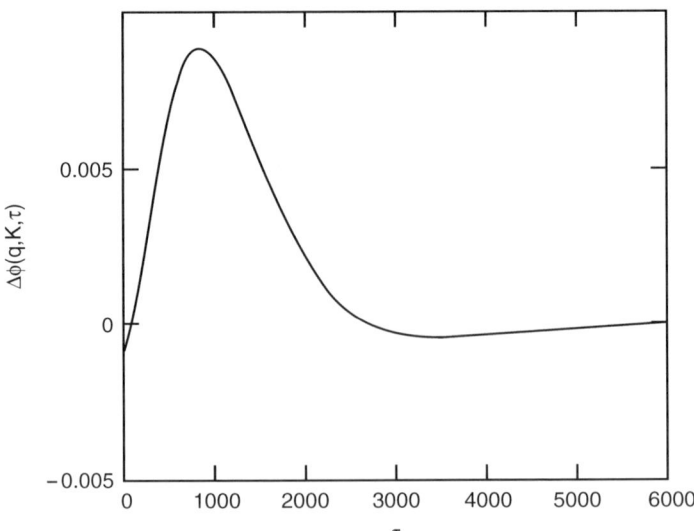

Figure 5. Deviation of the fully Markovian from the Non-Markovian solution, as a function of time. $K = 22.5$, $q = 0.004$. $\Delta\varphi(q, K; \tau) = \varphi(q, K; \tau) - \varphi_M(q, K; \tau)$.

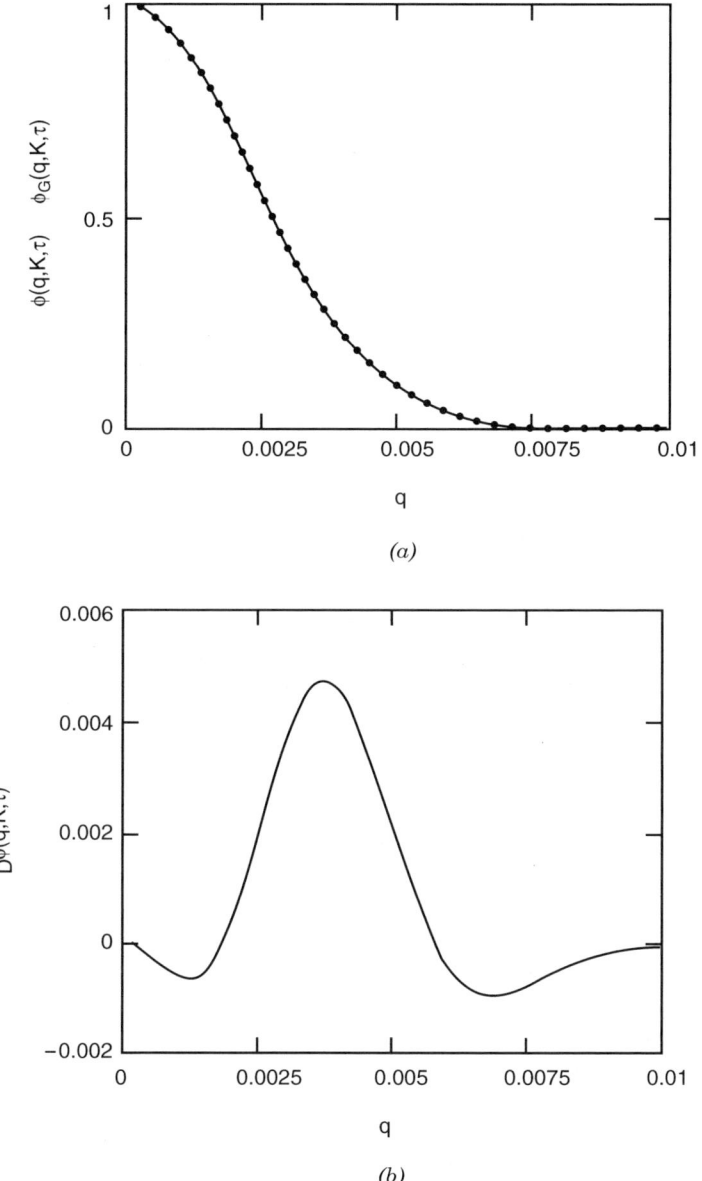

Figure 6. (a) Non-Markovian and Gaussian density profiles at $\tau = 1000$. $K = 22.5$. (b) Deviation of the Gaussian from the Non-Markovian solution at $\tau = 1000$. $K = 22.5$. $D\varphi(q, K; \tau) = \varphi(q, K; \tau) - \varphi_G(q, K; \tau)$.

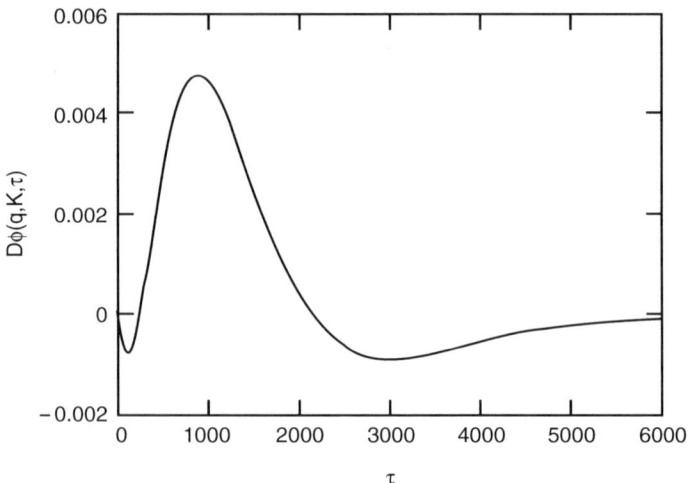

Figure 7. Deviation of the Gaussian density profile from the Non-Markovian solution, $D\varphi(q, K; \tau)$, as a function of time. $K = 22.5$; $q = 0.004$.

solution from the corresponding Gaussian is small, but visible. This is better visualized by plotting the difference $D\varphi(q; \tau) = \varphi(q; \tau) - \varphi_G(q; \tau)$ (Fig. 6b). The relatively important deviation has a nonnegligible maximum.

We now plot the deviation of the non-Markovian solution from the Gaussian diffusive profile $D\varphi(q; \tau)$ at the position of the maximum deviation $q = 0.004$ as a function of time (Fig. 7).

Comparing this figure with Fig. 5 (corresponding to the same value of q), we note that the deviation of the Gaussian has everywhere an opposite sign compared to the Markovian solution. Both deviations eventually go to zero; this takes, however, a very long time ($\tau = 6000$ in the present case); this time actually depends on q: Large values of q relax faster than the small ones. The following interesting conclusion thus follows from this discussion: *Over a long range of intermediate times* ($1000 < \tau < 10000$) *the asymptotic Markovian density profile is non-Gaussian. The "pure" diffusive regime (36) only sets in after a time much longer than the relaxation time.*

VI. CONCLUSIONS

In this work we presented a first step toward the construction of a *kinetic theory of chaotic dynamical systems described by iterative maps*. This theory follows as closely as possible the methodology of the kinetic theory of continuous-time dynamical systems developed in the framework of nonequilibrium statistical mechanics (see, e.g., Refs. 7 and 10). A closed equation for the density profile due to Bandtlow and Coveney [8] is taken as a starting point. Particular attention is devoted to the transition from this non-Markovian equation to a Markovian approximation, corresponding to the usual kinetic equation. This transition appears whenever there exist two widely separated time scales: a short *memory time* τ_M and a long *relaxation time* τ_R. Two levels of markovianization are described. The *zero-Markovian approximation* consists of neglecting altogether all memory effects: It is shown to correspond to the Vlasov approximation of the usual kinetic theory. The *full Markovian approximation* introduces a description formally similar to the collision operator of kinetic theory, although the role of collisions is played here by the intrinsic stochasticity. The evolution tends asymptotically, for times much longer than the relaxation time, toward an irreversible process of diffusive type.

The general properties described above are illustrated explicitly in the case of the Chirikov–Taylor standard map in the diffusive regime, where all calculations can be done analytically. The conditions of validity of the Markovian approximation are thus verified. The non-Markovian equation is solved explicitly in this regime. It can therefore be compared in detail with the Markovian approximations. The zero-Markovian approximation is shown to correspond to the quasilinear approximation; it appears to be inadequate for the asymptotic description of the evolution. Memory effects must necessarily be retained even in the asymptotic regime, through the construction of the fully Markovian evolution operator.

The transition from the non-Markovian to the Markovian regime is described in detail. It was also pointed out that the latter regime is not necessarily a purely diffusive one, described by a Gaussian density profile. Only for times very much longer than the relaxation time does the Markovian asymptotic solution tend towards a diffusive one.

References

1. B. V. Chirikov, *Phys. Rep.* **52**, 265 (1979).
2. A. J. Lichtenberg and M. A. Lieberman, *Regular and Stochastic Motion*, Springer, New York, 1983.
3. L. E. Reichl, *The Transition to Chaos*, Springer, New York, 1992.
4. A. B. Rechester and R. B. White, *Phys. Rev. Lett.* **44**, 1586 (1980).
5. H. D. J. Abarbanel, *Physica* **D4**, 89 (1981).
6. H. H. Hasegawa and W. C. Saphir, in *Aspects of Nonlinear Dynamics*, I. Antoniou and F. Lambert, eds., Springer, Berlin, 1991.
7. R. Balescu, *Statistical Dynamics: Matter out of Equilibrium*, Imperial College Press, 1997.
8. O. F. Bandtlow and P. V. Coveney, *J. Phys. A: Math. Gen.* **27**, 7939 (1994).
9. I. Prigogine and P. Résibois, *Physica* **27**, 629 (1961).
10. R. Balescu, *Equilibrium and Nonequilibrium Statistical Mechanics*, Wiley, New York, 1975.
11. R. Balescu, *J. Stat. Phys.* **98**, 1169 (2000).

THERMODYNAMICS OF A SIMPLE HAMILTONIAN CHAOTIC SYSTEM

HIROSHI H. HASEGAWA

Department of Mathematical Science, Ibaraki University, Mito, Japan; and Center for Statistical Mechanics, University of Texas, Austin, Texas, U.S.A.

CONTENTS

I. Introduction
II. Thermodynamics of the Cat Map
   A. Hamiltonian
   B. External Operations
   C. Time Evolution of Probability Density
   D. Work
   E. "The Second Law" of Thermodynamics
   F. Recover of the Second Law in Large System
III. Conclusions and Remarks
Acknowledgments
References

I. INTRODUCTION

A simple non-Hamiltonian chaotic system, such as the baker map, is a nice model to investigate the foundations of thermodynamics [1–3]. Because of such system's simplicity, we can use them to directly connect a time reversible dynamics with irreversible thermodynamics. In order to understand irreversibility, it was quite useful to see how a unique time scale such as the relaxation time, which characterizes irreversibility, appears in a time reversible dynamical system [2].

Dynamical Systems and Irreversibility: A Special Volume of Advances in Chemical Physics, Volume 122, Edited by Ioannis Antoniou. Series Editors I. Prigogine and Stuart A. Rice.
ISBN 0-471-22291-7. © 2002 John Wiley & Sons, Inc.

In this chapter we will investigate a simple Hamiltonian chaotic system as a small thermodynamics one [4,5]. To understand thermodynamic laws [6], we need real energy. We consider thermodynamics as a general theory, which describes how a system responds to slow external operations. We will introduce external operations such as isothermal and fixed-volume transformations in our system and study how the system responds.

Our study is motivated by a pioneer work by Sekimoto and Sasa [7]. They derived the thermodynamic laws of a system governed by the Langevin equation with slow external operations. In the Langevin equation, white noise plays a role of heatbath. The correlation time is negligible compared to the time scale of the external operations. They showed that the work is given as the difference between the initial and final free energy in the quasi-static isothermal process. They presented a beautiful derivation of the second law, the positive excess heat production, using an expansion with respect to the slow external operation.

In the next section we will construct thermodynamics of the cat map.

In Section II.A, we start with a time-dependent Hamiltonian, which governs the motion of a periodically kicked particle between two walls. The Poincaré map can be scaled as a generalized standard map. Because of the chaotic dynamics, the motion of the particle is irregular. We interpret the irregular motion as coming from thermal noise from a virtual heatbath. We introduce a

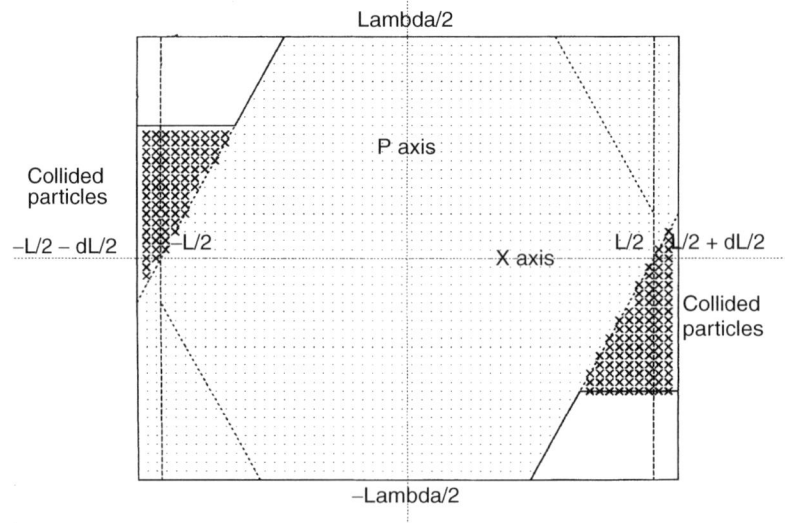

Figure 1. Phase space of the operated system.

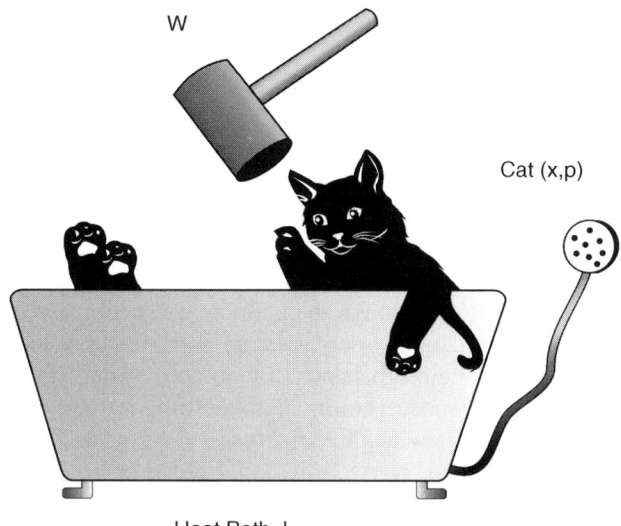

Figure 2. Thermodynamics of the cat map.

harmonic oscillator as the heatbath with which the system conserves total energy.

Because of the mixing property, the probability density physically approaches equilibrium. We define temperature as the average kinetic energy of the particle. The temperature is determined by the scale of the momentum.

In Section II.B, we will introduce the external isothermal operation and the fixed-volume one. In the former the walls move while the scale of the momentum stays fixed. In the latter the scale of the momentum changes without the walls moving.

In Section II.C, the time evolution of the probability density will be written as the repeated operation of a pair of transfer operators, with one describing an isothermal transformation and the other describing the chaotic map.

In Section II.D, we will calculate the work for slow isothermal process.

In Section II.E, following the argument of Sekimoto–Sasa, we derive the inequality corresponding to the 2nd law of thermodynamics. The inequality may lose the direct connection to the non-existence of perpetual motion of the second kind.

In Section II.F, we will discuss how the ordinary second law of thermodynamics can be recovered in large systems.

In Section III, we will conclude this chapter and comment on extension of thermodynamics into the region of weak chaos.

II. THERMODYNAMICS OF THE CAT MAP

A. Hamiltonian

We consider a periodically kicked particle between two walls at $X = \pm L/2$. The time evolution of the particle is governed by the following time-dependent Hamiltonian:

$$\mathcal{H}(t) = \frac{P^2}{2m} + V(X) \sum_{n=-\infty}^{\infty} \delta(t - n\tau) \qquad (1)$$

The potential depends on time through the periodic δ-function.

Adding a pair of action-angle variables (J, α) corresponding to a harmonic oscillator, we can rewrite the original non-autonomous system as an extended autonomous one. In the extended Hamiltonian, the time in the time-dependent potential is replaced by the new angle variable,

$$H = \mathcal{H}(\alpha) + J \qquad (2)$$

As a result, the total energy is conserved.

We interpret the virtual harmonic oscillator as a heatbath. When the system is chaotic, the particle moves irregularly. This irregular motion causes a random energy transfer between the particle and the heatbath. Although our heatbath includes only a few degrees of freedom, the detailed dynamics inside of the heatbath are not important. The random energy transfer between the system and the heatbath is what is essential for thermodynamics.

As we will discuss later, the walls play an important role in an isothermal operation. At the walls the particle collides elastically such that P changes to $-P$. To avoid the frequent flips of the sign at each collision, we choose a symmetric initial distribution function, $\rho_0(P, X) = \rho_0(-P, -X)$. Suppose we choose one symmetric pair of particles, (P, X) and $(-P, -X)$. When one particle changes the sign of its momentum, the other changes at the same time. We can reinterpret that as one particle keeping its momentum but jumping instantaneously from the right-hand wall to the left-hand wall. That is equivalent to the well-known periodic boundary condition. The periodic boundary condition means that $X_{n+1} + L/2$ should be taken mod L.

We choose a periodic boundary condition with respect to momentum at $P = \pm \Lambda/2$, when the particle is kicked.

The Poincaré map of our system is

$$P_{n+1} = P_n - V'(X_n) \qquad (3)$$
$$X_{n+1} = X_n + P_{n+1}\tau/m \qquad (4)$$
$$J_{n+1} = J_n - \frac{P_{n+1}^2}{2m} + \frac{P_n^2}{2m} \qquad (5)$$
$$\alpha_{n+1} = \alpha_n + 1 \qquad (6)$$

where the subscript n means the value just before the nth kick.

We insist that $P_{n+1} + \Lambda/2$ should be taken mod Λ just before a kick. Then, the kinetic energy reduces less than $(\Lambda/2)^2/2m$ before the kick. This means heatbath absorbs energy. It is one of the important roles of our heatbath. Between two consecutive kicks, we do not insist on any boundary condition with respect to momentum, because there is no connection with the heatbath.

We choose $V'(X) = -\Lambda f(2X/L)/2$ and $\tau = m\Lambda/L$. Then the map of the particle can be scaled as the generalized standard map,

$$p_{n+1} = p_n + f(x_n) \qquad (7)$$

$$x_{n+1} = x_n + p_{n+1} \qquad (8)$$

We have introduced scaled variables defined as $p_n = 2P_n/\Lambda$ and $x_n = 2X_n/L$. The boundary conditions imply that both $p_n + 1/2$ and $x_n + 1/2$ should be taken mod 1. Although we will consider varying the momentum scale Λ or system size L as external operation, the map of the scaled variables is kept the same.

For $V(X) = -\Lambda X^2/(2L)$ so $f(x) = x$, the map becomes a typical ideal chaotic map, the cat map. The phase space is uniformly chaotic. The natural invariant measure is uniform for unit square, $-1/2 < x_n < 1/2$ and $-1/2 < p_n < 1/2$. Because of the mixing property, the expectation value of the energy function approaches an equilibrium value,

$$\Theta = \frac{1}{L}\int_{-L/2}^{L/2} dX \, \frac{1}{\Lambda}\int_{-\Lambda/2}^{\Lambda/2} dP \, \frac{P^2}{2m} = \frac{1}{3}\frac{(\Lambda/2)^2}{2m} \qquad (9)$$

We define the temperature, Θ, as the average kinetic energy of the particle.

B. External Operations

We consider two typical external operations, one isothermal and the other at fixed volume.

In the isothermal process, we vary slowly the system size L while we keep the momentum scale Λ and the mass m so that the temperature remains at Θ. For the isothermal expansion/contraction, we move the two walls to outer/inner symmetrically. It makes negative/positive work by external operation. Because the average energy of the particle is kept the same, the energy of the heatbath decreases/increases.

In the fixed-volume process, the total energy is kept the same, because there is no external work. For the fixed-volume heating/cooling, the temperature increases/decreases and the energy transfers from/to the heatbath to/from the particle. By combining the isothermal contraction and expansion and also the fixed-volume heating and cooling, we can make a Carnot cycle in this simple chaotic system.

We vary the system size/the momentum scale in the isothermal/fixed volume operation. As we mentioned above, the map of the scaled variables is kept the same under these operations. We keep the basic dynamical properties the same under these operations. As an example, we will look in detail at the isothermal expansion. We consider the operation after the nth map. Before the operation, the particle has the scaled momentum and position (p_n, x_n). The momentum and position are scaled back as $(P_n, X_n) = (p_n \Lambda/2, x_n L_n/2)$ where L_n is the system size after the nth map.

We slowly vary the size of the system from L_n to $L_{n+1} = L_n + \delta L$ within a time interval δt. During the operation, the particle is moving freely between two moving walls and may collide elastically with the right or left wall. The two walls move out symmetrically. The velocity of the right-hand wall is given as $V_w = \delta L/2\delta t > 0$.

After the operation, the particle moves to $(\tilde{P}_n, \tilde{X}_n) = (P_n, X_n + P_n \delta t/m)$ without collision. If $X_n + P_n \delta t/m > L_{n+1}/2$, the particle elastically collides with the right-hand wall, and then the particle changes both momentum and position as $(\tilde{P}_n, \tilde{X}_n) = (-P_n + 2mV_w, L_{n+1} - X_n - P_n \delta t/m)$. On the other hand, if $X_n + P_n \delta t/m < -L_{n+1}/2$, then $(\tilde{P}_n, \tilde{X}_n) = (-P_n - 2mV_w, -L_{n+1} - X_n - P_n \delta t/m)$. After one collision, the particle loses kinetic energy as $\delta E(P_n) = -2V_w(|P_n| - mV_w)$. After the operation, the new scaled variables are given as $(\tilde{p}_n, \tilde{x}_n) = (2\tilde{P}_n/\Lambda, 2\tilde{X}_n/L_{n+1})$. We assume that the time interval of the operation is so short that there are no more than two collisions during one operation.

We consider the quasi-static isothermal process. We consider a pair of two consecutive operations. The former is isothermal expansion and the latter is the cat map. We repeat the pair N times. We make the quasi-static isothermal process by taking the limit of N going to infinity while keeping $\Delta L \equiv N\delta L$ constant.

We also keep the operation time δt finite in the limit of N going to infinity. This means that the duration of collision is finite on the analogy of the gas system [5]. During the operation, the system does not approach to equilibrium, because of the free motion. There is a time lag to start approaching to equilibrium. We will discuss the limit of the zero operation time later.

C. Time Evolution of Probability Density

Now we consider the time evolution of the symmetric probability density, $\rho_n(p, x)$, in the space of the scaled variables,

$$\rho_n(p, x) = U_{\text{map}} U_{\text{op}}(L_{n-1}) \cdots U_{\text{map}} U_{\text{op}}(L_0) \rho_0(p, x) \qquad (10)$$

where U_{map} is the Frobenius–Perron operator of the generalized standard map,

$$U_{\text{map}} \rho(p, x) = \rho(p - f(x - p), x - p) \qquad (11)$$

and $U_{op}(L_n)$ governs the time evolution of the probability density during the nth isothermal operation. If $4mV_w/\Lambda - 1 < p < 2mV(x-1)L_{n+1}/(2\Lambda\delta L) + 2mV_w/\Lambda$, the particle collides at right-hand wall, and then

$$U_{op}\rho(p,x) = \frac{L_{n+1}}{L_n}\rho\left(-p + \frac{4mV_m}{\Lambda}, \frac{L_{n+1}}{L_n}(x-1) + \frac{\Lambda\delta t}{mL_n}\left(p - \frac{4mV_m}{\Lambda}\right)\right) \quad (12)$$

If $1 - 4mV_w/\Lambda > p > 2mV(x+1)L_{n+1}/(2\Lambda\delta L) - 2mV_w/\Lambda$, the particle collides at left-handed wall, and then

$$U_{op}\rho(p,x) = \frac{L_{n+1}}{L_n}\rho\left(-p - \frac{4mV_m}{\Lambda}, \frac{L_{n+1}}{L_n}(x+1) + \frac{\Lambda\delta t}{mL_n}\left(p + \frac{4mV_m}{\Lambda}\right)\right) \quad (13)$$

If $2mV(x+1)L_{n+1}/(2\Lambda\delta L) - 2mV_w/ > 2mV(x-1)L_{n+1}/(2\Lambda\delta L) + 2mV_w/\Lambda$, the particle does not collide, and then

$$U_{op}\rho(p,x) = \frac{L_{n+1}}{L_n}\rho\left(p, \frac{L_{n+1}}{L_n}x + \frac{\Lambda\delta t}{mL_n}p\right) \quad (14)$$

If $p < 4mV_w/\Lambda - 1$ and $p < 2mV(x-1)L_{n+1}/(2\Lambda\delta L) + 2mV_w/\Lambda$ or if $p > 1 - 4mV_w/\Lambda$ and $p > 2mV(x+1)L_{n+1}/(2\Lambda\delta L) - 2mV_w/\Lambda$, two new spaces appear because of the expansion, and

$$U_{op}\rho(p,x) = 0 \quad (15)$$

We choose the uniform invariant density as the initial symmetric density, $\rho_0 = 1$.

D. Work

We will estimate the work in the isothermal process. A quasi-static isothermal process is realized in the limit of $N \to \infty$ while keeping ΔL and NV_w constant. In the case of strong chaos, the probability density is almost equilibrium, $\rho_n(p,x) \sim 1$, in the expectation of a smooth observable.

The total work during the isothermal operations is given as the sum of the energy loss of the particle in the N isothermal operations,

$$\Delta W = \sum_{n=0}^{N-1}\langle\delta E_n\rangle_n \quad (16)$$

where

$$\langle \cdot \rangle_n = \int_{-1/2}^{1/2} dp \int_{-1/2}^{1/2} dx \cdot \rho_n(p,x) \tag{17}$$

$$\delta E_n(p,x) = -V_w(|p|\Lambda - 2mV_w)\left[\theta\left(x + p\frac{L_\Lambda}{L_n} - \frac{L_{n+1}}{L_n}\right) + \theta\left(-x - p\frac{L_\Lambda}{L_n} - \frac{L_{n+1}}{L_n}\right)\right] \tag{18}$$

where $L_\Lambda = \Lambda \delta t/m$.

For large N, we treat $\delta U_{\text{op}} \equiv U_{\text{op}} - 1$ as a small perturbation in the expectation for a smooth observable. Using the formal expansion

$$\rho_n(p,x) = \rho_0(p,x) + \sum_{i=0}^{n-1} U_{\text{map}}^{n-i} \delta U_{\text{op}}(L_i)\rho_0(p,x) + \cdots \tag{19}$$

we obtain

$$\Delta W = \sum_{n=0}^{N-1} \langle \delta E_n \rangle_0 + \sum_{n=1}^{N-1}\sum_{i=0}^{n-1} \langle \delta E_n U_{\text{map}}^{n-i} \delta U_{\text{op}}(L_i)\rangle_0 + \cdots \tag{20}$$

After an easy integration, we obtain the first term as

$$\Delta W^{(0)} \equiv \sum_{n=0}^{N-1} \langle \delta E_n \rangle_0 = -\frac{2}{3\Lambda m}(\Lambda/2 - mV_w)^3 \delta L \sum_{n=0}^{N-1} \frac{1}{L_n} \tag{21}$$

The last sum is related to the Boltzmann entropy for large N,

$$\delta L \sum_{n=0}^{N-1} \frac{1}{L_n} = \log\left(\frac{L_N}{L_0}\right) + \frac{\Delta L^2}{2N(L_N)L_0} + O\left(\frac{1}{N^2}\right) \tag{22}$$

The phase volume of the equilibrium density for the initial and final states is $A_i = \Lambda L_0$ and $A_f = \Lambda L_N$, respectively. The first term of the average is the change of the Boltzmann entropy, $\Delta S = \log(A_f) - \log(A_i)$.

$$\Delta W^{(0)} = -2\Theta\left[\Delta S - \frac{6mV_w}{\Lambda}\Delta S + \frac{\Delta L^2}{2NL_NL_0}\right] + O\left(\frac{1}{N^2}\right) \tag{23}$$

We can understand why the Boltzman entropy appears in the work $\Delta W^{(0)}$. Since the phase volume increases during the isothermal expansion, the empty space appears after the operation

$$\chi_n(p,x) \equiv \left[1 - \frac{L_n}{L_{N+1}} U_{\text{op}}(L_i) 1\right] \quad (24)$$

After the mixing, the empty space is filled by the probability density and has the average energy $\Theta \langle \chi_n \rangle_0$, where $\langle \chi_n \rangle_0$ is the area of the empty space. Therefore, the loss of the kinetic energy of the particle can be related to the area of the empty space:

$$\langle \delta E_n \rangle_0 = \left\langle \left(\Theta - \frac{p^2 \Lambda^2}{8m}\right) \chi_n \right\rangle_0 \sim \left(\Theta - \frac{\Lambda^2}{8m}\right) \langle \chi_n \rangle_0 \quad (25)$$

Here we define

$$\delta \tilde{E}_n(p,x) \equiv \left(\Theta - \frac{p^2 \Lambda^2}{8m}\right) \chi_n(p,x) \quad (26)$$

This function will play an important role in an inequality in Section II.E.

From our definition, the temperature is same as the internal energy. In the limit $N \to \infty$, the work is given as change of the free energy, $F \equiv \Theta - 2\Theta S$,

$$\lim_{N \to \infty} \Delta W^{(0)} = \Delta F \quad (27)$$

Note that there are $O(1/N)$ correction terms in Eq. (23). As we mentioned before, these contributions are caused by the free motion of the particle during the operations.

E. "The Second Law" of Thermodynamics

Now we consider the excess heat production in Eq. (20),

$$\Delta W^{(1)} \equiv \sum_{n=1}^{N-1} \sum_{i=0}^{n-1} \langle \delta E_n U_{\text{map}}^{n-i} \delta U_{\text{op}}(L_i) \rangle_0 \quad (28)$$

In the case of the Langevin equation [7], the excess heat production satisfies the second law of thermodynamics, $\Delta W^{(1)} > 0$. The Fokker–Planck equation derived from the Langevin equation includes the small perturbation δE_n. This makes a bilinear form with respect to δE_n and guarantees the inequality.

By contrast, the time evolution operator of the probability density includes δU_{op} instead of δE_n in our system. Because $\delta E_n(p,x)$ is a different function from $U_{\text{op}}(L_i)1$, we cannot expect an inequality for the excess heat production.

If $\delta E_n(p,x)$ were $\delta \tilde{E}_n(p,x)$ defined in Eq. (26), we would have an inequality. Replacing $\delta \tilde{E}_n(p,x)$ by $\delta E_n(p,x)$, we introduce a modified work,

$$\Delta \tilde{W} = \sum_{n=1}^{N-1} \langle \delta \tilde{E}_n \rangle_n \tag{29}$$

As we have shown in Eq. (25), $\langle \delta \tilde{E}_n \rangle_0 = \langle \delta E_n \rangle_0$, the modified work is given as the change of the free energy in the quasi-static limit, $\Delta \tilde{W}^{(0)} = \Delta W^{(0)} \to \Delta F$.

Furthermore, we can derive the following inequality by neglecting $O(1/N^2)$

$$\Delta \tilde{W} - \Delta W(0) = \sum_{n=1}^{N-1} \sum_{i=0}^{n-1} \langle \delta \tilde{E}_n U_{\text{map}}^{n-i} \delta U_{\text{op}}(L_i) \rangle_0$$

$$= \sum_{n=1}^{N-1} \sum_{i=0}^{n-1} \left(\frac{p^2 \Lambda^2}{8m} - \Theta \right)^{-1} \langle \delta \tilde{E}_n U_{\text{map}}^{n-i} \delta \tilde{E}_i \rangle_0 \tag{30}$$

$$= \frac{1}{4\Theta} \left\langle \left| \sum_{n=1}^{N} \delta \tilde{\mathscr{E}}_n \right|^2 \right\rangle_0 \geq 0 \tag{31}$$

where

$$\delta \tilde{\mathscr{E}}_n \equiv U_{\text{map}}^{-n} \delta \tilde{E}_n U_{\text{map}}^n \tag{32}$$

The modified inequality may not be related to the nonexistence of perpetual motion of the second kind. Because we are considering the small dynamical system, the fluctuations of the momentum and position are the same order of the system size. This means that we can be the Maxwell demon. If there exists an inequality in the scale of the Maxwell demon, the inequality cannot be equivalent to the nonexistence of perpetual motion of the second kind.

We roughly estimate $\Delta \tilde{W}^{(1)}$. For convenience, we assume only one dominant decay mode in the correlation function in Eq. (30). Then, we rewrite

$$\Delta \tilde{W}^{(1)} \sim \sum_{n=1}^{N-1} \sum_{i=0}^{n-1} \frac{1}{4\Theta} \langle \delta \tilde{E}_n | \gamma_R \rangle e^{-\gamma(n-i)} \langle \gamma_L | \delta \tilde{E}_i \rangle_0$$

$$\sim \frac{1}{4\Theta} \langle \delta \tilde{E}_{N/2} | \gamma_R \rangle \frac{N}{1 - e^{-\gamma}} \langle \gamma_L | \delta \tilde{E}_{N/2} \rangle_0 \tag{33}$$

where $\gamma_R(p,x)/\gamma_L(p,x)$ is right/left eigenfunctional of the Frobenius–Perron operator.

For $\gamma \ll 1$, the modified excess heat production is proportional to the lifetime $1/\gamma$ as

$$\Delta \tilde{W}^{(1)} \sim \frac{1}{N\gamma} \tag{34}$$

From the perturbation theory, the modified excess heat production $\Delta \tilde{W}^{(1)}$ comes from deviation from the equilibrium state. The lifetime of the excited state is important factor to determine it. The decay constant is similar to the Kolmogorov–Sinai entropy that is equivalent to the positive Lyapunov exponent in a small Hamiltonian chaotic system. It is important that the modified excess heat production is greater for the smaller Lyapunov exponent. This means that it increases in the region of weak chaos.

F. Recover of the Second Law in Large System

Now we will discuss the limit of zero operation time $\delta t \to 0$. We consider the scaling $\delta t \to \epsilon \delta t$, where ϵ is an infinitesimal positive number. For this scaling, $\delta L \to \epsilon \delta L$ and $N \to N/\epsilon$. Therefore, $O(1/N)$ contributions are suppressed as $O(\epsilon/N)$. In order to avoid this trivial result, we have to scale the decay constant $\gamma \to \epsilon \gamma$ in Eq. (34).

We naturally obtain the scaling of the decay constant by multiplying system size as $L \to [1/\sqrt{\epsilon}]_G L$, where the Gauss symbol $[R]_G$ means the greatest integer less than R. By multiplying system size as $N_{\text{cell}} L$, the system becomes multi-cat map, called the sawteeth map. The extended map has diffusion modes. The least decay constant has expected scaling as $\gamma = D\pi^2/N_{\text{cell}}^2 \sim D\pi^2/\epsilon$, where the diffusion coefficient is $D = 1/24$.

By extending the small system to a large one, we keep the nontrivial contribution $\Delta \tilde{W}^{(1)}$. For the large system, does the ordinary second law recover? The answer is yes. Because the diffusion modes are associated with intercell motion of the particle, details of intracell motion do not affect the diffusion modes. The difference between $\delta \tilde{E}_n(p,x)$ and $\delta E_n(p,x)$ vanishes in the limit of zero operation time and infinite system size. Therefore, the inequality means the nonexistence of perpetual motion of the second kind.

The excess heat production is proportional to the variance of the total energy distribution in Eq. (31). For the Gaussian distribution, we can interpret the variance as information $\langle \Delta E^2 \rangle = -\langle \log e^{-\Delta E^2} \rangle$. It is quite important that the excess heat production can be interpret as scaled loss of information of the total energy distribution. We will report the details in our forthcoming paper [8].

III. CONCLUSIONS AND REMARKS

In this chapter we constructed thermodynamics of the simple Hamiltonian chaotic system with a Poincaré map that can be scaled as the cat map. In the cat map, the ordinary thermodynamics almost worked. The work became a change of free energy in a quasi-static isothermal process. An important difference appeared in the second law of thermodynamics. The inequality lost the direct connection to the nonexistence of perpetual motion of the second kind.

In our treatment we chose the finite operation time, which is corresponding the finite duration of collision. When we took the limit of the zero operation time, the contributions from the decay modes associate with Kolmogorov–Sinai entropy vanished. Nontrivial contribution can survive for the multiple system. In the large system, the second law recover and the excess heat production became the loss of information of the total energy distribution.

We consider the thermodynamics as a general theory that describes how a system responds to slow external operation. There is no reason to restrict it in the region of the strong chaos. We will extend it into the region of the weak chaos, in which the correlation decays in power law. As we encountered in the previous arguments, new rules may lose the connection of the nonexistence of perpetual motion of the second kind. But the rules can characterize the complicated phase structure in the weak chaos.

The weak chaos is an other possibility to obtain nontrivial contributions in the limit of the zero operation even for small system. The long time correlation makes nontrivial contribution to the excess heat production. Actually, we have found anomalous behavior of the excess heat production in the standard map. We will report it in our forthcoming paper [9].

Acknowledgments

I thank Prof. I. Prigogine and Prof. K. Kitahara for their support and encouragement. I acknowledge comments on our manuscript from D. Driebe, K. Nelson, C.-B. Li, and Y. Ohtaki. This work was supported by the Engineering Research Program of the Office of Basic Energy Sciences at the U.S. Department of Energy, Grant DE-FG03-94ER14465, the Robert A. Welch Foundation Grant F-0365, the European Commission Contract ESPRT Project 28890 (CTIAC), Grant in Aid for Scientific Research from the Japanese Ministry of Education, Science and Culture (Hakken-Kagaku) and SVBL program in Ibaraki University.

References

1. P. Gaspard, *Chaos, Scattering, and Statistical Mechanics*, Cambridge University Press, Cambridge, 1998.
2. D. Driebe, *Fully Chaotic Maps and Broken Time Symmetry*, Kluwer Academic Publishers, Dordrecht, 1999.
3. J. R. Dorfman, *An Introduction to Chaos in Nonequilibrium Statistical Mechanics*, Cambridge University Press, Cambridge, 1999.
4. V. Berdichevsky, *Thermodynamics of Chaos and Order*, Pitman Monographs and Surveys in Pure and Applied Mathematics, Vol. 90, Addison-Wesley Longman, Reading, MA, 1997.
5. R. Balescu, *Statistical Dynamics: Matter out of Equilibrium*, Imperial College Press, London, 1997.
6. D. Kondepudi and I. Prigogine, *Modern Thermodynamics: From Heat Engines to Dissipative Structure*, John Wiley & Sons, New York, 1998.
7. K. Sekimoto, *J. Phys. Soc. Japan* **66**, 1234 (1997); K. Sekimoto and S. Sasa, *J. Phys. Soc. Japan* **66**, 3326 (1997); S. Sasa and T. Kanatsu, *Phys. Rev. Lett.* **82**, 912 (1999).
8. H. H. Hasegawa, C.-B. Li, and Y. Ohtaki, submitted to *Phys. Lett.*
9. Y. Ohtaki and H. H. Hasegawa, in preparation.

HARMONIC ANALYSIS OF UNSTABLE SYSTEMS

I. ANTONIOU

International Solvay Institutes for Physics and Chemistry, Free University of Brussels, Brussels, Belgium; and Theoretische Natuurkunde, Free University of Brussels, Brussels, Belgium

Z. SUCHANECKI

International Solvay Institutes for Physics and Chemistry, Free University of Brussels, Brussels, Belgium; Theoretische Natuurkunde, Free University of Brussels, Brussels, Belgium; and Institute of Mathematics, University of Opole, ul. Oleska 48, Opole, Poland

CONTENTS

I. Introduction
II. Generalized Spectral Decompositions and Probabilistic Extension of the Dynamics of Unstable Systems
    A. The Renyi Maps
    B. The Tent Maps
    C. The Logistic Map
    D. The Baker's Transformations
III. Time Operator and Shift Representation of the Evolution
    A. Time Operator for the Renyi Map
    B. Time Operator for the Cusp Map
Acknowledgments
References

Dynamical Systems and Irreversibility: A Special Volume of Advances in Chemical Physics, Volume 122, Edited by Ioannis Antoniou. Series Editors I. Prigogine and Stuart A. Rice. ISBN 0-471-22291-7. © 2002 John Wiley & Sons, Inc.

I. INTRODUCTION

The idea of using operator theory for the study of dynamical systems is due to Koopman and was extensively used thereafter in statistical mechanics and ergodic theory because the dynamical properties are reflected in the spectrum of the density evolution operators. The evolution of a dynamical system on the phase space Ω is described by the evolution group $\{S_t\}$, with t real for flows and integer for cascades. The phase space is endowed with a σ-algebra S of measurable subsets of Ω and a probability measure μ. Usually μ is the equilibrium measure; that is, S_t preserve the measure μ:

$$\mu(S_t^{-1}(A)) = \mu(A) \qquad \text{for all } A \in S$$

The evolution of dynamical systems can be classified according to different ergodic properties which correspond to various degrees of irregular behavior such as ergodicity, mixing, exact systems, and Kolmogorov systems. The ergodic properties of classic dynamical systems can be conveniently studied in the Hilbert space formulation of dynamics due to Koopman [1,2]. We consider the Hilbert space $L^2 = L^2(\Omega, S, \mu)$ of square integrable functions on Ω. The transformations S_t induce the Koopman evolution operators V_t acting on the functions $f \in L^2$ as follows:

$$V_t f(\omega) = f(S_t \omega), \qquad \omega \in \Omega$$

The adjoint operator

$$U_t = V_t^\dagger$$

is called the Frobenius–Perron operator [2].

As S_t preserve the measure μ, the operators V_t are isometries. If S_t are, in addition, invertible, then V_t are unitary. In the case of flows the self-adjoint generator of V_t is known as the Liouville operator:

$$V_t = e^{iLt}$$

In the case of Hamiltonian flows the Liouville operator is given by the Poisson bracket associated with the Hamiltonian function H:

$$Lf = i\{H, f\}$$

The work of the Brussels group and collaborators over the last 8 years [3–23] has demonstrated that for unstable systems, classic or quantum, there exist spectral decompositions of the evolution in terms of resonances and resonance states which appear as eigenvalues and eigenprojections of the evolution operator.

These new spectral decompositions in terms of resonances include the correlation decay rates, the lifetimes, and the Lyapunov times. In this sense they provide a natural representation for the evolution of unstable dynamical systems. The decompositions acquire meaning in suitable dual pairs of functional spaces beyond the conventional Hilbert space frameworks. For invertible systems the reversible evolution group, once extended to the functional space, splits into two distinct semigroups. Irreversibility emerges therefore naturally as the selection of the semigroup corresponding to the future observations. The resonances are the singularities of the extended resolvent of the evolution operator, while the resonance states are the corresponding Riesz projections computed as the residues of the extended resolvent at the singularities [10,11].

The spectral decomposition of evolution operators employing the methods of functional analysis is a new tool for the probabilistic analysis of dynamical systems. The canonically conjugate approach in terms of the time operator has also been developed in Brussels during the last 20 years [3,24] and proved recently to be even more powerful tool for the probabilistic analysis of complex systems.

Time operators in dynamical systems were introduced in 1978 by B. Misra and I. Prigogine [3,24] as self-adjoint operators T satisfying the canonical commutation relation with the Koopman operators V_t:

$$TV_t = V_t T + tV_t \qquad (1)$$

Dynamical systems with time operators are intrinsically irreversible because they admit Lyapunov operators as operator functions of the time operator $M = M(T)$, where M is any decreasing function [3,24]. The typical example is Kolmogorov systems where the spectral projections of the time operator are averaging projections on the refining K-partitions.

The idea behind the analysis of the evolution semigroup $\{V_t\}$ on a Hilbert space $\mathscr{H}$ through the time operator T is to decompose T in terms of a complete orthonormal set of eigenvectors $\varphi_{n,\alpha}$, known also as age eigenfunctions:

$$T\varphi_{n,\alpha} = n\varphi_{n,\alpha}$$

$$T = \sum_n n \sum_\alpha |\varphi_{n,\alpha}\rangle\langle\varphi_{n,\alpha}|$$

$$\sum_{n,\alpha} |\varphi_{n,\alpha}\rangle\langle\varphi_{n,\alpha}| = I$$

The Koopman operator V_t shifts the eigenvectors $\varphi_{n,\alpha}$:

$$V_t \varphi_{n,\alpha} = \varphi_{n+t,\alpha}$$

The index n labels the age and α the multiplicity of the spectrum of the time operator. As a result the eigenvectors $\varphi_{n,\alpha}$ of the time operator provide a shift representation of the evolution:

$$f = \sum_{n,\alpha} a_{n,\alpha} \varphi_{n,\alpha} \Rightarrow V_t f = \sum_{n,\alpha} a_{n,\alpha} \varphi_{n+t,\alpha} = \sum_{n,\alpha} a_{n-t,\alpha} \varphi_{n,\alpha}$$

The knowledge of the eigenvectors of T amounts therefore to a probabilistic solution of the prediction problem for the dynamical system described by the semigroup $\{V_t\}$. The spaces $\mathcal{N}_n$ spanned by the eigenvectors $\varphi_{n,\alpha}$ are called age eigenspaces or spaces of innovation at time (stage) n, as they correspond to the new information or detail brought at time n.

The relation between the spectral and shift representation of dynamical systems is like the canonical relation between the position and momentum representations in quantum mechanics. However, the shift representations can be derived for a substantially a wider class of systems than dynamical systems associated with maps. This includes evolution semigroups associated with some classes of stochastic processes such as stationary, Markov or self similar processes. We present below some selected results concerning generalized spectral decompositions and time operator obtained in recent years [6–22, 25–32].

II. GENERALIZED SPECTRAL DECOMPOSITIONS AND PROBABILISTIC EXTENSION OF THE DYNAMICS OF UNSTABLE SYSTEMS

The ergodic properties of classic dynamical systems: ergodicity, mixing, exactness, and Kolmogorov property can be expressed as spectral properties of the corresponding Koopman or Frobenius–Perron evolution operators. The spectral approach allows us to study and classify in a unified way dynamical systems in terms of operator theory and functional analysis.

The drawback of this approach is that the spectra of Frobenius–Perron or Koopman operators of unstable systems considered on the space L^2 of square integrable densities do not contain characteristic time scales of irreversible changes such as decay rates of the correlation functions. In order that isolated point spectra corresponding to the decay rates of the correlation functions emerge, the Frobenius–Perron operator has to be restricted to suitable locally convex subspaces of L^2. For example, for expanding maps, as the differentiability of the domain functions increases, the essential spectrum (i.e., the spectrum excluding isolated point spectra) of the Frobenius–Perron operator decreases from unit disk to a smaller one and isolated point spectra appear in the annulus between the two disks [33,34]. The logarithms of these isolated point eigenvalues of the Frobenius–Perron operator are known as the Pollicot–Ruelle resonances [35,36].

In some particular cases it is possible to restrict the Frobenius–Perron operator to a dense subspace in such a way that its spectrum becomes discrete. In such a case the Hilbert space L^2 can be replaced by the dual pair $(\Phi, \Phi^\times)$, where Φ is an invariant with respect to U locally convex subspace of L^2 and $\Phi^\times$ is its topological dual. Here the Koopman operator V is extended to the space $\Phi^\times$. Moreover, in many cases it is possible to obtain a generalized spectral decomposition of V (and also of U) in the form

$$V = \sum_i z_i |\varphi_i)(F_i| \qquad (2)$$

where z_i are the eigenvalues of U, and $|\varphi_i)$ and $(F_i|$ are a biorthogonal family of the corresponding eigenvectors which are elements of Φ and $\Phi^\times$, respectively. Following Dirac's notation [37] we denote by the bras (| and kets |) the linear and antilinear functionals, respectively. Formula (2) has to be understood as follows:

$$(\phi|Vf) = \sum_i z_i (\phi|\varphi_i)(F_i|f) = (U\phi|F)$$

for any state ϕ and observable f from a suitable dual pair. This procedure is referred to as rigging, and the triple

$$\Phi \subset L^2 \subset \Phi^\times$$

is called a *rigged Hilbert space* (see Ref. 38 for details).

Summarizing for the reader's convenience, a dual pair $(\Phi, \Phi^\times)$ of linear topological spaces constitutes a rigged Hilbert space for the linear endomorphism V of the Hilbert space $\mathcal{H}$ if the following conditions are satisfied:

1. Φ is a dense subspace of $\mathcal{H}$.
2. Φ is complete and its topology is stronger than the one induced by $\mathcal{H}$.
3. Φ is stable with respect to the adjoint $V^\dagger$ of V, i.e. $V^\dagger \Phi \subset \Phi$.
4. The adjoint $V^\dagger$ is continuous on Φ.

The extension V_{ext} of V to the dual $\Phi^\times$ of Φ is then defined in the standard way as follows:

$$(\phi|V_{\text{ext}}f) = (V^\dagger \phi|f)$$

for every $\phi \in \Phi$. The choice of the test function space Φ depends on the specific operator V and on the physically relevant questions to be asked about the system.

Let us present now some of the generalized spectral decompositions of the Frobenius–Perron operators.

A. The Renyi Maps

The β-adic Renyi map S on the interval $[0,1)$ is the multiplication, modulo 1, by the integer $\beta \geq 2$:

$$S: [0,1) \to [0,1) \qquad x \mapsto Sx = \beta x \pmod{1}$$

The Koopman operator admits the following generalized spectral decomposition [8,9]:

$$V = \sum_{n=0}^{\infty} \frac{1}{\beta^n} |\tilde{B}_n)(B_n| \tag{3}$$

where $B_n(x)$ is the n-degree Bernoulli polynomial defined by the generating function:

$$\frac{ze^{zx}}{e^z - 1} = \sum_{n=0}^{\infty} \frac{B_n(x)}{n!} z^n \tag{4}$$

and

$$|\tilde{B}_n) = \begin{cases} |1), & n = 0 \\ |\frac{(-1)^{(n-1)}}{n!} \{\delta^{(n-1)}(x-1) - \delta^{(n-1)}(x)\}) & n = 1, 2, \ldots \end{cases}$$

Formula (3) defines a spectral decomposition for the Koopman and Frobenius–Perron operators in the following sense:

$$(\rho|Vf) = \sum_{n=0}^{\infty} \frac{1}{\beta^n} (\rho|\tilde{B}_n)(B_n|f)$$

for any density function ρ and observable f in the appropriate pair $(\Phi, \Phi^\times)$.

The generalized spectral decomposition of the Renyi map has no meaning in the Hilbert space L^2. The derivatives $\delta^{(n)}(x)$ of Dirac's delta function that appear as right eigenvectors of the Koopman operator V of the Renyi map are outside L^2. In order to give meaning to these formal eigenvectors the Koopman operator has to be extended to a suitable rigged Hilbert space. In the case of the Renyi map, various riggings exist [19]. For example, we can consider the restrictions of the Frobenius–Perron operator to a series of test function spaces such as the space $\mathscr{P}$ of polynomials in one variable, the Banach space $\mathscr{E}_c$ of entire functions of exponential type c, and others [19]. Among them we looking for such space that gives the tightest rigging i.e. such for which the test function space is the (set-theoretically) largest possible within a chosen family of test function spaces, such that the physically relevant spectral decomposition is meaningful. In the case of the Renyi map the tight rigging is realized in a natural way by the space $\tilde{\mathscr{E}}_{2\pi} \equiv \bigcup_{c<2\pi} \mathscr{E}_c$, where $\mathscr{E}_c$ is the space of entire functions of exponential type less than c, with the inductive limit topology (see Ref. 19 for details).

B. The Tent Maps

The family of tent maps is defined by

$$T_m : [0, 1) \to [0, 1)$$

$$T_m = \begin{cases} m\left(x - \frac{2n}{m}\right) & \text{for } x \in \left[\frac{2n}{m}, \frac{2n+1}{m}\right) \\ m\left(\frac{2n+2}{m} - x\right) & \text{for } x \in \left[\frac{2n+1}{m}, \frac{2n+2}{m}\right) \end{cases}$$

where $m = 2, 3, \ldots, n = 0, 1, \ldots, \left[\frac{m-1}{2}\right]$ and $[y]$ denotes the integer part of real number y. The case $m = 2$ corresponds to the well known tent map. The absolutely continuous invariant measure is the Lebesgue measure dx for all maps T_m. The Frobenius–Perron operator for the tent maps has the form

$$U_T \rho(x) = \frac{1}{m} \left\{ \sum_{n=0}^{\left[\frac{m-1}{2}\right]} \rho\left(\frac{2n+x}{m}\right) + \sum_{n=0}^{\left[\frac{m-2}{2}\right]} \rho\left(\frac{2n+2-x}{m}\right) \right\}$$

The spectrum consists of the eigenvalues [21]

$$z_i = \frac{1}{m^{i+1}} \left\{ \left[\frac{m-1}{2}\right] + 1 + (-1)^i \left(\left[\frac{m-2}{2}\right] + 1 \right) \right\}$$

which means that for the even tent maps, $m = 2, 4, \ldots$, the eigenvalues are

$$z_i = \begin{cases} \frac{1}{m^i}, & i \text{ even} \\ 0, & i \text{ odd} \end{cases} \tag{5}$$

and for the odd tent maps, $m = 3, 5, \ldots$, the eigenvalues are

$$z_i = \begin{cases} \frac{1}{m^i}, & i \text{ even} \\ \frac{1}{m^{i+1}}, & i \text{ odd} \end{cases} \tag{6}$$

The eigenvectors of the tent maps can be expressed in terms of the Bernoulli and Euler polynomials. For even tent maps $m = 2, 4, \ldots$

$$f_i(x) = \begin{cases} B_i\left(\frac{x}{2}\right), & i = 0, 2, 4, \ldots \\ \frac{i+1}{2m^{i+1}} E_i(x), & i = 1, 3, 5, \ldots \end{cases}$$

For odd tent maps $m = 3, 5, \ldots$

$$f_i(x) = \begin{cases} 1, & i = 0 \\ E_i(x), & i = 1, 3, 5, \ldots \\ B_i(x) + E_{i-1}(x), & i = 2, 4, \ldots \end{cases}$$

The Bernoulli polynomials are defined by the generating function (4). The Euler polynomials are defined by the generating function

$$\frac{2e^{xt}}{e^t + 1} = \sum_{n=0}^{\infty} E_n(x) \frac{t^n}{n!}, \qquad |t| < \pi$$

The left eigenvectors are given by the following formulas:
For even tent maps $m = 2, 4, \ldots$

$$F_i(x) = \begin{cases} m^i \tilde{B}_i(x), & i = 0, 2, 4, \ldots \\ m^{i+1}\left(\frac{2}{i+1}\tilde{E}_i(x) + \tilde{B}_{i+1}(x)\right), & i = 1, 3, 5, \ldots \end{cases}$$

For odd tent maps $m = 3, 5, \ldots$

$$F_i(x) = \begin{cases} 1, & i = 0 \\ \tilde{E}_i(x), & i = 1, 3, 5, \ldots \\ \tilde{B}_i(x) + \tilde{E}_{i-1}(x), & i = 2, 4, \ldots \end{cases}$$

The expressions $\tilde{B}_i(x)$, $\tilde{E}_i(x)$ are given by the following formulas as in the case of the Renyi maps:

$$\tilde{B}_i(x) = \begin{cases} 1, & i = 0 \\ \frac{(-1)^{i-1}}{i!}\{\delta^{(i-1)}(x-1) - \delta^{(i-1)}(x)\}, & i = 1, 2, \ldots \end{cases}$$

$$\tilde{E}_i(x) = (-1)^i 2(i!)\{\delta^{(i)}(x-1) + \delta^{(i)}(x)\}, \qquad i = 0, 1, \ldots$$

From Eqs. (5) and (6) we observe that the spectrum of the symmetric tent maps $m = 2, 4, \ldots$ does not contain the odd powers of $1/m$. This is a general property of symmetric maps and is due to the fact that the antisymmetric eigenfunctions are in the null space of the Frobenius–Perron operator.

Summarizing, the spectra of the tent maps T_m depend upon m but the eigenvectors depend only on the evenness of m.

C. The Logistic Map

The logistic map in the case of fully developed chaos is defined by

$$S(x) = 4x(1-x) \quad \text{for } x \in [0, 1]$$

The logistic map is a typical example of exact system. The invariant measure for the logistic map is [2] $d\nu(x) = \dfrac{1}{\pi\sqrt{x(1-x)}} dx$.

The spectral decomposition of the logistic map can be obtained from the spectral decomposition of the dyadic tent map through the well-known topological equivalence of these transformations [22]. The transformation $g: [0, 1) \to [0, 1)$ defined by

$$g(x) = \frac{1}{\pi} \arccos(1 - 2x)$$

defines a topological equivalence between the logistic map and the dyadic tent map $T: [0, 1) \to [0, 1)$

$$T(x) = \begin{cases} 2x, & \text{for } x \in [0, \tfrac{1}{2}) \\ 2(1-x), & \text{for } x \in [\tfrac{1}{2}, 1) \end{cases}$$

expressed through the formula

$$S = g^{-1} \circ T \circ g$$

The transformation g transforms the Lebesgue measure, which is the invariant measure of the tent map to the invariant measure of the logistic map.

The transformation G intertwines the Koopman operator V of the logistic map with the Koopman operator V_T of the tent map:

$$V = G V_T G^{-1}$$

The intertwining transformations G and G^{-1}, when suitably extended, map the eigenvectors of V_T onto the eigenvectors of V. Therefore

$$V = \sum_{n=0}^{+\infty} z_n G|\Phi_n)(\varphi_n|G^{-1}$$

$$V = \sum_{n=0}^{+\infty} z_n |F_n)(f_n| \tag{7}$$

with $z_n = \frac{1}{2^{2n}}$, $F_n(x) = 2^{2n}\tilde{B}_{2n}\left(\frac{1}{\pi}\arccos(1-2x)\right)$, $f_n(x) = B_{2n}\left(\frac{1}{2\pi}\arccos(1-2x)\right)$. In formula (7) the bras and kets correspond to the invariant measure of the logistic map. The meaning of the spectral decomposition of the logistic map is inherited from the meaning of the spectral decomposition of the tent map in terms of the dual pair of polynomials; that is it can be understood in the sense

$$(\rho|Vf) = \sum_{n=0}^{+\infty} \frac{1}{2^{2n}} \left(\rho \left| 2^{2n} \tilde{B}_{2n} \left(\frac{1}{\pi} \arccos(1-2x) \right) \right. \right) \left(B_{2n} \left(\frac{1}{2\pi} \arccos(1-2x) \right) \middle| f \right)$$

for any state ρ in the space $\mathscr{P}_{\left(\frac{1}{\pi}\arccos(1-2x) \right)}$, and any observable f in the anti-dual space $^\times\mathscr{P}_{\left(\frac{1}{\pi}\arccos(1-2x) \right)}$.

D. The Baker's Transformations

The β-adic, $\beta = 2, 3, \ldots$, baker's transformation B on the unit square $Y = [0, 1) \times [0, 1)$ is a two-step operation: (1) Squeeze the 1×1 square to a $\beta \times 1/\beta$ rectangle and (2) cut the rectangle into $\beta(1 \times 1/\beta)$-rectangles and pile them up to form another 1×1 square:

$$(x, y) \mapsto B(x, y) = \left(\beta x - r, \frac{y+r}{\beta} \right) \quad \left(\text{for } \frac{r}{\beta} \leq x < \frac{r+1}{\beta}, \ r = 0, \ldots \beta - 1 \right) \tag{8}$$

The invariant measure of the β-adic baker transformation is the Lebesgue measure on the unit square. The Frobenius–Perron and Koopman operators are unitary on the Hilbert space $L^2 = L_x^2 \otimes L_y^2$ of square integrable densities over the unit square and has countably degenerate Lebesgue spectrum on the unit circle plus the simple eigenvalue 1 associated with the equilibrium (as is the case for all Kolmogorov automorphisms).

The Koopman operator V has a spectral decomposition involving Jordan blocks, which was obtained [7,23] using a generalized iterative operator method based on the subdynamics decomposition:

$$V = |F_{00})(f_{00}| + \sum_{\nu=1}^{\infty} \left\{ \sum_{r=0}^{\nu} \frac{1}{\beta^\nu} |F_{\nu,r})(f_{\nu,r}| + \sum_{r=0}^{\nu-1} |F_{\nu,r+1})(f_{\nu,r}| \right\} \tag{9}$$

The vectors $|F_{\nu,r})$ and $(f_{\nu,r}|$ form a Jordan basis:

$$(f_{\nu,r}| V = \begin{cases} \frac{1}{\beta^\nu} \{(f_{\nu,r}| + (f_{\nu,r+1}|\} & (r = 0, \ldots, \nu-1) \\ \frac{1}{\beta^\nu} (f_{\nu,r}| & (r = \nu) \end{cases}$$

$$V |F_{\nu,r}) = \begin{cases} \frac{1}{\beta^\nu} \{ |F_{\nu,r}) + |F_{\nu,r-1}) \} & (r = 1, \ldots, \nu) \\ \frac{1}{\beta^\nu} |F_{\nu,r}) & (r = 0) \end{cases}$$

$$(f_{\nu,r}| F_{\nu',r'}) = \delta_{\nu\nu'} \delta_{rr'}, \quad \sum_{\nu=0}^{\infty} \sum_{r=0}^{\nu} |F_{\nu,r})(f_{\nu,r}| = I$$

While the Koopman operator V is unitary in the Hilbert space L^2 and thus has spectrum on the unit circle $|z| = 1$ in the complex plane, the spectral decomposition (9) includes the numbers $1/\beta^\nu < 1$ which are not in the Hilbert space spectrum. The spectral decomposition (9) also shows that the Frobenius–Perron operator has Jordan-block parts despite the fact that it is diagonalizable in the Hilbert space. As the left and right principal vectors contain generalized functions, the spectral decomposition (9) has no meaning in the Hilbert space L^2.

The principal vectors $f_{\nu,j}$ and $F_{\nu,j}$ are linear functionals over the spaces $L_x^2 \otimes \mathscr{P}_y$ and $\mathscr{P}_x \otimes L_y^2$, respectively [7].

III. TIME OPERATOR AND SHIFT REPRESENTATION OF THE EVOLUTION

Time operators appear naturally for chaotic systems because the associated Koopman operators are shift operators on the Hilbert space $L^2 \ominus [1]$. In particular the Koopman operator is a unilateral shift for exact endomorphisms and a bilateral shift for Kolmogorov automorphisms. It is therefore natural to consider time operators associated with shifts in general.

Let us recall first the basic notions concerning shifts. A linear continuous operator V on a Hilbert space $\mathscr{H}$ is called a *shift* if there exists a sequence $\{\mathscr{N}_n | n = 0, 1, 2, \ldots\}$, enumerated by the set of all integers or by the set of all positive integers, of closed linear subspaces of $\mathscr{H}$ such that

1. $\mathscr{N}_n$ is orthogonal to $\mathscr{N}_m$ if $m \neq n$.
2. $\mathscr{H} = \bigoplus_n \mathscr{N}_n$.
3. For any n, V isometrically maps $\mathscr{N}_n$ onto $\mathscr{N}_{n+1}$.

V is called *unilateral shift* if $n = 0, 1, 2, \ldots$ and *bilateral shift* if $n \in \mathbb{Z}$.

The number $m = \dim \mathscr{N}_0$ is called the *multiplicity* of the (bilateral or unilateral) shift. We shall call $\mathscr{N} \equiv \mathscr{N}_0$ the *generating space* or the *innovation generator* of the shift V and call $\mathscr{N}_n = V^n(\mathscr{N})$ the *space of innovations* at time (stage) n.

Let V be a unilateral shift of multiplicity $m \in \mathscr{N} \cup \{\infty\}$ on a Hilbert space $\mathscr{H}$. An orthonormal basis $\{\chi_n^\alpha | n = 0, 1, \ldots, 1 \leq \alpha < m + 1\}$ is called a *generating basis* for V iff $V\chi_n^\alpha = \chi_{n+1}^\alpha$ for all n, α (or equivalently, $V^\dagger \chi_n^\alpha = \chi_{n-1}^\alpha$ if $n \geq 1$ and $V^\dagger \chi_0^\alpha = 0$).

Let V be a bilateral shift of multiplicity $m \in \mathscr{N} \cup \{\infty\}$ on a Hilbert space $\mathscr{H}$. An orthonormal basis $\{\chi_n^\alpha | n \in \mathbb{Z}, 1 \leq \alpha < m + 1\}$ is called a *generating basis* for V iff $V\chi_n^\alpha = \chi_{n+1}^\alpha$ for all n, α (or equivalently, $V^\dagger \chi_n^\alpha = \chi_{n-1}^\alpha$ for all n, α).

Let W_t, where $t \in [0, \infty)$, $t \in \mathbb{R}$, $t \in \mathbb{Z}$ or $t = 0, 1, 2, \ldots$, be a semigroup flow or cascade of continuous linear operators on a Hilbert space $\mathscr{H}$. A linear operator T on $\mathcal{H}$ is called *a time operator for the semigroup W_t* if T has dense domain D_T, $W_t(D_T) \subset D_T$ for all $t \geq 0$ and $TW_t f = W_t(T+t)f$ for all $f \in D_T$.

A. Time Operator for the Renyi Map [25]

We consider here the 2-adic Renyi map, that is, the transformation $S : [0,1] \to [0,1]$:

$$Sx = 2x \pmod{1}$$

The dynamical system is $([0,1], \{\mathcal{B}, dx, \{S_n\}_{n\in\mathbb{N}}\}$, where $\mathcal{B}$ is the Borel σ-algebra of the interval $[0,1]$, dx symbolizes the Lebesgue measure, and $S_n \stackrel{df}{=} \underbrace{S \circ S \circ \cdots \circ S}_{n-\text{times}}$. The Koopman operator of the Renyi map is

$$Vf(x) = f(Sx) = \begin{cases} f(2x), & \text{for } x \in [0, \tfrac{1}{2}] \\ f(2x-1), & \text{for } x \in [\tfrac{1}{2}, 1] \end{cases}$$

As the first step in the construction of the time operator we define the functions

$$\varphi_1(x) = \mathbb{1}_{[0,1]}(2x) - \mathbb{1}_{[0,1]}(2x-1)$$

and

$$\varphi_{n+1}(x) = V^n \varphi_1(x), \quad \text{for } n = 1, 2, \ldots$$

Then, for a given ordered set of integers $\mathbf{n}$, $\mathbf{n} = \{n_1, \ldots, n_k\}$ $n_1 < \cdots < n_k$, define the function

$$\varphi_{\mathbf{n}}(x) \stackrel{df}{=} \varphi_{n_1}(x) \ldots \varphi_{n_k}(x)$$

The Koopman operator acts as a shift on $\varphi_{\mathbf{n}}$, i.e. $V\varphi_{\mathbf{n}} = \varphi_{\mathbf{n}+1}$, where $\mathbf{n} + 1 = \{n_1 + 1, \ldots, n_m + 1\}$. Moreover, the functions $\varphi_{\mathbf{n}}$, where $\mathbf{n}$ runs over all ordered subsets of $\mathbb{N}$, together with the constant $\equiv 1$, form an orthonormal basis in $L^2_{[0,1)}$.

Theorem. *Each vector $\varphi \in \mathcal{H}$ has the following expansion in the basis $\{\varphi_{\mathbf{n}}\}$:*

$$\varphi = \sum_{\tau=1}^{\infty} \sum_{k=1}^{2^{\tau-1}} a_{\tau,k} \varphi_{\mathbf{n}_\tau^k} \tag{10}$$

where $\mathbf{n}_\tau^k$ denotes the set $\{n_1, \ldots, n_i\}$ with fixed $n_i = \tau$ while k runs through all $2^{\tau-1}$ possible choices of integers $n_1 < \cdots < n_{i-1} < \tau$ (N can be $+\infty$). The operator T acts on φ of the form (10) as follows:

$$T\varphi = \sum_{\tau=1}^{N} \sum_{k=1}^{2^{\tau-1}} \tau a_{\tau,k} \varphi_{\mathbf{n}_\tau^k}$$

Moreover, T satisfies

$$TV^n = V^n T + nV^n, \qquad n = 1, 2, \ldots$$

B. Time Operator for the Cusp Map [29]

In this section we present a Time Operator for the so-called cusp map

$$F: [-1, 1] \to [-1, 1], \qquad \text{where} \quad F(x) = 1 - 2\sqrt{|x|}$$

which is an approximation of the Poincaré section of the Lorenz attractor [39]. The absolutely continuous invariant measure of the cusp map has density

$$\rho(x) = \frac{1-x}{2}$$

The Koopman operator of the cusp map is the operator V acting on $L^2 = L^2([-1, 1], \mu)$, where μ is the measure with density ρ defined by the above formula, is

$$Vf(x) = f(1 - 2\sqrt{|x|})$$

The generating space $\mathcal{N}$ of the Koopman operator regarded as a shift coincides with the space of functions:

$$\{f \in L_2 : f(x) = (1+x)g(x), \qquad \text{where} \quad g \text{ is odd}\}$$

The set of functions

$$\chi^\alpha(x) = \sqrt{\frac{2(1+x)}{1-x}} \sin \pi \alpha x, \qquad \alpha = 1, 2, \ldots$$

is an orthonormal basis in the space $\mathcal{N}$. We have the following theorem.

Theorem. *The set* $\chi_n^\alpha = V^n \chi^\alpha = \chi^\alpha(S^n(x))$, $\alpha = 1, 2, \ldots, n = 0, 1, \ldots$, *is a generating basis for the Koopman operator V of the cusp map acting on $L_2 \ominus \mathbf{1}$.*

The operator with eigenvectors χ_n^α and eigenvalues n is the natural time operator of the shift V:

$$T\chi_n^\alpha = n\chi_n^\alpha; \qquad T = \sum_{n=0}^\infty n \sum_{\alpha=1}^\infty |\chi_n^\alpha\rangle\langle\chi_n^\alpha|$$

Remark. Although the spectral representation is formally equivalent to the shift representation, in the case of the cusp map, the spectral analysis seems to be quite involved and even not possible for analytic functions [30]. However, the shift representation of the cusp map is quite simple and explicit, allowing for probabilistic predictions.

Acknowledgments

We are grateful to Professor Ilya Prigogine for numerous insightful discussions on time operators and subdynamics which were the origins of the harmonic analysis of dynamical systems.

This work received financial support from the Belgian Government through the Interuniversity Attraction Poles.

References

1. B. Koopman, *Proc. Natl. Acad. Sci. USA* **17**, 315–318 (1931).
2. A. Lasota and M. Mackey, *Probabilistic Properties of Deterministic Systems*, Cambridge University Press, U.K., 1985.
3. I. Prigogine, *From Being to Becoming*, Freeman, San Francisco, 1980.
4. T. Petrosky, I. Prigogine, and S. Tasaki, *Physica* **A173**, 175–242 (1991).
5. T. Petrosky and I. Prigogine, *Physica* **A175**, 146–209 (1991).
6. I. Antoniou and I. Prigogine, *Physica* **A192**, 443–464 (1993).
7. I. Antoniou and S. Tasaki, *Physica* **A190**, 303–329 (1992).
8. I. Antoniou and S. Tasaki, *J. Phys.* **A26**, 73–94 (1993).
9. I. Antoniou and S. Tasaki, *Int. J. Quantum Chemistry* **46**, 425–474 (1993).
10. I. Antoniou, L. Dmitrieva, Yu. Kuperin, and Yu. Melnikov, *Comput. Math. Appl.* **34**, 399–425 (1997).
11. O. F. Bandtlow, I. Antoniou, and Z. Suchanecki, *Comput. Math. Appl.* **34**, 95–102 (1997).
12. S. Tasaki, Z. Suchanecki, and I. Antoniou, *Phys. Lett. A* **179**, 103–110 (1993).
13. S. Tasaki, I. Antoniou, and Z. Suchanecki, *Chaos Solitons and Fractals* **4**, 227–254 (1994).
14. S. Tasaki, I. Antoniou, and Z. Suchanecki, *Phys. Lett. A* **179**, 97–102 (1993).
15. I. Antoniou and Z. Suchanecki, *Advances in Chemical Physics*, Vol. 99, I. Prigogine and S. Rice, eds., Wiley, New York, 1997, pp. 299–332.
16. I. Antoniou and Z. Suchanecki, in *Nonlinear, Deformed and Irreversible Systems*, H.-D. Doebner, V. K. Dobrev, and P. Nattermann, eds., World Scientific, Singapore, 1995, pp. 22–52.
17. I. Antoniou and Z. Suchanecki, *Found. Phys.* **24**, 1439–1457 (1994).
18. Z. Suchanecki, I. Antoniou, S. Tasaki, and O. F. Bandtlow, *J. Math. Phys.* **37**, 5837–5847 (1996).
19. I. Antoniou, Yu. Melnikov, S. Shkarin, and Z. Suchanecki, *Chaos Solitons and Fractals* **11**, 393–421 (2000).

20. I. Antoniou, Bi Qiao, and Z. Suchanecki, *Chaos Solitons and Fractals* **8**, 77–90 (1997).
21. I. Antoniou and Bi Qiao, *Phys. Lett. A* **215**, 280–290 (1996).
22. I. Antoniou and Bi Qiao, *Nonlinear World* **4**, 135–143 (1997).
23. D. Driebe, *Fully Chaotic Maps with Broken Time Symmetry*, Kluwer Academic Publishers, Boston 1997.
24. B. Misra, *Proc. Natl. Acad. USA* **75**, 1627–1631 (1978).
25. I. Antoniou and Z. Suchanecki, *Chaos Solitons and Fractals* **11**, 423–435 (2000).
26. I. Antoniou and K. Gustafson, *Chaos, Solitons and Fractals* **11**, 443–452 (2000).
27. I. Antoniou, V. A. Sadovnichii, and S. A. Shkarin, *Physica* **A269**, 299–313 (1999).
28. I. Antoniou, I. Prigogine, V. A. Sadovnichii, and S. A. Shkarin, *Chaos, Solitons and Fractals* **11**, 465–477 (2000).
29. I. Antoniou, *Chaos Solitons and Fractals* **12**, 1619–1627 (2001).
30. I. Antoniou, S. A. Shkarin, and E. Yarevsky, Spectral Properties of the Cusp Families (in preparation).
31. I. Antoniou and Z. Suchanecki, Time operator and approximation (to appear).
32. I. Antoniou and Z. Suchanecki, Time operator and stochastic processes (to appear).
33. M. Pollicott, *Ann. Math.* **131**, 331 (1990).
34. D. Ruelle, *Commun. Math. Phys.* **125**, 239 (1990); *Publ. Math. IHES* **72**, 175 (1989).
35. M. Pollicott, M *Invent. Math.* **81**, 413 (1986); *Invent. Math.* **85**, 147 (1985).
36. D. Ruelle, *Phys. Rev. Lett.* **56**, 405–407 (1986); *J. Stat. Phys.* **44**, 281 (1986).
37. P. A. M. Dirac, *The Principles of Quantum Mechanics*, Clarendon Press, Oxford, 1958.
38. A. Bohm and M. Gadella, *Dirac Kets, Gamow Vectors and Gelfand Triplets*, Springer Lecture Notes on Physics, No. 348, Berlin, 1989.
39. E. Ott, *Rev. Mod. Phys.* **53**, 655 (1981).

PROPERTIES OF PERMANENT AND TRANSIENT CHAOS IN CRITICAL STATES

P. SZÉPFALUSY

Department of Physics of Complex Systems, Eötvös University, Budapest, Hungary; and Research Institute for Solid State Physics and Optics, Budapest, Hungary

CONTENTS

I. Introduction and Summary
Acknowledgment
References

I. INTRODUCTION AND SUMMARY

Fully chaotic dissipative systems are considered and changes of characteristics are followed when a control parameter is varied in such a way that these changes are smooth; in particular the chaotic state is maintained. Besides the permanent chaos, equal weight is given to the transient one, and the features of the two types of chaotic motions are compared.

In the case of the transient chaos, one can specify a region Ω in the phase space, where the chaotic motion takes place [1–3]. After leaving this region the trajectory finally settles to an attractor (which can be simple or a chaotic one). The residence time in Ω depends on the initial condition; its average is given by the inverse of the escape rate, which is related to the conditionally invariant measure, a basic concept in the field of transient chaos [1–3]. This measure is invariant if after each step it is renormalized by a factor fixed by the escape rate. The physically relevant conditionally invariant measures are those which are smooth (absolutely continuous with respect to the Lebesgue measure) along the

---

Dynamical Systems and Irreversibility: A Special Volume of Advances in Chemical Physics, Volume 122, Edited by Ioannis Antoniou. Series Editors I. Prigogine and Stuart A. Rice.
ISBN 0-471-22291-7. © 2002 John Wiley & Sons, Inc.

expanding directions, and in the following it will be always assumed that this requirement is fulfilled. (Such a measure can be called Sinai–Ruelle–Bowen (SRB) type of conditionally invariant measure in analogy with the SRB measure for permanent chaos [4].) The maximal invariant set S in Ω, which is a fractal with rich multifractal properties [2], is called the repeller. Note that the time spent in Ω is longer if the initial point is closer to S. Consequently, the transiently chaotic motion can be arbitrarily long, which makes this type of chaos also experimentally relevant. The natural measure on the repeller is obtained by restricting the conditionally invariant measure to the invariant set S [5].

Scenarios and properties are demonstrated by using a family of one-dimensional maps with one increasing and one decreasing monotonic branch. They act on the two preimages I_0 and I_1 of the interval I chosen as [0, 1]. In the case of permanent chaos $I = I_0 \cup I_1$, while in the case of transient chaos the trajectory is regarded as escaped, if it gets outside the $I_0 \cup I_1$ interval. The maps are complete in the sense of symbolic dynamics based on a bipartition with elements I_0, I_1.

For maps close to the piecewise linear one, perturbation expansion is applied to calculate the spectrum of the Frobenius–Perron operator and the entropies [2, 6,7]. By changing further the control parameter, one can arrive at the weakly intermittent state in the case of permanent chaos. Here the Kolmogorov–Sinai entropy is still positive (despite of the fact that the fixed point at the origin becomes marginally unstable, the slope there is unity); the Rényi entropies for $q > 1$ are, however, zero [2,8]. (Weak intermittent states are, of course, not restricted to $1D$ maps. One can identify such a state also in the Lorenz model [9].)

The analogous state of transient chaos is achieved when the logarithm of the slope of the map at the origin becomes equal to the escape rate. When approaching this state the conditionally invariant measure remains smooth (like the invariant measure in the weakly intermittent state), but the natural measure highly degenerates; it is entirely concentrated in the fixed point at the origin, which remains, however, now a strictly unstable fixed point. As a consequence, not only the $q > 1$ Rényi entropies, but also the Kolmogorov–Sinai entropy, become zero [10–12]. One has to emphasize that the repeller preserves its fractal nature.

The states obtained this way representing border states of chaos are called critical in the permanent and in the transient chaos as well, since phase transition-like phenomena can be associated with them within the framework of the thermodynamic formalism of chaotic systems [2,8,10,11]. The corresponding value of the control parameter is referred to as its critical value. It is worth noting that while the Kolmogorov–Sinai entropy (the measure of chaoticity) is decreasing when approaching toward the critical state, the complexity is increasing [13].

Finally, a further essential difference between the two types of chaos in their critical states has to be stressed. Namely, in the transient case, when the control parameter has arrived at its critical value, a second important conditional invariant measure emerges, whose properties are very different from the first one discussed above [12]. The two conditionally invariant measures have large basins of attraction, a fact that makes both of them physically relevant. The density of the second conditionally invariant measure is zero at the left endpoint of the interval I [11,12]. Its basin of attraction is constituted by functions also sharing this property. In particular, an initial distribution restricted to the inside of the interval I will approach the second conditionally invariant measure. The basin of attraction of the density of the first one is formed by functions having finite values at the left endpoint of the interval I. The situation can be understood by investigating the eigenfunctions of the operator adjoint to the Frobenius–Perron operator [11,12]. The two measures produce different escape rates. Interesting consequences concerning transient diffusion processes taking place in a chain of such maps can be drawn [12]. Namely, the diffusion coefficient can have a finite jump, when the control parameter reaches its citical value.

Acknowledgment

The author would like to thank the organizers of the XXI Solvay Conference for their invitation and for the kind hospitality.

References

1. Ch. Beck and F. Schlögl, *Thermodynamics of Chaotic Systems*, Cambridge University Press, 1993.
2. A. Csordás, G. Györgyi, P. Szépfalusy, and T. Tél, *Chaos* **3**, 31 (1993).
3. T. Tél, in *STATPHYS 19*, Hao Bai-Lin, ed., World Scientific, Singapore, 1996, pp. 346–362.
4. J.-P. Eckmann and D. Ruelle, *Rev. Mod. Phys.* **57**, 617 (1985).
5. H. Kantz and P. Grassberger, *Physica* **D17**, 75 (1985).
6. G. Györgyi and P. Szépfalusy, *Acta Phys. Hung.* **64**, 33 (1988).
7. A. Csordás and P. Szépfalusy, *Phys. Rev.* **A38**, 2582 (1988).
8. P. Szépfalusy, T. Tél, A. Csordás, and Z. Kovács, *Phys. Rev.* **A36**, 3525 (1987).
9. Z. Kaufmann and P. Szépfalusy, *Phys. Rev.* **A40**, 2615 (1989).
10. A. Németh and P. Szépfalusy, *Phys. Rev.* **E52**, 1544 (1995).
11. H. Lustfeld and P. Szépfalusy, *Phys. Rev.* **E53**, 5882 (1996).
12. Z. Kaufmann, H. Lustfeld, A. Németh, and P. Szépfalusy, *Phys. Rev. Lett.* **78**, 4031 (1997).
13. P. Szépfalusy, *Phys. Scripta* **T25**, 226 (1986).

FROM COUPLED DYNAMICAL SYSTEMS TO BIOLOGICAL IRREVERSIBILITY

KUNIHIKO KANEKO

Department of Pure and Applied Sciences, College of Arts and Sciences, University of Tokyo, Tokyo, Japan

CONTENTS

I. Introduction
II. High-Dimensional Chaos Revisited
III. Prevalence of Milnor Attractors
IV. Chaotic Itinerancy
V. Collective Dynamics
VI. Cell Differentiation and Development as Dynamical Systems
VII. Scenario for Cell Differentiation
VIII. Dynamical Systems Representations of Cell Differentiation
IX. Toward Biological Irreversibility Irreducible to Thermodynamics
X. Discussion: Toward Phenomenology Theory of Development Process
Acknowledgments
References

I. INTRODUCTION

How should we understand the origin of biological irreversibility?

As an empirical fact, we know that the direction from the alive to the dead is irreversible. At a more specific level, we know that in a multicellular organism with a developmental process, there is a definite temporal flow. Through the developmental process, the multipotency (i.e., the ability to create different types of cells) decreases. Initially, the embryonic stem cell has totipotency and has the potentiality to create all types of cells in the organism. Then a stem cell can create a limited variety of cells, having multipotency. This hierarchical loss

---

Dynamical Systems and Irreversibility: A Special Volume of Advances in Chemical Physics, Volume 122, Edited by Ioannis Antoniou. Series Editors I. Prigogine and Stuart A. Rice.
ISBN 0-471-22291-7. © 2002 John Wiley & Sons, Inc.

of multipotency terminates at a determined cell, which can only replicate its own type, in the normal developmental process. The degree of determination increases in the normal course of development. How can one understand such irreversibility?

Of course this question is not easy to answer. However, it should be pointed out that:

1. It is very difficult to imagine that this irreversibility is caused by a set of specific genes. The present irreversibility is too universal to be attributed to characteristics of a few molecules.
2. It is also impossible to simply attribute this irreversibility to the second law of thermodynamics. One can hardly imagine that the entropy, even if it were possible to be defined, suddenly increases at the death, or successively increases at the cell differentiation process. Furthermore, it should be generally very difficult to define a thermodynamic entropy to a highly nonequilibrium system such as a cell.

Then what strategy should we choose?

A biological system contains always sufficient degrees of freedom—say, a set of chemical concentrations in a cell, whose values change in time. Then, one promising strategy for the study of a biological system lies in the use of dynamical systems [1]. By setting a class of dynamical systems, we search for universal characteristics that are robust against microscopic and macroscopic fluctuations.

A biological unit, such as a cell, has always some internal structure that can change in time. As a simple representation, the unit can be represented by a dynamical system. For example, consider a representation of a cell by a set of chemical concentrations. A cell, however, is not separated from the outside world completely. For example, isolation by a biomembrane is flexible and incomplete. In this way, the units, represented by dynamical systems, interact with each other through the external environment. Hence, we need a model consisting of the interplay between inter-unit and intra-unit dynamics. For example, the complex chemical reaction dynamics in each unit (cell) is affected by the interaction with other cells, which provides an interesting example of "intra–inter dynamics." In the "intra–inter dynamics," elements having internal dynamics interact with each other. This type of intra–inter dynamics is not necessarily represented only by the perturbation of the internal dynamics by the interaction with other units, nor is it merely a perturbation of the interaction by adding some internal dynamics.

As a specific example of the scheme of intra–inter dynamics, we will mainly discuss the developmental process of a cell society accompanied by cell differentiation. Here, the intra–inter dynamics consists of several biochemical reaction processes. The cells interact through the diffusion of chemicals or their active signal transmission.

If N cells with k degrees of freedom exist, the total dynamics is represented by an Nk-dimensional dynamical system (in addition to the degrees of freedom of the environment). Furthermore, the number of cells is not fixed in time, but they are born by division (and die) in time.

After the division of a cell, if two cells remained identical, another set of variables would not be necessary. If the dynamical system for chemical state of a cell has orbital instability (such as chaos), however, the orbits of chemical dynamics of the (two) daughters will diverge. Hence, the number of degrees of freedom, Nk, changes in time. This increase in the number of variables is tightly connected with the internal dynamics. It should also be noted that in the developmental process, in general, the initial condition of the cell states is chosen so that their reproduction continues. Thus, a suitable initial condition for the internal degrees of freedom is selected through interaction.

Now, to study a biological system in terms of dynamical systems theory, it is first necessary to understand the behavior of a system with internal degrees of freedom and interaction [4]. This is the main reason why I started a model called Coupled Map Lattice [5] (and later Globally Coupled Map [6]) about 2 decades ago. Indeed, several discoveries in GCM seem to be relevant to understand some basic features in a biological system. GCM has provided us some novel concepts for nontrivial dynamics between microscopic and macroscopic levels, while the dynamic complementarity between a part and the whole is important to study biological organization [7]. In the present chapter we briefly review the behaviors of GCM in Section II, and in Sections III–V we discuss some recent advances about dominance of Milnor attractors, chaotic itinerancy, and collective dynamics. Then we will switch to the topic of development and differentiation in an interacting cell system. After presenting our model based on dynamical systems in Section VI, we give a basic scenario discovered in the model, and interpret cell differentiation in terms of dynamical systems. Then, the origin of biological irreversibility is discussed in Section IX. Discussion toward the construction of phenomenology theory of development is given in Section X.

II. HIGH-DIMENSIONAL CHAOS REVISITED

The simplest case of global interaction is studied as the "globally coupled map" (GCM) of chaotic elements [6]. A standard example is given by

$$x_{n+1}(i) = (1-\epsilon)f(x_n(i)) + \frac{\epsilon}{N}\sum_{j=1}^{N} f(x_n(j)) \qquad (1)$$

where n is a discrete time step and i is the index of an element ($i = 1, 2, \ldots, N$=system size), and $f(x) = 1 - ax^2$. The model is just a mean-field-theory-type extension of coupled map lattices (CML) [5].

Through the interaction, elements tend to oscillate synchronously, while chaotic instability leads to destruction of the coherence. When the former tendency wins, all elements oscillate coherently, while elements are completely desynchronized in the limit of strong chaotic instability. Between these cases, elements split into clusters in which they oscillate coherently. Here a cluster is defined as a set of elements in which $x(i) = x(j)$ [6]. Attractors in GCM are classified by the number of synchronized clusters k and the number of elements for each cluster N_i. Each attractor is coded by the clustering condition $[k, (N_1, N_2, \ldots, N_k)]$. Stability of each clustered state is analyzed by introducing the split exponent [6,9].

An interesting possibility in the clustering is that it provides a source for diversity. In clustering it should be noted that identical chaotic elements differentiate spontaneously into different groups: Even if a system consists of identical elements, they split into groups with different phases of oscillations. Hence a network of chaotic elements gives a theoretical basis for isologous diversification and provides a mechanism for the origin of diversity and complexity in biological networks [10,11].

In a globally coupled chaotic system in general, the following phases appear successively with the increase of nonlinearity in the system (a in the above logistic map case) [6]:

1. **Coherent phase:** Only a coherent attractor ($k = 1$) exists.
2. **Ordered phase:** All attractors consist of a few ($k = o(N)$) clusters.
3. **Partially ordered phase:** Attractors with a variety of clusterings coexist, while most of them have many clusters ($k = O(N)$).
4. **Turbulent phase:** Elements are completely desynchronized, and all attractors have N clusters.

The above clustering behaviors have universally been confirmed in a variety of systems.

In the partially ordered (PO) phase, there are a variety of attractors with a different number of clusters and a different type of partitions $[N_1, N_2, \ldots, N_k]$. The clustering here is typically inhomogeneous: The partition $[N_1, N_2, \ldots, N_k]$ is far from equal partition. Often this clustering is hierarchical as for the number of elements, and as for the values. For example, consider the following idealized clustering: First split the system into two equal clusters. Take one of them and split it again into two equal clusters, but leave the other without split. By repeating this process, the partition is given by $[N/2, N/4, N/8, \ldots]$. In this case, the difference of the values of $x_n(i)$ is also hierarchical. The difference between the values of $x_n(i)$ decreases as the above process of partition is iterated. Although the above partition is too much simplified, such hierarchical structure in partition and in the phase space is typically observed in the PO phase. The partition is organized as an inhomogeneous tree structure, as in the spin glass model [8].

We have also measured the fluctuation of the partitions, using the probability Y that two elements fall on the same cluster. In the PO phase, this Y value fluctuates by initial conditions, and the fluctuation remains finite even if the size goes to infinity [12,13]. It is noted that such remnant fluctuation of partitions is also seen in spin glass models [8].

III. PREVALENCE OF MILNOR ATTRACTORS

In the partially ordered (PO) phase, there coexist a variety of attractors depending on the partition [12]. To study the stability of an attractor against perturbation, we introduce the return probability $P(\sigma)$, defined as follows [14]: Take an orbit point $\{x(i)\}$ of an attractor in an N-dimensional phase space, and perturb the point to $x(i) + \frac{\sigma}{2} rnd_i$, where rnd_i is a random number taken from $[-1, 1]$, uncorrelated for all elements i. Check if this perturbed point returns to the original attractor via the original deterministic dynamics (1). By sampling over random perturbations and the time of the application of perturbation, the return probability $P(\sigma)$ is defined as (# of returns)/(# of perturbation trials). As a simple index for robustness of an attractor, it is useful to define σ_c as the largest σ such that $P(\sigma) = 1$. This index measures what we call the *strength* of an attractor.

The strength σ_c gives a minimum distance between the orbit of an attractor and its basin boundary. In contrast with our naive expectation from the concept of an attractor, we have often observed "attractors" with $\sigma_c = 0$; that is, $P(+0) \equiv \lim_{\delta \to 0} P(\delta) < 1$. If $\sigma_c = 0$ holds for a given state, it cannot be an "attractor" in the sense with asymptotic stability, since some tiny perturbations kick the orbit out of the "attractor." The attractors with $\sigma_c = 0$ are called Milnor attractors [15,16]. In other words, Milnor attractor is defined as an attractor that is unstable by some perturbations of arbitrarily small size, but globally attracts orbital points. The basin of attraction has a positive Lebesgue measure. (The basin is riddled here [17,18].) Because it is not asymptotically stable, one might, at first sight, think that it is rather special, and it appears only at a critical point like the crisis in the logistic map [15]. To our surprise, the Milnor attractors are rather commonly observed around the PO phase in our GCM. The strength and basin volume of attractors are not necessarily correlated. Attractors with $\sigma_c = 0$ often have a large basin volume.

Still, one might suspect that such Milnor attractors must be weak against noise. Indeed, by a very weak noise with the amplitude σ, an orbit at a Milnor attractor is kicked away, and if the orbit is reached to one of attractors with $\sigma_c > \sigma$, it never comes back to the Milnor attractor. Rather, an orbit kicked out from a Milnor attractor is often found to stay in the vicinity of it [16]. The orbit comes back to the original Milnor attractor before it is kicked away to other attractors with $\sigma_c > \sigma$. Furthermore, by a larger noise, orbits sometimes are more attracted to Milnor attractors. Such attraction is possible, since Milnor attractors here have global attraction in the phase space, in spite of their local instability.

Prevalence of Milnor attractors gives us reason to suspect the computability of our system. Once the digits of two variables $x(i)$ and $x(j)$ agree down to the lowest bit, the values never split again, even though the state with the synchronization of the two elements may be unstable. As long as digital computation is adopted, it is always possible that an orbit is trapped to such unstable state. In this sense a serious problem is cast in numerical computation of GCM in general.[1]

Existence of Milnor attractors may lead us to suspect the correspondence between a (robust) attractor and memory, often adopted in neuroscience (and theoretical cell biology). It should be mentioned that Milnor attractors can provide dynamic memory [4,19] allowing for an interface between outside and inside, external inputs, and internal representation.

IV. CHAOTIC ITINERANCY

Besides the above *static* complexity, *dynamic* complexity is more interesting at the PO phase. Here orbits make itinerancy over ordered states with partial synchronization of elements, via highly chaotic states. This dynamics, called chaotic itinerancy (CI), is a novel universal class in high-dimensional dynamical systems. Our CI consists of a quasi-stationary high-dimensional state, exits to "attractor ruins" with low effective degrees of freedom, residence therein, and chaotic exits from them. In the CI, an orbit successively itinerates over such "attractor ruins," involving ordered motion with some coherence among elements. The motion at "attractor ruins" is quasi-stationary. For example, if the effective degrees of freedom is two, the elements split into two groups, in each of which elements oscillate almost coherently. The system is in the vicinity of a two-clustered state, which, however, is not a stable attractor, but keeps attraction to its vicinity globally within the phase space. After staying at an attractor ruin, an orbit exits from it due to chaotic instability, and it shows a high-dimensional chaotic motion without clear coherence. This high-dimensional

---

[1] Indeed, in our simulations we have often added a random floating at the smallest bit of $x(i)$ in the computer, to partially avoid such computational problems.

state is again quasi-stationary, although there are some holes connecting to the attractor ruins from it. Once the orbit is trapped at a hole, it is suddenly attracted to one of attractor ruins—that is, ordered states with low-dimensional dynamics.

This CI dynamics has independently been found in a model of neural dynamics (by Tsuda [19]), optical turbulence [20], and GCM. It provides an example of successive changes of relationships among elements.

Note that the Milnor attractors satisfy the condition of the above ordered states constituting chaotic itinerancy. Some Milnor attractors we have found keep global attraction, which is consistent with the observation that the attraction to ordered states in chaotic itinerancy occurs globally from a high-dimensional chaotic state. Attraction of an orbit to precisely a given attractor requires infinite time, and before the orbit is really settled to a given Milnor attractor, it may be kicked away.[2] When Milnor attractors that lose the stability ($P(0) < 1$) keep global attraction, the total dynamics can be constructed as the successive alternations to the attraction to, and escape from, them. If the attraction to robust attractors from a given Milnor attractor is not possible, the long-term dynamics with the noise strength $\to +0$ is represented by successive transitions over Milnor attractors. Then the dynamics is represented by transition matrix over Milnor attractors. This matrix is generally asymmetric: Often, there is a connection from a Milnor attractor A to a Milnor attractor B, but not from B to A. The total dynamics is represented by the motion over a network, given by a set of directed graphs over Milnor attractors.

In general, the "ordered states" in CI may not be exactly Milnor attractors but can be weakly destabilized states from Milnor attractors. Still, the attribution of CI to Milnor attractor network dynamics is expected to work as one ideal limit.[3]

As already discussed about the Milnor attractor, computability of the switching over Milnor attractor networks has a serious problem. In each event of switching, which Milnor attractor is visited next after the departure from a Milnor attractor may depend on the precision. In this sense, the order of visits to Milnor attractors in chaotic itinerancy may not be undecidable in a digital computer. In other words, motion at a macroscopic level may not be decidable from a microscopic level. With this respect, it may be interesting to note that there are similar statistical features between (Milnor attractor) dynamics with a riddled basin and undecidable dynamics of a universal Turing machine [23].

---

[2] This problem is subtle computationally, since any finite precision in computation may have a serious influence on whether the orbit remains at a Milnor attractor or not.

[3] The notion of chaotic itinerancy is rather broad, and some of CI may not be explained by the Milnor attractor network. In particular, chaotic itinerancy in a Hamiltonian system [21,22] may not fit directly with the present correspondence.

V. COLLECTIVE DYNAMICS

If the coupling strength ϵ is small enough, oscillation of each element has no mutual synchronization. In this turbulent phase, $x(i)$ takes almost random values almost independently, and the number of degrees of freedom is proportional to the number of elements N; that is, the Lyapunov dimension increases in proportion to N. Even in such cases, the macroscopic motion shows some coherent motion distinguishable from noise, and there remains some coherence among elements, even in the limit of $N \to \infty$. As a macroscopic variable we adopt the mean field,

$$h_n = \frac{1}{N}\sum_{i=1}^{N} f(x_n(i)) \qquad (2)$$

In almost all the parameter values, the mean field motion shows some dynamics that is distinguishable from noise, ranging from torus-like to higher-dimensional motion. This motion remains even in the thermodynamic limit [24].

This remnant variation means that the collective dynamics $h_n(i)$ keeps some structure. One possibility is that the dynamics is low-dimensional. Indeed in some system with a global coupling, the collective motion is shown to be low-dimensional in the limit of $N \to \infty$ (see Refs. 25 and 26). In the GCM equation [Eq. (1)], with the logistic or tent map, low-dimensional motion is not detected generally, although there remains some collective motion in the limit of $N \to \infty$. The mean field motion in GCM is regarded to be infinite-dimensional, even when the torus-like motion is observed [27–30]. Then it is important to clarify the nature of this mean-field dynamics.

It is not so easy to examine the infinite-dimensional dynamics, directly. Instead, Shibata, Chawanya, and the author have first made the motion low-dimensional by adding noise, and then they studied the limit of noise $\to 0$. To study this effect of noise, we have simulated the model

$$x_{n+1}(i) = (1-\epsilon)f(x_n(i)) + \frac{\epsilon}{N}\sum_{j=1}^{N} f(x_n(j)) + \sigma \eta_n^i \qquad (3)$$

where η_n^i is a white noise generated by an uncorrelated random number homogeneously distributed over $[-1, 1]$.

The addition of noise can destroy the above coherence among elements. In fact, the microscopic external noise leads the variance of the mean field distribution to decrease with N [24,31]. This result also implies decrease of the mean field fluctuation by external noise.

Behavior of the above equation in the thermodynamic limit $N \to \infty$ is represented by the evolution of the one-body distribution function $\rho_n(x)$ at time step n directly. Because the mean field value

$$h_n = \int f(x)\rho_n(x)\,dx \tag{4}$$

is independent of each element, the evolution of $\rho_n(x)$ obeys the Perron–Frobenius equation given by

$$\rho_{n+1}(x) = \int dy \frac{1}{\sqrt{2\pi}\sigma} e^{-\frac{(F_n(y)-x)^2}{2\sigma^2}} \rho_n(y) \tag{5}$$

with

$$F_n(x) = (1-\epsilon)f(x) + \epsilon h_n \tag{6}$$

By analyzing the above Perron–Frobenius equation [32], it is shown that the dimension of the collective motion increases as $log(1/\sigma^2)$, with σ as the noise strength. Hence in the limit of $\sigma \to 0$, the dimension of the mean field motion is expected to be infinite. Note that the mean field dynamics (at $N \to \infty$) is completely deterministic, even under the external noise.

With the addition of noise, high-dimensional structures in the mean-field dynamics are destroyed successively; and the bifurcation from high-dimensional to low-dimensional chaos, and then to torus, proceeds with the increase of the noise amplitude. With a further increase of noise to $\sigma > \sigma_c$, the mean field goes to a fixed point through Hopf bifurcation. This destruction of the hidden coherence leads to a strange conclusion. Take a globally coupled system with a desynchronized and highly chaotic state, and add noise to the system. Then the dimension of the mean field motion gets lower with the increase of noise.

The appearance of low-dimensional "order" through the destruction of small-scale structure in chaos is also found in noise-induced order [33]. Note, however, that in a conventional noise-induced transition [34], the ordered motion is still stochastic, since the noise is added into a low-dimensional dynamical system. On the other hand, the noise-induced transition in the collective dynamics occurs after the thermodynamic limit is taken. Hence the low-dimensional dynamics induced by noise is truly low-dimensional. When we say a torus, the Poincaré map shows a curve without thickness by the noise, since the thermodynamic limit smears out the fluctuation around the tours. Also, it is interesting to note that a similar mechanism of the destruction of hidden coherence is observed in quantum chaos.

This noise-induced low-dimensional collective dynamics can be used to distinguish high-dimensional chaos from random noise. If the irregular behavior is originated in random noise, (further) addition of noise will result in an increase of the fluctuations. If the external application of noise leads to the decrease of fluctuations in some experiment, it is natural to assume that the irregular dynamics there is due to high-dimensional chaos with a global coupling of many nonlinear modes or elements.

VI. CELL DIFFERENTIATION AND DEVELOPMENT AS DYNAMICAL SYSTEMS

Now we come back to the problem of cell differentiation and development. A cell is separated from environment by a membrane, whose separation, however, is not complete. Some chemicals pass through the membrane; and through this transport, cells interact with each other. When a cell is represented by a dynamical system the cells interact with each other and with the external environment. Hence, we need a model consisting of the interplay between inter-unit and intra-unit dynamics. Here we will mainly discuss the developmental process of a cell society accompanied by cell differentiation, where the intra–inter dynamics consist of several biochemical reaction processes. Cells interact through the diffusion of chemicals or their active signal transmission, while they divide into two when some condition is satisfied with the chemical reaction process in it. (See Fig. 1 for schematic representation of our model.)

We have studied several models [2,3,35–38] with (a) internal (chemical) dynamics of several degrees of freedom, (b) cell–cell interaction type through the medium, and (c) the division to change the number of cells.

As for the internal dynamics, an autocatalytic reaction among chemicals is chosen. Such autocatalytic reactions are necessary to produce chemicals in a cell, required for reproduction [39]. Autocatalytic reactions often lead to nonlinear oscillation in chemicals. Here we assume the possibility of such oscillation in the intracellular dynamics [40,41]. As the interaction mechanism, the diffusion of chemicals between a cell and its surroundings is chosen.

To be specific, we mainly consider the following model here. First, the state of a cell i is assumed to be characterized by the cell volume and a set of functions $x_i^{(m)}(t)$ representing the concentrations of k chemicals denoted by $m = 1,\ldots,k$. The concentrations of chemicals change as a result of internal biochemical reaction dynamics within each cell and cell–cell interactions communicated through the surrounding medium.

For the internal chemical reaction dynamics, we choose a catalytic network among the k chemicals. The network is defined by a collection of triplets (l, j, m) representing the reaction from chemical m to l catalyzed by j. The rate of increase of $x_i^l(t)$ (and decrease of $x_i^m(t)$) through this reaction is given by $x_i^{(m)}(t)(x_i^{(j)}(t))^\alpha$, where α is the degree of catalyzation ($\alpha = 2$ in the simulations considered presently). Each chemical has several paths to other chemicals, and

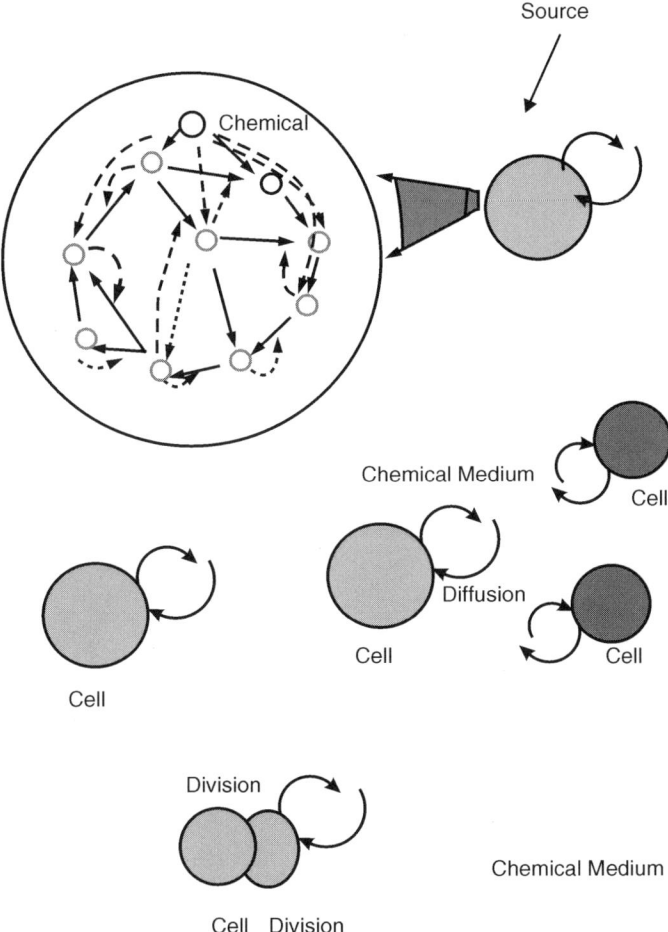

Figure 1. Schematic representation of our model.

thus a complex reaction network is formed. Thus, the change in the chemical concentrations through all such reactions is determined by the set of all terms of the above type for a given network. (These reactions can include genetic processes.)

Cells interact with each other through the transport of chemicals out of and into the surrounding medium. As a minimal case, we consider only indirect cell–cell interactions through diffusion of chemicals via the medium. The transport rate of chemicals into a cell is proportional to the difference in chemical concentrations between the inside and the outside of the cell and is given by $D(X^{(l)}(t) - x_i^{(l)}(t))$, where D denotes the diffusion constant, and $X^{(l)}(t)$ is the concentration of the chemical at the medium. The diffusion of a chemical species through cell membrane should depend on the properties of this species. In this model, we consider the simple case in which there are two types of chemicals; one that can penetrate the membrane and one that cannot. For simplicity, we assume that all the chemicals capable of penetrating the membrane have the same diffusion coefficient, D. With this type of interaction, corresponding chemicals in the medium are consumed. To maintain the growth of the organism, the system is immersed in a bath of chemicals through which (nutritive) chemicals are supplied to the cells.

As chemicals flow out of and into the environment, the cell volume changes. The volume is assumed to be proportional to the sum of the quantities of chemicals in the cell, and thus is a dynamical variable. Accordingly, chemicals are diluted as a result of the increase of the cell volume.

In general, a cell divides according to its internal state, for example, as some products, such as DNA or the membrane, are synthesized, accompanied by an increase in cell volume. Again, considering only a simple situation, we assume that a cell divides into two when the cell volume becomes double the original. At each division, all chemicals are almost equally divided, with random fluctuations.

Of course, each result of simulation depends on the specific choice of the reaction network. However, the basic feature of the process to be discussed does not depend on the details of the choice, as long as the network allows for the oscillatory intracellular dynamics leading to the growth in the number of cells. Note that the network is not constructed to imitate an existing biochemical network. Rather, we try to demonstrate that important features in a biological system are a natural consequence of a system with internal dynamics, interaction, and reproduction. From the study we try to extract a universal logic underlying a class of biological systems.

VII. SCENARIO FOR CELL DIFFERENTIATION

From several simulations of the models starting from a single cell initial condition, we have shown that cells undergo spontaneous differentiation as the number is increased (see Fig. 2 for schematic representation): The first differentiation starts with the clustering of the phase of the oscillations, as discussed in globally coupled maps (see Fig. 2a). Then, the differentiation comes to the stage that the average concentrations of the biochemicals over the cell cycle become different. The composition of biochemicals as well as the rates of catalytic reactions and transport of the biochemicals become different for each group.

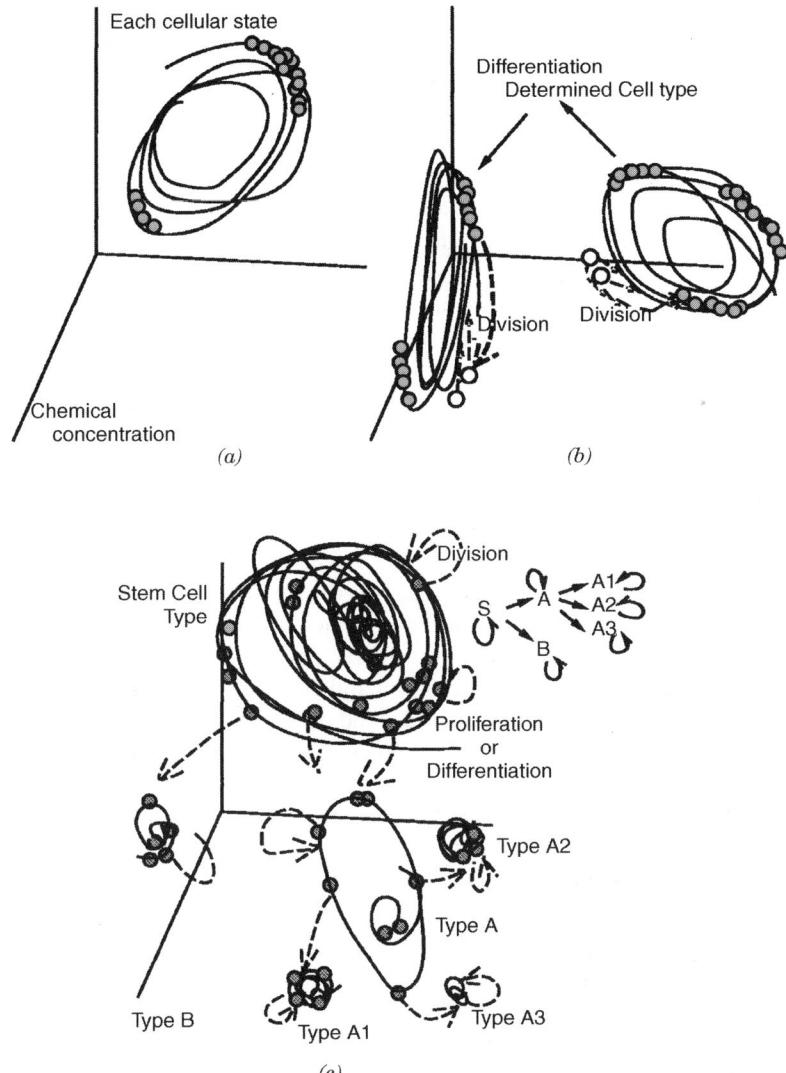

Figure 2. Schematic representation of cell differentiation process, plotted in the phase space of chemical concentrations.

After the formation of cell types, the chemical compositions of each group are inherited by their daughter cells. In other words, chemical compositions of cells are recursive over divisions. The biochemical properties of a cell are inherited by its progeny; in other words, the properties of the differentiated cells are stable, fixed, or determined over the generations (see Fig. 2b). After several divisions, such an initial condition of units is chosen to give the next generation of the same type as its mother cell.

The most interesting example here is the formation of stem cells, schematically shown in Fig. 2c [37]. This cell type, denoted as S here, either reproduces the same type or forms different cell types, denoted for example as type A and type B. Then after division events $S \to S, A, B$ occur. Depending on the adopted chemical networks, the types A and B replicate, or switch to different types. For example, $A \to A, A1, A2, A3$ is observed in some network. This hierarchical organization is often observed when the internal dynamics have some complexity, such as chaos.

The differentiation here is "stochastic," arising from chaotic intracellular chemical dynamics. The choice for a stem cell either to replicate or to differentiate seems stochastic as far as the cell type is concerned. Because such stochasticity is not due to external fluctuation but is a result of the internal state, the probability of differentiation can be regulated by the intracellular state. This stochastic branching is accompanied by a regulative mechanism. When some cells are removed externally during the developmental process, the rate of differentiation changes so that the final cell distribution is recovered.

In some biological systems such as the hematopoietic system, stem cells either replicate or differentiate into different cell type(s). This differentiation rule is often hierarchical [42,43]. The probability of differentiation to one of the several blood cell types is expected to depend on the interaction. Otherwise, it is hard to explain why the developmental process is robust. For example, when the number of some terminal cells decreases, there should be some mechanism to increase the rate of differentiation from the stem cell to the differentiated cells. This suggests the existence of interaction-dependent regulation of the differentiation ratio, as demonstrated in our results.

Microscopic Stability

The developmental process is stable against molecular fluctuations. First, intracellular dynamics of each cell type are stable against such perturbations. Then, one might think that this selection of each cell type is nothing more than a choice among basins of attraction for a multiple attractor system. If the interaction were neglected, a different type of dynamics would be interpreted as a different attractor. In our case, this is not true, and cell–cell interactions are necessary to stabilize cell types. Given cell-to-cell interactions, the cell state is stable against perturbations on the level of each type of intracellular dynamics.

Next, the number distribution of cell types is stable against fluctuations. Indeed, we have carried out simulations of our model, by adding a noise term, considering finiteness in the number of molecules [3,44]. The obtained cell type, as well as the number distribution, is hardly affected by the noise as long as the noise amplitude is not too large.[4]

Macroscopic Stability

Each cellular state is also stable against perturbations of the interaction term. If the cell type number distribution is changed within some range, each type of cellular dynamics keeps its identity. Hence discrete, stable types are formed through the interplay between intracellular dynamics and interaction. The recursive production is attained through the selection of initial conditions of the intracellular dynamics of each cell, so that it is rather robust against the change of interaction terms as well.

The macroscopic stability is clearly shown in the spontaneous regulation of differentiation ratio. How is this interaction-dependent rule formed? Note that depending on the distribution of the other cell types, the orbit of the internal cell state is slightly deformed. For a stem cell case, the rate of the differentiation or the replication (e.g., the rate to select an arrow among $S \rightarrow S, A, B$) depends on the cell-type distribution. For example, when the number of "A" type cells is reduced, the orbit of an "S"-type cell is shifted toward the orbits of "A," with which the rate of switch to "A" is enhanced. The information of the cell-type distribution is represented by the internal dynamics of "S"-type cells, and it is essential to the regulation of differentiation rate [37].

It should be stressed that our dynamical differentiation process is always accompanied by this kind of regulation process, without any sophisticated programs implemented in advance. This autonomous robustness provides a novel viewpoint to the stability of the cell society in multicellular organisms.

VIII. DYNAMICAL SYSTEMS REPRESENTATIONS OF CELL DIFFERENTIATION

Because each cell state is realized as a balance between internal dynamics and interaction, one can discuss which part is more relevant to determine the stability of each state. In one limiting case, the state is an attractor as internal dynamics [45], which is sufficiently stable and not destabilized by cell–cell interaction. In this case, the cell state is called "determined," according to the terminology in cell biology. In the other limiting case, the state is totally governed by the interaction; and by changing the states of other cells, the cell state in concern is destabilized. In this case, each cell state is highly dependent on the environment or other cells.

---

[4] When the noise amplitude is too large, distinct types are no longer formed. Cell types are continuously distributed. In this case the division speed is highly reduced, since the differentiation of roles by differentiated cell types is destroyed.

Each cell type in our simulation generally lies between these two limiting cases. To see such intra–inter nature of the determination explicitly, one effective method is a transplantation experiment. Numerically, such an experiment is carried out by choosing determined cells (obtained from the normal differentiation process) and putting them into a different set of surrounding cells, to set distribution of cells so that it does not appear through the normal course of development.

When a differentiated and recursive cell is transplanted to another cell society, the offspring of the cell keeps the same type, unless the cell-type distribution of the society is strongly biased. When a cell is transplanted into a biased society, differentiation from a "determined" cell occurs. For example, a homogeneous society consisting only of one determined cell type is unstable, and some cells start to switch to a different type. Hence, the cell memory is preserved mainly in each individual cell, but suitable intercellular interactions are also necessary to keep it.

Because each differentiated state is not an attractor, but is stabilized through the interaction, we propose to define *partial attractor*, to address attraction restricted to the internal cellular dynamics. Tentative definition of this partial attractor is as follows:

1. **Internal stability.** Once the cell–cell interaction is specified (i.e., the dynamics of other cells), the state is an attractor of the internal dynamics. In other words, it is an attractor when the dynamics is restricted only to the variables of a given cell.
2. **Interaction stability.** The state is stable against change of interaction term, up to some finite degree. With the change of the interaction term of the order ϵ, the change in the dynamics remains of the order of $O(\epsilon)$.
3. **Self-consistency.** For some distribution of units of cellular states satisfying 1 and 2, the interaction term continues to satisfy condition 1 and 2.

We tentatively call a state satisfying 1–3 a partial attractor. Each determined cell type we found can be regarded as a partial attractor. To define the dynamics of stem cell in our model, however, we have to slightly modify the condition of 2 to a "Milnor-attractor" type. Here, small perturbation to the interaction term (by the increase of the cell number) may lead the state to switch to a differentiated state. Hence, instead of 2, we set the condition:

2′. For some change of interaction with a finite measure, some orbits remain to be attracted to the state.

So far we have discussed the stability of a state by fixing the number of cells. In some case, condition 3 may not be satisfied when the system is developed from a single cell following the cell division rule. As for developmental process, the condition has to be satisfied for a restricted range of cell distribution realized by the evolution from a single cell. Then we need to add the condition:

4. **Accessibility.** The distribution (3) is satisfied from an initial condition of a single cell and with the increase of the number of cells.

Cell types with determined differentiation observed in our model is regarded as a state satisfying 1–4, while the stem cell type is regarded as a state satisfying 1, 2′, 3, and 4.

In fact, as the number is increased, some perturbations to the interaction term is introduced. In our model, the stem-cell state satisfies 2 up to some number; but with the further increase of number, the condition 2 is no more satisfied and is replaced by 2′. Perturbation to the interaction term due to the cell number increase is sufficient to bring about a switch from a given stem-cell dynamics to a differentiated cell. Note again that the stem-cell type state with weak stability has a large basin volume when started from a single cell.

IX. TOWARD BIOLOGICAL IRREVERSIBILITY IRREDUCIBLE TO THERMODYNAMICS

In the normal development of cells, there is clear irreversibility, resulting from the successive loss of multipotency.

In our model simulations, this loss of multipotency occurs irreversibly. The stem-cell type can differentiate to other types, while the determined type that appears later only replicates itself. In a real organism, there is a hierarchy in determination, and a stem cell is often over a progenitor over only a limited range of cell types. In other words, the degree of determination is also hierarchical. In our model, we have also found such hierarchical structure. So far, we have found only up to the second layer of hierarchy in our model with the number of chemicals $k = 20$.

Here, dynamics of a stem-type cell exhibit irregular oscillations with orbital instability and involve a variety of chemicals. Stem cells with these complex dynamics have a potential to differentiate into several distinct cell types. Generally, the differentiated cells always possess simpler cellular dynamics than do the stem cells—for example, fixed-point dynamics and regular oscillations.

Although we have not yet succeeded in formulating the irreversible loss of multipotency in terms of a single fundamental quantity (analogous to thermodynamic entropy), we have heuristically derived a general law describing the change of the following quantities in all of our numerical experiments, using a variety of reaction networks [44,46]. As cell differentiation progresses through development, we encounter the following:

I. Stability of intracellular dynamics increases.
II. Diversity of chemicals in a cell decreases.
III. Temporal variations of chemical concentrations decrease, by realizing less chaotic motion.

The degree of statement I could be determined by a minimum change in the interaction to switch a cell state, by properly extending the "attractor strength" in Section III. Initial undifferentiated cells spontaneously change their state even without the change of the interaction term, while stem cells can be switched by a tiny change in the interaction term. The degree of determination is roughly measured as the minimum perturbation strength required for a switch to a different state.

The diversity of chemicals (statement II) can be measured, for example, by $S = -\sum_{j=1}^{k} p(j) \log p(j)$, with $p(j) = \langle x(j)/\sum_{m=1}^{k} x(m) \rangle$, with $\langle \cdots \rangle$ as temporal average. Loss of multipotency in our model is accompanied by a decrease in the diversity of chemicals and is represented by the decrease of this diversity S.

The tendency (III) is numerically confirmed by the subspace Kolmorogorov–Sinai (KS) entropy of the internal dynamics for each cell. Here, this subspace KS entropy is measured as a sum of positive Lyapunov exponents, in the tangent space restricted only to the intracellular dynamics for a given cell. Again, this exponent decreases through the development.

X. DISCUSSION: TOWARD PHENOMENOLOGY THEORY OF DEVELOPMENTAL PROCESS

In the present chapter, we have first surveyed some recent progress in coupled dynamical systems—in particular, globally coupled maps. Then we discuss some of our recent studies on the cell differentiation and development, based on coupled dynamical systems with some internal degrees of freedom and the potentiality to increase the number of units (cells). Stability and irreversibility of the developmental process are demonstrated by the model, and they are discussed in terms of dynamical systems.

Of course, results based on a class of models are not sufficient to establish a theory to understand the stability and irreversibility in development of multicellular organisms. We need to unveil the logic that underlies such models and real development universally. Although mathematical formulation is not yet established, supports are given to the following conjecture:

Assume a cell with internal chemical reaction network whose degrees of freedom is large enough and which interacts each other through the environment. Some chemicals are transported from the environment and converted to other chemicals within a cell. Through this process the cell volume increases and the cell is divided. Then, for some chemical networks, each chemical state of a cell remains to be a fixed point. In this case, cells remain identical, where the competition for chemical resources is higher, and the increase of the cell number is suppressed. On the other hand, for some reaction networks, cells differentiate and the increase in the cell number is not suppressed. The differentiation of cell types forms a hierarchical rule. The initial cell types have large chemical diversity and show irregular temporal change of chemical concentrations. As the number of cells increases and the differentiation progresses, irreversible loss of multipotency is observed. This differentiation process is triggered by instability of some states by cell–cell interaction, while the realized states of cell types and the number distribution of such cell types are stable against perturbations, following the spontaneous regulation of differentiation ratio.

When we recall the history of physics, the most successful phenomenological theory is nothing but thermodynamics. To construct a phenomenology theory for development, or generally a theory for biological irreversibility, comparison with the thermodynamics should be relevant. Some similarity between the phenomenology of development and thermodynamics is summarized in Table I.

As mentioned, both the thermodynamics and the development phenomenology have stability against perturbations. Indeed, the spontaneous regulation in a stem cell system found in our model is a clear demonstration of stability against perturbations, similar to the Le Chatelier–Braun principle. The irreversibility in thermodynamics is defined by suitably restricting possible operations, as formulated by adiabatic process. Similarly, the irreversibility in a multicellular organism has to be suitably defined by introducing an ideal developmental process. Note that in some experiments such as cloning from somatic cells in animals [47], the irreversibility in normal development can be reversed.

The last question that should be addressed here is the search for macroscopic quantities to characterize each (differentiated) cellular 'state'. Although thermodynamics is established by cutting the macroscopic out of microscopic

TABLE I
Comparison Between Development Phenomenology and Thermodynamics

Parameter	Development Phenomenology	Thermodynamics
Stability	Cellular and ensemble level	Macroscopic
Stability against perturbation	Regulation of differentiation ratio	Le Chatelier–Braun principle
Irreversibility	Loss of multipotency	Second law
Quantification of irreversibility	Some pattern of gene expression?	Entropy
Cycle	Somatic clone cycle?	Carnot cycle

levels, in a cell system, it is not yet sure if such macroscopic quantities can be defined, by separating a macroscopic state from the microscopic level. At the present stage, there is no definite answer. Here, however, it is interesting to recall recent experiments of tissue engineering. By changing the concentrations of only three control chemicals, Ariizumi and Asashima [48] have succeeded in constructing all tissues from *Xenopus* undifferentiated cells (animal cap). Hence there may be some hope that a reduction to a few variables characterizing macroscopic "states" may be possible.

Construction of phenomenology for development charactering its stability and irreversibility is still at the stage "waiting for Carnot"; but following our results based on coupled dynamical systems models and some of recent experiments, I hope that such phenomenology theory will be realized in the near future.

Acknowledgments

The author is grateful to T. Yomo, C. Furusawa, and T. Shibata for discussions. The work is partially supported by Grant-in-Aids for Scientific Research from the Ministry of Education, Science, and Culture of Japan.

References

1. For pioneering work on development from dynamical systems, see A. M. Turing, The chemical basis of morphogenesis. *Philos. Trans. R. Soc. B* **237**, 5 (1952).
2. K. Kaneko and T. Yomo, *Bull. Math. Biol.* **59**, 139–196 (1997).
3. K. Kaneko and T. Yomo, *J. Theor. Biol.* 243–256 (1999).
4. K. Kaneko and I. Tsuda, *Complex Systems: Chaos and Beyond—A Constructive Approach with Applications in Life Sciences*, Springer, Berlin, 2000 (based on K. Kaneko and I. Tsuda, *Chaos Scenario for Complex Systems*, Asakura, 1996, in Japanese).
5. K. Kaneko, *Prog. Theor. Phys.* **72**, 480–486 (1984); K. Kaneko, ed., *Theory and Applications of Coupled Map Lattices*, Wiley, New York, 1993.
6. K. Kaneko, *Physica* **41D**, 137–172 (1990).
7. K. Kaneko, *Complexity* **3**, 53–60 (1998).
8. M. Mezard, G. Parisi, and M. A. Virasoro, eds., *Spin Glass Theory and Beyond*, World Scientific Publishers, Singapore, 1988.
9. K. Kaneko, *Physica* **77D**, 456 (1994).
10. K. Kaneko, *Physica* **75D**, 55 (1994).
11. K. Kaneko, *Artificial Life* **1**, 163 (1994).
12. K. Kaneko, *J. Phys. A* **24**, 2107 (1991).
13. A. Crisanti, M. Falcioni, and A. Vulpiani, *Phys. Rev. Lett.* **76**, 612 (1996); S. C. Manruiba and A. Mikhailov, *Europhys. Lett.* **53**, 451–457 (2001).
14. K. Kaneko, *Phys. Rev. Lett.* **78**, 2736–2739 (1997); *Physica* **124D**, 308–330 (1998).
15. J. Milnor, *Commun. Math. Phys.* **99**, 177 (1985); **102**, 517 (1985).
16. P. Ashwin, J. Buescu, and I. Stuart, *Phys. Lett. A* **193**, 126 (1994); *Nonlinearity* **9**, 703 (1996).
17. J. C. Sommerer and E. Ott, *Nature* **365**, 138 (1993); E. Ott et al., *Phys. Rev. Lett.* **71**, 4134 (1993).
18. Y.-C. Lai and R. L. Winslow, *Physica* **74D**, 353 (1994).

19. I. Tsuda, *World Futures* **32**, 167 (1991); *Neural Networks* **5**, 313 (1992).
20. K. Ikeda, K. Matsumoto, and K. Ohtsuka, *Prog. Theor. Phys. Suppl.* **99**, 295 (1989).
21. T. Konishi and K. Kaneko, *J. Phys. A* **25**, 6283 (1992).
22. K. Shinjo, *Phys. Rev. B* **40**, 9167 (1989).
23. A. Saito and K. Kaneko, *Physica* **155D**, 1 (2001).
24. K. Kaneko, *Phys. Rev. Lett.* **65**, 1391 (1990); *Physica* **55D**, 368 (1992).
25. A. S. Pikovsky and J. Kurths, *Phys. Rev. Lett.* **72**, 1644 (1994).
26. T. Shibata and K. Kaneko, *Europhys. Lett.* **38**, 417–422 (1997).
27. S. V. Ershov and A. B. Potapov, *Physica* **86D**, 532 (1995); *Physica* **106D**, 9 (1997).
28. T. Chawanya and S. Morita, *Physica* **116D**, 44 (1998).
29. N. Nakagawa and T. S. Komatsu, *Phys. Rev. E* **57**, 1570 (1998).
30. T. Shibata and K. Kaneko, *Phys. Rev. Lett.* **81**, 4116–4119 (1998).
31. G. Perez et al., *Phys. Rev.* **45A**, 5469 (1992); S. Sinha et al., *Phys. Rev.* **46A**, 3193 (1992).
32. T. Shibata, T. Chawanya, and K. Kaneko, *Phys. Rev. Lett.* **82**, 4424–4427 (1999).
33. K. Matsumoto and I. Tsuda, *J. Stat. Phys.* **31**, 87 (1983).
34. W. Horsthemke and R. Lefever, *Noise-Induced Transitions*, H. Haken, ed., Springer, Berlin, 1984.
35. K. Kaneko and T. Yomo, *Physica* **75D**, 89–102 (1994).
36. K. Kaneko, *Physica* **103D**, 505–527 (1997).
37. C. Furusawa and K. Kaneko, *Bull. Math. Biol.* **60**, 659–687 (1998).
38. C. Furusawa and K. Kaneko, *Artificial Life* **4**, 79–93 (1998).
39. M. Eigen and P. Schuster, *The Hypercycle*, Springer, Berlin, 1979.
40. B. Goodwin, *Temporal Organization in Cells*, Academic Press, London, 1963.
41. B. Hess and A. Boiteux, *Annu. Rev. Biochem.* **40**, 237–258 (1971).
42. B. Alberts, D. Bray, J. Lewis, M. Raff, K. Roberts, and J. D. Watson, *Mol. Biol. Cell* **1989** (1994).
43. M. Ogawa, *Blood* **81**, 2844 (1993).
44. C. Furusawa and K. Kaneko, *J. Theor. Biol.* **209**, 395–416 (2001).
45. For a pioneering work to relate cell differentiation with multiple attractors, see S. Kauffman, *J. Theor. Biol.* **22**, 437 (1969).
46. K. Kaneko and C. Furusawa, *Physica* **280A**, 23–33 (2000); C. Furusawa and K. Kaneko, *Phys. Rev. Lett.* **84**, 6130–6133 (2000).
47. J. B. Gurdon, R. A. Laskey, and O. R. Reeves, *J. Embryol. Exp. Morphol.* **34**, 93–112 (1975); K. H. S. Campbell, J. McWhir, W. A. Ritchie, and I. Wilmut, *Nature* **380**, 64–66 (1996).
48. T. Ariizumi and M. Asashima, *Int. J. Dev. Biol.* **45**, 273–279 (2001).

PART TWO

TRANSPORT AND DIFFUSION

IRREVERSIBILITY IN REVERSIBLE MULTIBAKER MAPS—TRANSPORT AND FRACTAL DISTRIBUTIONS

S. TASAKI*

Department of Physics, Nara Women's University, Nara, Japan; and Institute for Fundamental Chemistry, Kyoto, Japan

CONTENTS

I. Introduction
II. Multibaker Map with Energy Coordinate
III. States and Their Evolution
    A. Measures and Time Evolution
    B. Forward Time Evolution
        1. Steady States
        2. Decay Modes
    C. Backward Time Evolution and Time Reversal Symmetry
    D. Unidirectional Evolution of Measures and Time Reversal Symmetry
IV. Description with Broken Time Reversal Symmetry
    A. Complex Spectral Theory
    B. Multibaker Map Revisited
    C. Condition for Description with Broken Time Reversal Symmetry
V. Measure Selection and Macroscopic Properties
VI. Conclusions
Acknowledgments
References

---

*Present address: Department of Applied Physics and Advanced Institute for Complex Systems, Waseda University, Tokyo, Japan.

Dynamical Systems and Irreversibility: A Special Volume of Advances in Chemical Physics, Volume 122, Edited by Ioannis Antoniou. Series Editors I. Prigogine and Stuart A. Rice.
ISBN 0-471-22291-7. © 2002 John Wiley & Sons, Inc.

I. INTRODUCTION

Consistent incorporation of irreversibility into conservative dynamical systems is one of important problems in physics. Prigogine and his collaborators have been studying this problem [1–5] and recently proposed a formalism, called the complex spectral theory [6–14], which explicitly represents irreversible time evolution of states for conservative systems as a superposition of decaying terms. The early stage of this theory was the subdynamics theory [3–5], where solutions of the Liouville equation are decomposed into components (or subdynamics) each obeying Markovian time evolution. The decomposition is carried out in such a way that the lowest-order approximation satisfies the Markovian master equation obtained in the van Hove limit [4]. Later the subdynamic decomposition was shown to be equivalent to the eigenvalue problem of the Liouville operator corresponding to complex eigenvalues, the imaginary parts of which are relaxation rates for the forward time evolution [6].[1] Because the Liouville operator of a conservative system is Hermitian in a suitable Hilbert space, it is inevitable to change mathematical settings in order for the Liouville operator to have complex eigenvalues. One of the possible settings is the Gelfand triple (i.e., the rigged Hilbert space), where admissible dynamical variables and/or states are restricted and the eigenmodes corresponding to complex eigenvalues are defined as Schwartz' distributions [8]. In many cases, the test function space of the Gelfand triple is invariant only for the forward time evolution; as a result, the evolution becomes a forward semigroup on the space of generalized functions.

Here we remark that the complex spectral theory applied to quantum unstable systems such as the Friedrichs model gives the same results as the contour deformation theory of Sudarshan et al. [15], the rigged Hilbert space approach of Bohm et al. [16], and the complex scaling approach of Combes et al. [17]. In the contour deformation theory, the continuous spectra of the Hamiltonian are analytically continued to complex values and the unstable states are obtained as usual eigenstates of the analytically continued Hamiltonian. In the Bohm's rigged Hilbert space approach, the completeness relation of the energy eigenstates is represented as an energy integral of a product of the S-matrix element and analytical functions and unstable states are obtained from the residues of the S-matrix poles in the second Riemann sheet. Such unstable states are justified as generalized vectors with the aid of suitable rigged Hilbert spaces. In the complex scaling approach, a family of one real-parameter unitary deformation of the Hamiltonian is introduced, the parameter is analytically continued to complex values, and unstable states are

---

[1]Although it was not explicitly explained, one-dimensional subdynamics decomposition of George and Mayné [5] solves the eigenvalue equation of the Liouville operator.

obtained as usual eigenfunctions of the analytically continued non-Hermitian Hamiltonian.

On the other hand, stimulated by the recent progress of the dynamical systems theory, a new approach is developed in classical statistical mechanics [18–42]. This dynamical systems approach mainly deals with hyperbolic systems. Roughly speaking, a hyperbolic system is a system where nearby trajectories of any trajectory either exponentially deviate from or exponentially approach to the original trajectory, or, a system where every trajectory is exponentially unstable along some directions and exponentially stable along the other directions. Because of the exponential instability of every trajectory, a hyperbolic system is chaotic. Note that a hyperbolic system can be either conservative or dissipative. Indeed, there are two directions in the dynamical systems approach: In one direction, dissipative and time-reversal symmetric systems, thermostated systems, are investigated; and in the other direction, open conservative systems are studied. Because we are interested in the conservative systems, the dissipative case is not discussed.

The dynamical systems approach is believed to be sufficiently generic because typical systems are considered to satisfy the Gallavotti–Cohen hypothesis [26,27], which asserts that the microdynamics of an N-body system for large N is of hyperbolic character. And, so far, several new results have been obtained such as relations among Kolmogorov–Sinai entropy, Lyapunov exponents and transport coefficients (escape-rate formalism [18,19,22]), simple symmetry of a large-deviation distribution of the entropy production in non-equilibrium steady states (fluctuation theorem [25,26]), and fractal distributions describing nonequilibrium steady states [23,25–27,31].

While the complex spectral theory asserts that the state evolution becomes semigroup, the dynamical systems approach deals with both forward and backward time evolutions. Then, it is interesting to investigate conditions, which lead to a description with broken time-reversal symmetry. This point will be discussed later.

Note that the dynamical systems approach is different from the conventional ergodic theory because, in the latter, an invariant measure is fixed from the beginning, while in the former, invariant measures are not fixed. As is well known in the dynamical systems theory [43], a hyperbolic system may admit uncountably many invariant (even ergodic) measures, and the selection of a "physical measure" is one of important problems. So far, two candidates are proposed for the physical measure [43]: the Kolmogorov measure and the Sinai–Ruelle–Bowen measure [44]. The former is a zero noise limit of the invariant measure of the given system perturbed by a random noise. The latter describes the time averages of observables on motions where initial data are randomly sampled with respect to the Lebesgue measure. It should be noted that both criteria of selecting physical measures are nondynamical because the noise in the former is of external origin and the Lebesgue measure in the latter is not derived from dynamics. This observation seems to imply that the statistical behavior of a given system cannot be derived from its dynamics alone, contrary to reductionists' view. The problem of measure selection, would be irrelevant for physicists if the choice of an invariant measure did not affect observable phenomena. However, as will be shown later, this is not the case in general.

In this report, using a reversible area-preserving hyperbolic multibaker map, the appearance of irreversible state evolution in reversible conservative systems is illustrated, and several related problems will be discussed.

A multibaker map [19,31,32] is a lattice extension of the conventional baker transformation, which exhibits a deterministic diffusion. The multibaker maps and their generalizations are extensively used to illustrate transport properties of hyperbolic systems including the fractality of nonequilibrium steady states [31, 33], the problem of irreversible entropy production [33,34,36–38,41,42], two or more kinds of transport [37,38,42], fluctuation theorem [40], and new relations between fractality of the steady states and transport coefficients [41]. For the details of recent developments, see the review by Tél and Vollmer [39]. Here, we deal with the multibaker map with energy coordinate introduced in Ref. 37.

The report is organized as follows. In Section II we describe the multibaker map with energy coordinate. The evolution of statistical ensembles starting from smooth initial distributions are discussed in Section III. The nonequilibrium steady state obtained from the forward time evolution and the decay modes are shown to have fractality along the contracting direction, and their properties are consistent with thermodynamics. As a result of the time reversal symmetry, there exists another steady state with anti-thermodynamical properties. The states are shown to evolve unidirectionally from the anti-thermodynamical steady state to the thermodynamically normal steady state in a way consistent with dynamical reversibility. In Section IV, after briefly reviewing the complex spectral theory, we show that a description with broken time-reversal symmetry naturally arises from the requirement of including the (fractal) steady states into a class of admissible initial states. Problem of measure selection is discussed in Section V. The last section is devoted to the summary.

II. MULTIBAKER MAP WITH ENERGY COORDINATE

The multibaker map with energy coordinate introduced in Ref. 37 is a caricature of the periodic Lorentz gas and was constructed based on the following observation: For the $2d$ periodic Lorentz gas, the dimension of the phase space is four: two for the position and two for the momentum of a moving particle. And each trajectory of a particle can be fully determined by the states at collisions, which are specified by the name of a scatterer, the scattering position of the particle on the scatterer (θ), the direction of the particle velocity just before the scattering (ψ), and the kinetic energy of the particle. When one considers the motion of particles having the same total energy, the kinetic energy can be omitted and the dynamics of the Lorentz gas is fully described by a map defined on an array of (θ, ψ)-rectangles, and this map resembles the usual multibaker map [19,39]. On the other hand, in order to describe the motion of particles with different energies, it is necessary to use a map defined on an array of pillars, where the vertical direction corresponds to the kinetic-energy axis and the horizontal section represents a (θ, ψ)-rectangle. A multibaker map with energy coordinate introduced in Ref. 37 mimics this map.

The phase space is a chain Γ of three-dimensional cells as shown in Fig. 1a:

$$\Gamma = \{(n, x, y, E) | n \in \mathbf{Z}, \ E \in \mathbf{R}^+, \quad 0 < x \leq a_E, \ 0 < y \leq a_E\} \quad (1)$$

where $\mathbf{Z}$ and $\mathbf{R}^+$ stand for the sets of integers and of positive real numbers, respectively, and a_E is a positive function of E. The variables n, x, y, and E correspond respectively to the name of a scatterer, θ, ψ (or ψ, θ), and the kinetic energy, and a_E^2 is the area of a constant "kinetic energy" section of each cell. As shown in Ref. 37, the form of a_E^2 is fixed to be $a_E^2 \propto e^{2E}$ by three conditions: (i) conservation of the total energy, (ii) invertibility of the map, and (iii) independence of the transition rates from the "kinetic" energy E.

In case corresponding to a system under constant external field, the multibaker map B_F is given by (cf. Fig. 1b)

$$B_F(n, x, y, E)$$
$$= \begin{cases} \left(n - 1, \frac{x}{l^+ e^F}, l^+ e^F y, E + F\right), & (0 < \frac{x}{a_E} \leq l^-) \\ \left(n, \frac{x - l^- a_E}{s}, sy + l^+ a_E, E\right), & (l^- < \frac{x}{a_E} \leq 1 - l^+) \\ \left(n + 1, \frac{x - (1 - l^+) a_E}{l^- e^{-F}}, \{l^- y + (1 - l^-) a_E\} e^{-F}, E - F\right), \\ \quad (1 - l^+ < \frac{x}{a_E} \leq 1) \end{cases} \quad (2)$$

In the above, F is the parameter corresponding to the applied field, $l^\pm \equiv 2l/(1 + e^{\pm 2F})$ is the transition rate from nth to $(n \pm 1)$th cells, and $s \equiv 1 - 2l$ to the self-transition rate with $l \in (0, 1/2]$ a real parameter.

Because we are interested in transports of open conservative systems, we consider a system where the multibaker chain of length $N + 1$ is embedded between infinitely extended "free chains" (Fig. 2). Then, (2) holds for $n \in [1, N - 1]$. The free motion is modeled by simple shifts, that is,

$$B_F(n, x, y, E) = \begin{cases} (n - 1, x, y, E), & (0 < \frac{x}{a_E} \leq l_f) \\ (n, x, y, E), & (l_f < \frac{x}{a_E} \leq 1 - l_f) \\ (n + 1, x, y, E), & (1 - l_f < \frac{x}{a_E} \leq 1) \end{cases} \quad (3)$$

where $l_f = l^+$ for $n \leq -2$ and $l_f = l^-$ for $n \geq N + 2$.

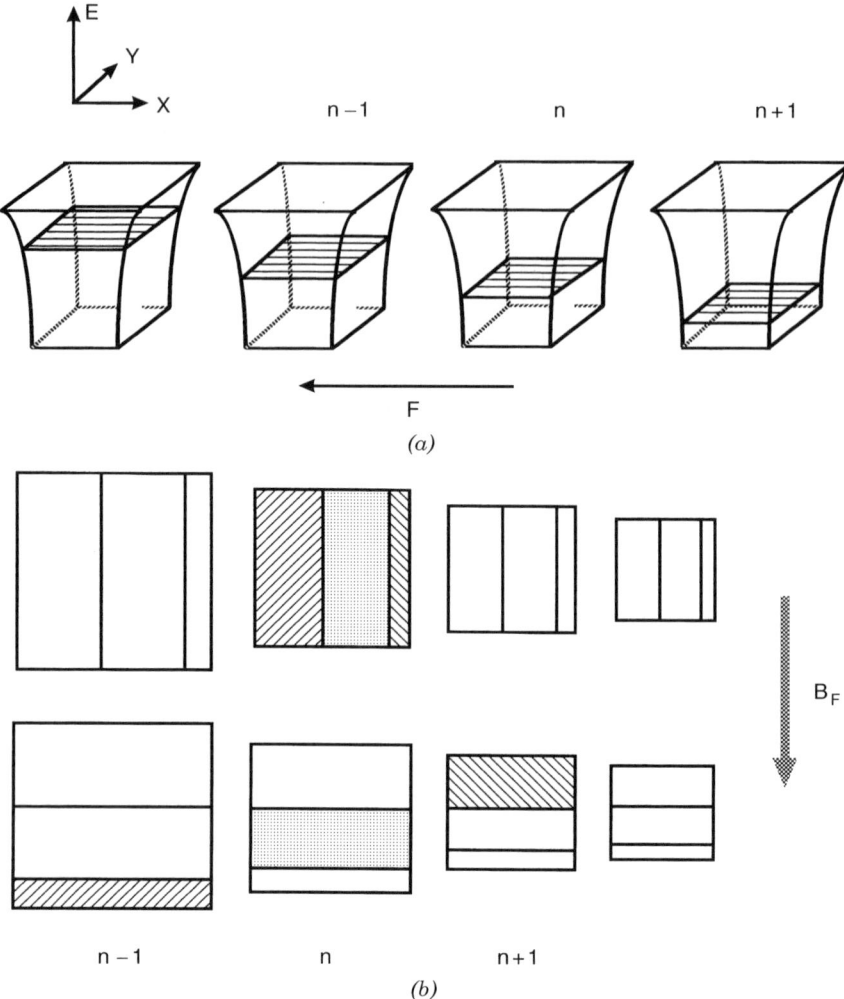

Figure 1. (a) Schematic representation of the phase space Γ. The sectional area at "kinetic energy" E depends on E. The arrow represents the applied field, and hatched squares correspond to a constant total energy surface. (b) Hyperbolic part of the multibaker map B_F on the constant total energy surface.

The sites $n = -1, 0, N, N+1$ are joint sites. The transformation rules for $n = -1$ and $n = N+1$ are mixtures of (2) and (3): $B_F(-1, x, y, E)$ for $1 - l^+ < x/a_E \leq 1$ and $B_F(N+1, x, y, E)$ for $0 < x/a_E \leq l^-$ are given by (2), and $B_F(-1, x, y, E)$ for $0 < x/a_E \leq 1 - l^+$ and $B_F(N+1, x, y, E)$ for $l^- < x/a_E \leq 1$ by (3). The transformation rule for the sites $n = 0$ and $n = N$

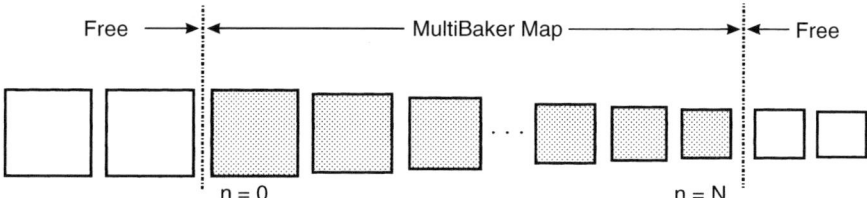

Figure 2. Overall view of the constant energy surface of the open multibaker map B_F. In the shaded cells the map is hyperbolic, and in the unshaded cells the map is a simple shift.

is partly given by (2); that is, $B_F(0, x, y, E)$ for $l^- < x/a_E \leq 1$ and $B_F(N, x, y, E)$ for $0 < x/a_E \leq 1 - l^+$ are given by (2). And the rest cases are given by

$$B_F(0, x, y, E) = \left(-1, l^+ e^F y, \frac{x}{l^- e^{-F}}, E + F\right), \quad \left(0 < \frac{x}{a_E} \leq l^-\right) \quad (4)$$

$$B_F(N, x, y, E) = \left(N + 1, \{l^- y + (1 - l^-)a_E\}e^{-F}, \frac{x - (1 - l^+)a_E}{l^+ e^F}, E + F\right),$$
$$\left(1 - l^+ < \frac{x}{a_E} \leq 1\right) \quad (5)$$

Equations (2)–(5) provide the transformation rule of the multibaker map with energy coordinate imbedded between free chains.

As clearly seen from (2), the phase space is stretched along the x axis and is contracted along the y axis, or the multibaker part is hyperbolic. Contrarily, the dynamics of the free parts is a simple shift operation and there is no phase space deformation. Hence, the present system is of chaotic scattering type.

Before closing this section, we remark that the map B_F is time reversal symmetric. Indeed, let

$$I(n, x, y, E) \equiv \begin{cases} (n, a_E - y, a_E - x, E) & \text{(for } 0 \leq n \leq N\text{)} \\ (n, a_E - x, a_E - y, E) & \text{(for } n \leq -1 \text{ or } n \geq N + 1\text{)} \end{cases} \quad (6)$$

Then it is an involution: $I^2 = I$ and $IB_F^t I = B_F^{-t}$.

III. STATES AND THEIR EVOLUTION

A. Measures and Time Evolution

We are interested in a situation where identical particles are distributed to the whole system with finite density. This can be described by the Poisson suspension [33,45] because the multibaker map is a toy model of the Lorentz gas which consists of particles without mutual interaction and admits an equilibrium state given by a Poisson suspension.

Each configuration of the system is specified by a list of particle coordinates $\zeta \equiv \{z_1, z_2, \ldots\}$ with $z_j = (n_j, x_j, y_j, E_j) \in \Gamma$ ($j = 1, 2, \ldots$), which is not ordered and where repetitions are permitted. Because the particle density is assumed to be finite everywhere, the number of particles included in any compact subset K of Γ should be finite, or only configurations ζ satisfying $\#(\zeta \cap K) < +\infty$ are allowed where $\#$ stands for the cardinal number of the subsequent set. The collection of such lists forms the phase space X called the configuration space.[2]

A probability measure on X is defined by specifying probability of finding n particles ($n = 0, 1, \ldots$) in any compact subset $K \subset \Gamma$. Here we consider the case where the number of particles in a compact set K obeys the Poisson distribution. More precisely, let $C_{K,m} = \{\zeta \in X | \#(\zeta \cap K) = m\}$ be a set of configurations with m particles in a set K; then the Poisson measure P is defined [45] on the minimal σ-algebra containing all sets of the form $C_{K,m}$ by putting

$$P(C_{K,m}) = \frac{\nu(K)^m}{m!} e^{-\nu(K)} \tag{7}$$

$$P(C_{K,m} \cap C_{K',m'}) = P(C_{K,m}) P(C_{K',m'}) \tag{8}$$

where $\nu(K)$ is a measure on Γ representing the average number of particles in a set K (hence it is not a probability measure) and the sets $K, K' \subset \Gamma$ are disjoint.

The multibaker map B_F induces dynamics of configurations via $\hat{B}_F^t \zeta \equiv \{B_F^t z_1, B_F^t z_2, \ldots\}$. The time evolution of the Poisson measure induced by the map $\hat{B}_F$ is given by

$$P_t(C_{K,m}) = P(\hat{B}_F^{-t} C_{K,m}) = \frac{\nu_t(K)^m}{m!} e^{-\nu_t(K)} \tag{9}$$

where $\nu_t(K) \equiv \nu(B_F^{-t} K)$ is the evolved measure on Γ.

When the initial measure ν_0 is absolutely continuous with respect to the Lebesgue measure with density ρ_0, it is convenient to consider the time evolution of the partially integrated distribution G_t at fixed total energy E, which is defined by

$$G_t(n, x, y, E) \equiv \int_0^y dy' \rho_0(B_F^{-t}(n, x, y', E - nF)) \tag{10}$$

Let $\tilde{G}_t$ be the partially integrated distribution function expressed in terms of the rescaled coordinates $\xi \equiv x/a_{E,n} \in [0, 1]$ and $\eta \equiv y/a_{E,n} \in [0, 1]$ with

---

[2] For technical reasons, lists of finite number of particle coordinates are also included in X [45].

$a_{E,n} = ae^E e^{-Fn}$; then the evolution equation of the partially integrated distribution [37] is

$$\tilde{G}_{t+1}(n, \xi, \eta, E) = l^- e^{-F} \tilde{G}_t\left(n + 1, l^- \xi, \frac{\eta}{l^+}, E\right) \quad (11)$$

for $\eta \in [0, l^+)$;

$$\tilde{G}_{t+1}(n, \xi, \eta, E) = l^- e^{-F} \tilde{G}_t(n + 1, l^- \xi, 1, E) + s \, \tilde{G}_t\left(n, s\xi + l^-, \frac{\eta - l^+}{s}, E\right) \quad (12)$$

for $\eta \in [l^+, 1 - l^-)$; and

$$\tilde{G}_{t+1}(n, \xi, \eta, E) = l^- e^{-F} \tilde{G}_t(n + 1, l^- \xi, 1, E) + s \, \tilde{G}_t(n, s\xi + l^-, 1, E) + l^+ e^F \tilde{G}_t\left(n - 1, l^+ \xi + 1 - l^+, \frac{\eta - 1 + l^-}{l^-}, E\right) \quad (13)$$

for $\eta \in [1 - l^-, 1]$.

Here we consider a class of initial distributions where particles are uniformly distributed with respect to the Lebesgue measure on the left chain $(-\infty, -1]$ with particle density (per energy) $\rho_-(E)$ and on the right chain $[N + 1, +\infty)$ with particle density $\rho_+(E)$. When one is interested in the state evolution only in the middle chain, this initial condition is equivalent to the flux boundary conditions [31]

$$\tilde{G}_t(-1, \xi, \eta, E) = \rho_-(E) a_E e^F \eta, \qquad \tilde{G}_t(N + 1, \xi, \eta, E) = \rho_+(E) a_E e^{-(N+1)F} \eta \quad (14)$$

B. Forward Time Evolution

1. Steady States

Equations (11)–(14) were shown to admit a unique stationary solution $\tilde{G}_\infty$ which does not depend on ξ [37]. In terms of the measure $\nu_{+\infty}$ represented by $\tilde{G}_\infty$ and three sets

$$\Gamma_n(E, \Delta) = \left\{(n, x, y, E' - nF) \middle| 0 \leq \frac{x}{a_{E',n}} < 1, 0 \leq \frac{y}{a_{E',n}} < 1, E < E' < E + \Delta\right\} \quad (15)$$

$$R_n(E, \Delta) = \left\{(n, x, y, E' - nF) \middle| 1 - l^+ \leq \frac{x}{a_{E',n}} < 1, 0 \leq \frac{y}{a_{E',n}} < 1, E < E' < E + \Delta\right\} \quad (16)$$

$$L_n(E, \Delta) = \left\{(n, x, y, E' - nF) \middle| 0 \leq \frac{x}{a_{E',n}} < l^-, 0 \leq \frac{y}{a_{E',n}} < 1, E < E' < E + \Delta\right\} \quad (17)$$

the particle distribution $\Pi_{+\infty}(n, E)$ per energy per site is given by

$$\Pi_{+\infty}(n, E) \equiv \lim_{\Delta \to 0} \nu_{+\infty}(\Gamma_n(E, \Delta))/\Delta$$
$$= a_E^2 \left\{ (\rho_+(E) - \rho_-(E)) \frac{e^{-2(N+1)F}(1 - e^{-2(n+1)F})}{1 - e^{-2(N+2)F}} + \rho_-(E)e^{-2nF} \right\} \quad (18)$$

and the corresponding flow $J_{n|n+1}^{+\infty}(E)$ from the nth to the $(n+1)$th sites is given by

$$J_{n|n+1}^{+\infty}(E) \equiv \lim_{\Delta \to 0} \{\nu_{+\infty}(R_n(E, \Delta)) - \nu_{+\infty}(L_{n+1}(E, \Delta))\}/\Delta$$
$$= l^+ \Pi_{+\infty}(n, E) - l^- \Pi_{+\infty}(n+1, E)$$
$$= -\frac{2l}{1 + e^{-2F}} (\rho_+(E) - \rho_-(E)) \frac{a_E^2 e^{-2(N+1)F}(1 - e^{-2F})}{1 - e^{-2(N+2)F}} \quad (19)$$

Obviously, when $\rho_+ \neq \rho_-$, the flow is nonvanishing.

When $F = 0$, the particle distribution $\Pi_{+\infty}(n, E)$ is linear in the site coordinate n:

$$\Pi_{+\infty}(n, E) = \frac{\Pi_{+\infty}(N+1, E) - \Pi_{+\infty}(-1, E)}{N + 2}(n + 1) + \Pi_{+\infty}(-1, E) \quad (20)$$

and Fick's law of diffusion holds:

$$J_{n|n+1}^{+\infty}(E) = -l \frac{\Pi_{+\infty}(N+1, E) - \Pi_{+\infty}(-1, E)}{N + 2} \quad (21)$$

This flow-density relation and, thus, the steady state $\tilde{G}_\infty$ are not time reversal symmetric. This is also the case for $F \neq 0$.

In terms of the particle distribution and the flow, the intracell distribution is given by

$$a_{E,n}\tilde{G}_{+\infty}(n, \eta, E) = \Pi_{+\infty}(n, E)\eta - \frac{J_{n|n+1}^{+\infty}(E)}{l} \varphi_n(\eta) \quad (22)$$

where φ_n is defined as the unique solution of a functional equation

$$\varphi_n(\eta) = \begin{cases} l^- \varphi_{n+1}\left(\frac{\eta}{l^+}\right) + \frac{l}{l^+} \eta, & (0 \leq \eta \leq l^+) \\ s\, \varphi_n\left(\frac{\eta - l^+}{s}\right) + l, & (l^+ \leq \eta \leq 1 - l^-) \\ l^+ \varphi_{n-1}\left(\frac{\eta - 1 + l^-}{l^-}\right) + \frac{l}{l^-}(1 - \eta), & (1 - l^- \leq \eta \leq 1) \end{cases} \quad (23)$$

with the boundary conditions $\varphi_{-1}(\eta) = \varphi_{N+1}(\eta) = 0$. As shown in Fig. 3, the function φ_n has a self-similar graph and the intracell distribution of the steady state $\tilde{G}_\infty$ is fractal [31,37]. However, the self-similarity is not complete. Indeed, the steady-state measure $\tilde{G}_\infty$ is absolutely continuous with respect to the Lebesgue measure with a density $\rho_{+\infty}(n, y, E - nF)$. The density takes $\rho_-(E)$ or

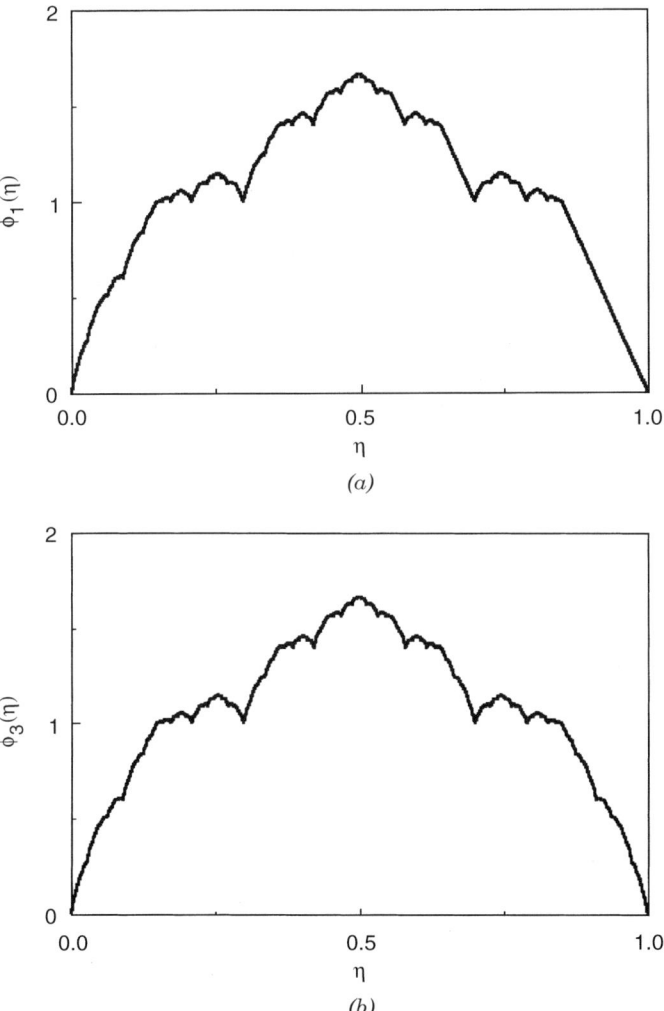

Figure 3. The fractal part $\varphi_n(\eta)$ of the partially integrated distribution $\tilde{G}_{+\infty}$ versus the rescaled intracell coordinate η for $l^+ = l^- = 0.3$ and $N = 8$. The functions φ_n ($n = 1, 3, 5,$ and 7) are shown, respectively, in (a), (b), (c), and (d).

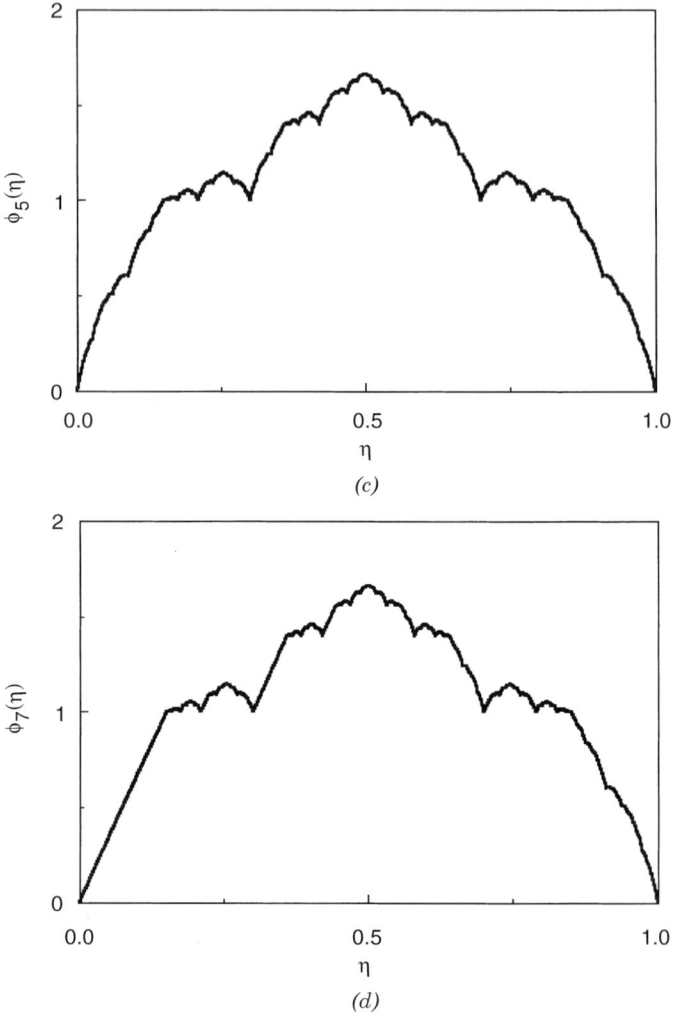

Figure 3. (*Continued*)

$\rho_+(E)$ on the complement of the stable manifold of a fractal set, called the fractal repeller, which has zero Lebesgue measure [33].

2. Decay Modes

The evolution equations (11)–(13) provide more information. With the aid of a method of Ref. 46 (see also Ref. 47), one can show that the partially integrated distribution of any initial measure, which is absolutely continuous with respect to the Lebesgue measure and possesses a density piecewise continuously differentiable in ξ, approaches the steady-state distribution $\tilde{G}_{+\infty}$ when $t \to +\infty$:

$$\tilde{G}_t(n,\xi,\eta,E) - \tilde{G}_{+\infty}(n,\eta,E) = \sum_{\substack{j=1 \\ |\kappa_j|>\lambda}}^{N+1} \kappa_j^t b_j(E)\gamma_j(n,\eta) + \delta\mathscr{G}_t(n,\xi,\eta,E) \quad (24)$$

where $\lambda = \max(1 - 2l, \sqrt{l^+l^-})$, the j-sum runs over all $j = 1,\ldots,N+1$ satisfying $|\kappa_j| > \lambda$, a function $\delta\mathscr{G}_t$ decays as $|\delta\mathscr{G}_t| = O(t^2\lambda^t)$ uniformly with respect to n, ξ and η, and the relaxation rates $\kappa_j < 1$ are given by

$$\kappa_j = 1 - 2l + 2\sqrt{l^+l^-}\cos\left(\frac{\pi j}{N+2}\right) \quad (25)$$

Those rates correspond to the Pollicott–Ruelle resonances [48,49].

The function $\gamma_j(n,\eta)$ describing intracell distribution of the jth decay mode is defined as the unique solution of a functional equation:

$$\gamma_j(n,\eta) = \begin{cases} \frac{\sqrt{l^+l^-}}{\kappa_j}\gamma_j(n+1, \frac{\eta}{l^+}), & (0 \leq \eta \leq l^+) \\ \frac{s}{\kappa_j}\gamma_j\left(n, \frac{\eta-l^+}{s}\right) + \frac{\sqrt{l^+l^-}}{\kappa_j}\sin\left(\frac{(n+2)\pi j}{N+2}\right), & (l^+ \leq \eta \leq 1-l^-) \\ \frac{\sqrt{l^+l^-}}{\kappa_j}\gamma_j\left(n-1, \frac{\eta-1+l^-}{l^-}\right) & \\ \quad + \frac{s}{\kappa_j}\sin\left(\frac{(n+1)\pi j}{N+2}\right) + \frac{\sqrt{l^+l^-}}{\kappa_j}\sin\left(\frac{(n+2)\pi j}{N+2}\right), & (1-l^- \leq \eta \leq 1) \end{cases} \quad (26)$$

and the coefficient $b_j(E)$ is determined by the initial measure:

$$b_j(E) = \frac{2}{N+2}\sum_{n=0}^{N}\int_0^1 d\xi \bar{\gamma}_j(n,\xi)\{\tilde{G}_0(n,\xi,1,E) - \tilde{G}_{+\infty}(n,1,E)\} \quad (27)$$

where $\bar{\gamma}_j(n,\xi) \equiv \gamma_j(n,1) - \gamma_j(n,1-\xi)$. The function $\gamma_j(n,\eta)$ is dipicted in Fig. 4, and its graph is fractal. As seen in the figure, faster decay modes have more singular distributions. The slowest decay mode (the $j = 1$ mode) is a conditionally invariant measure [50] for a chaotic repeller formed under the absorbing boundary condition: $\rho_+(E) = \rho_-(E) = 0$.

We remark that particle distribution per site (18) of the steady state $\nu_{+\infty}$ and relaxation rates (25) are identical to the corresponding quantities for a one-dimensional random walk shown in Fig. 5, where the transition rate to the left adjacent site (the right adjacent site) is $l^-(l^+)$. Moreover, the macroscopic distribution corresponding to the decay mode (26) is

$$\Pi_j(n,E) = a_{E,n}\int_0^1 d\xi\gamma_j(n,1) = a_E\left(\frac{l^+}{l^-}\right)^{n/2}\sin\left(\frac{\pi(n+1)j}{N+2}\right) \quad (28)$$

which is again equal to the decay mode of the random walk. In short, when $t \to +\infty$, the evolution of the measure is equivalent to a random walk, which is an irreversible stochastic process.

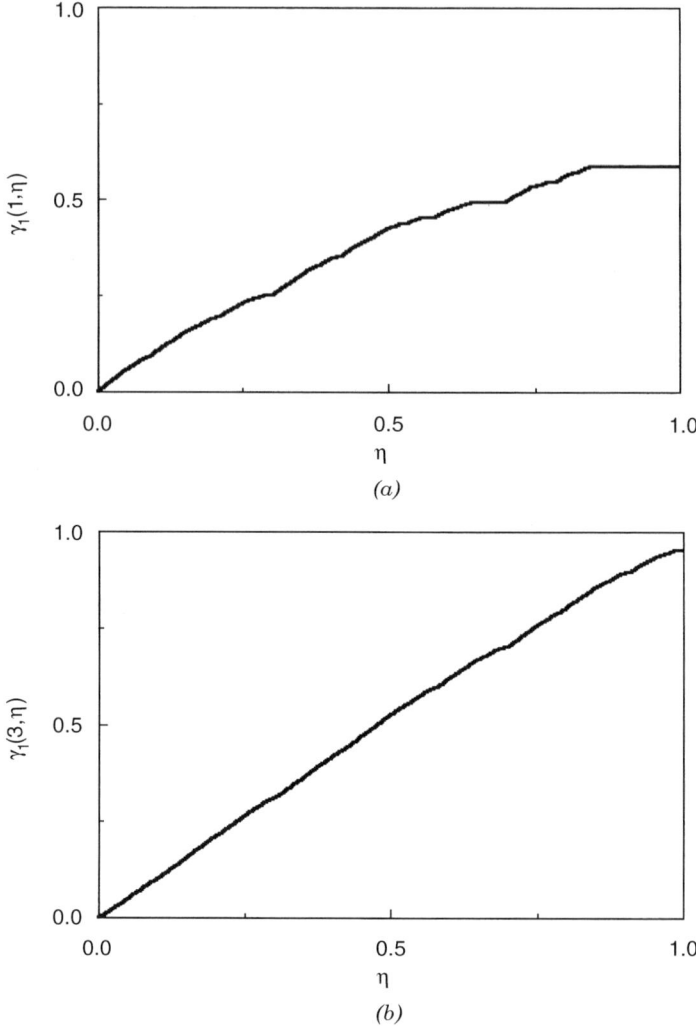

Figure 4. Intracell distributions of the jth decay modes $\gamma_j(n, \eta)$ versus the rescaled intracell coordinate η for $l^+ = l^- = 0.3$ and $N = 8$. The distributions $\gamma_1(n, \eta)$ ($n = 1, 3, 5$, and 7) of the first mode are shown in (a), (b), (c), and (d), and those $\gamma_3(n, \eta)$ ($n = 1, 3, 5, 7$) of the third mode are shown in (e), (f), (g), and (h).

Furthermore, along the line of thoughts of Gaspard [33] and Gilbert and Dorfman [36], the production of a relative coarse-grained entropy of $\tilde{G}_\infty$ with respect to the Lebesgue measure was shown to be positive and to reduce to the phenomenological expression in an appropriate scaling limit [37]. Thus, the steady state $\tilde{G}_\infty$ is irreversible in this sense as well.

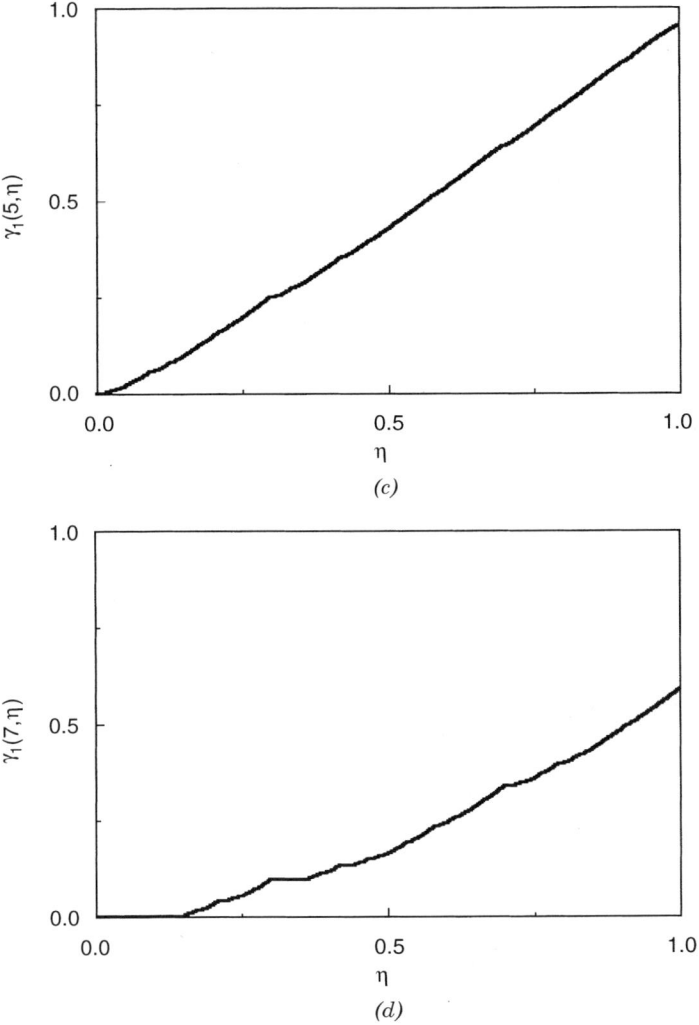

Figure 4. (*Continued*)

C. Backward Time Evolution and Time Reversal Symmetry

Now we study the backward time evolution. Then the state is conveniently described by another partially integrated distribution function:

$$H_t(n, x, y, E) \equiv \int_0^x dx' \rho_0\left(B_F^{-t}(n, x', y, E - nF)\right) \tag{29}$$

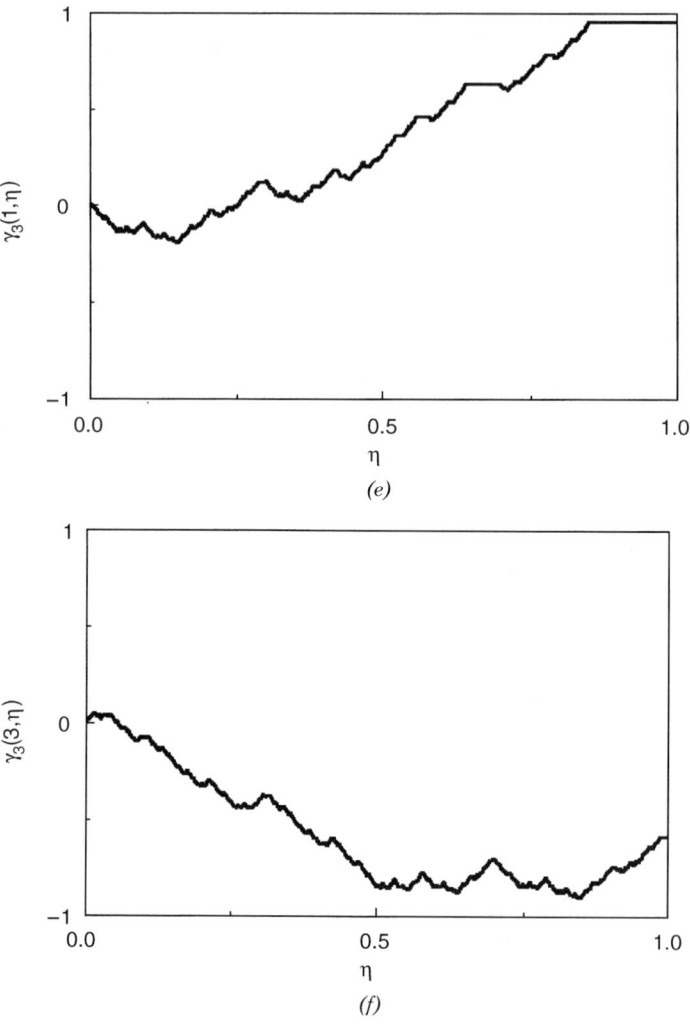

Figure 4. (*Continued*)

or its rescaled version $\tilde{H}_t(n, \xi, \eta, E) \equiv H_t(n, \xi a_{E,n}, \eta a_{E,n}, E)$. Because of the time reversal symmetry of the map B_F, $\tilde{H}_t$ for $0 \leq n \leq N$ is related to the partially integrated distribution $\tilde{G}_t$ via the time reversal operation I:

$$\tilde{H}_t(n, \xi, \eta, E) = a_{E,n} \int_0^\xi d\xi' \rho_0(B_F^{-t} I(n, (1-\eta)a_{E,n}, (1-\xi')a_{E,n}, E - nF))$$
$$= \{\tilde{G}_{-t}(n, 1-\eta, 1, E) - \tilde{G}_{-t}(n, 1-\eta, 1-\xi, E)\}_{\rho_0 \circ I} \quad (30)$$

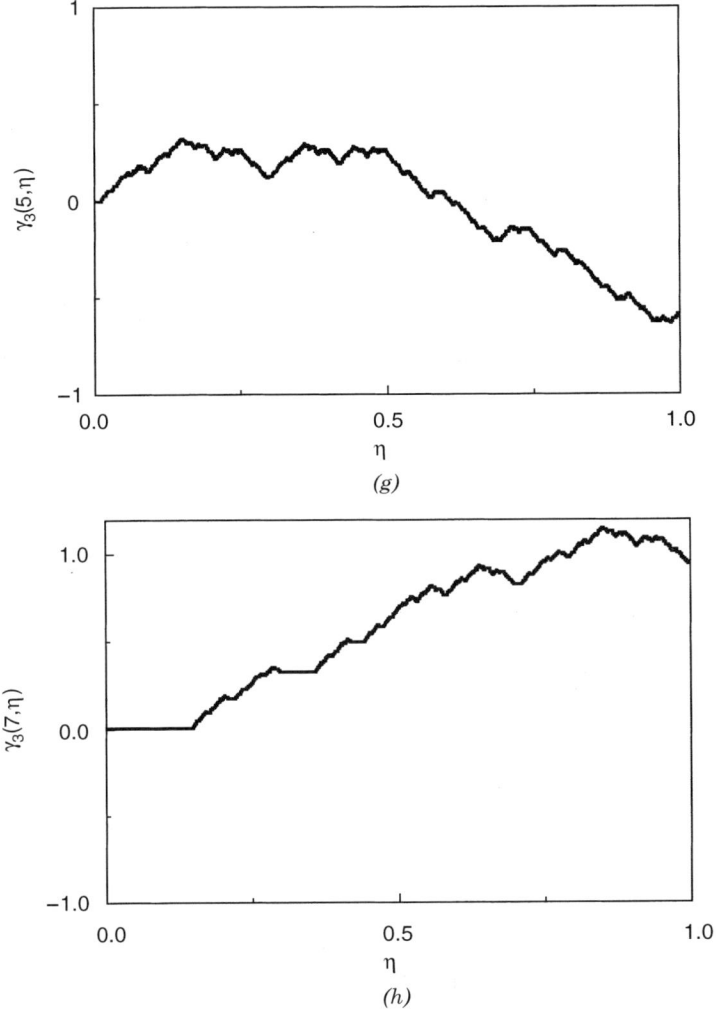

Figure 4. (*Continued*)

where the subscript $\rho_0 \circ I$ implies that the initial state of $\tilde{G}_{-t}$ has a density $\rho_0 \circ I$. Applying the previous result (24), if $\rho_0 \circ I(n, x, y, E)$ is piecewise continuously differentiable in x or $\rho_0(n, x, y, E)$ is piecewise continuously differentiable in y, one finds for $t < 0$

$$\tilde{H}_t(n, \xi, \eta, E) - \tilde{H}_{-\infty}(n, \xi, E) = \sum_{\substack{j=1 \\ |\kappa_j| > \lambda}}^{N+1} \kappa_j^{|t|} \bar{b}_j(E) \bar{\gamma}_j(n, \xi) + \delta \mathscr{H}_t(n, \xi, \eta, E) \quad (31)$$

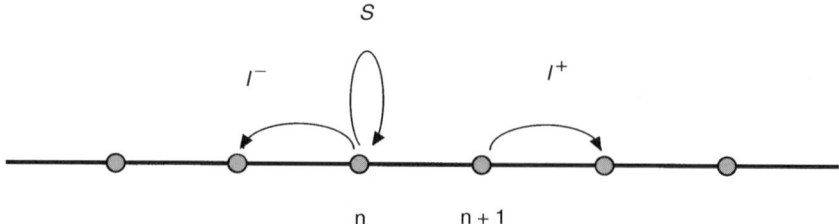

Figure 5. A random walk corresponding to the asymptotic evolution of measure ν_t ($t \gg 1$). The transition probabilities to the left and right adjacent sites are, respectively, l^- and l^+. The random walk has the same decay rate (25), and its stationary distribution and decay modes have the same expression as those (18) and (28) for the multibaker map B_F.

where $\delta\mathcal{H}_t = O(t^2 \lambda^{|t|})$, $\bar{\gamma}_j$ is defined just below (27), the steady-state $\tilde{H}_{-\infty}$ is defined by

$$a_{E,n}\tilde{H}_{-\infty}(n, \xi, E) = \Pi_{+\infty}(n, E)\xi + \frac{J_{n|n+1}^{+\infty}(E)}{l} \varphi_n(1-\xi) \qquad (32)$$

and the coefficient $\bar{b}_j(E)$ is given by

$$\bar{b}_j(E) = \frac{2}{N+2} \sum_{n=0}^{N} \int_0^1 d\gamma_j(n, \eta)\{\tilde{H}_0(n, 1, \eta, E) - \tilde{H}_{-\infty}(n, 1, E)\} \qquad (33)$$

In the above, we have used $\tilde{H}_{-\infty}(n, 1, E) = \tilde{G}_{+\infty}(n, 1, E) = \Pi_{+\infty}(n, E)/a_{E,n}$. Note that two steady states $\nu_{\pm\infty}$ are time-reversed states with each other. Equation (31) shows that when $t \to -\infty$, the measure converges to a steady-state measure $\nu_{-\infty}$ represented by $\tilde{H}_{-\infty}$. The functions $\bar{\gamma}_j(n, \xi)$ represent relaxation for $t \to -\infty$, or growing for $t \to +\infty$, and, thus, will be referred to as grow modes.

The particle distribution $\Pi_{-\infty}(n, E)$ per site per energy of the steady state $\nu_{-\infty}$ is equal to that of the state $\nu_{+\infty}$:

$$\Pi_{-\infty}(n, E) = \lim_{\Delta \to 0} \nu_{-\infty}(\Gamma_n(E, \Delta))/\Delta = \Pi_{+\infty}(n, E) \qquad (34)$$

Similarly, the macroscopic distribution $\bar{\Pi}_j(n, E)$ corresponding to the grow mode is equal to that of the decay mode (28): $\bar{\Pi}_j(n, E) = \Pi_j(n, E)$. The difference between the two steady states $\nu_{+\infty}$ and $\nu_{-\infty}$ as well as that between the decay and grow modes lie in the direction along which the distributions possess fractality. Indeed, the steady state $\nu_{+\infty}$ and decay modes are fractal along the contracting y direction, while the state $\nu_{-\infty}$ and grow modes are fractal along the expanding x direction.

The two steady states are different also in their flows. The flow $J^{-\infty}_{n|n+1}$ in the state $\nu_{-\infty}$ is opposite to the flow $J^{+\infty}_{n|n+1}$ in the state $\nu_{+\infty}$:

$$J^{-\infty}_{n|n+1}(E) \equiv \lim_{\Delta \to 0} \{\nu_{-\infty}(R_n(E,\Delta)) - \nu_{-\infty}(L_{n+1}(E,\Delta))\}/\Delta = -J^{+\infty}_{n|n+1}(E) \quad (35)$$

Because the state $\nu_{+\infty}$ has consistent properties with the second law of thermodynamics, the state $\nu_{-\infty}$ is anti-thermodynamical. For example, the anti-Fick law holds for $F = 0$:

$$J^{-\infty}_{n|n+1}(E) = l \frac{\Pi_{-\infty}(N+1,E) - \Pi_{-\infty}(-1,E)}{N+2} \quad (36)$$

which corresponds to a negative diffusion coefficient.

D. Unidirectional Evolution of Measures and Time Reversal Symmetry

We have shown that the measure evolves to the steady state $\nu_{+\infty}$ for $t \to +\infty$ and to the state $\nu_{-\infty}$ for $t \to -\infty$. More precisely, this is the case when the initial density $\rho_0(n,x,y,E)$ is piecewise continuously differentiable in both x and y; that is, the support $\text{supp}\rho_0$ consists of a finite number of sets with smooth boundaries: $\text{supp}\rho_0 = \cup_{m=1}^M D_m$ and ρ_0 is continuously differentiable in both x and y on the closure of each set D_m with bounded derivatives. Because the state $\nu_{+\infty}$ attracts the class of initial states for $t \to +\infty$, it behaves as an attracting fixed point in the space of states. On the other hand, the initial states evolve toward $\nu_{-\infty}$ for $t \to -\infty$, and this implies that the initial states deviate from $\nu_{-\infty}$ for $t \to +\infty$, or the state $\nu_{-\infty}$ behaves as a repelling fixed point. Thus, the states evolve unidirectionally from the repeller $\nu_{-\infty}$ to the attractor $\nu_{+\infty}$ (c.f. Fig. 6).

This unidirectional evolution is fully consistent with the time reversal symmetry of the map B_F. To clarify this point, we consider the following thought experiment. At $t = 0$, the system is prepared to be in a state ν_0, which is one of the initial states described before (14). Until time $t = t_1 (> 0)$, the system evolves according to the multibaker map B_F. At $t = t_1$, time reversal operation I is applied. After $t = t_1$, the system again evolves according to B_F. Let $\mathscr{B}_F$ and $\mathscr{I}$ be transformations of measures induced by the multibaker map B_F and the time reversal operation I:

$$\mathscr{B}_F \nu(K) \equiv \nu(B_F^{-1} K), \qquad \mathscr{I}\nu(K) \equiv \nu(I^{-1} K) \quad (37)$$

where $K \subset \Gamma$ is an arbitrary measurable set. Then the properties of B_F and I implies

$$\mathscr{I}\mathscr{B}_F\mathscr{I} = \mathscr{B}_F^{-1}, \qquad \mathscr{I}^2 = \mathscr{I}$$

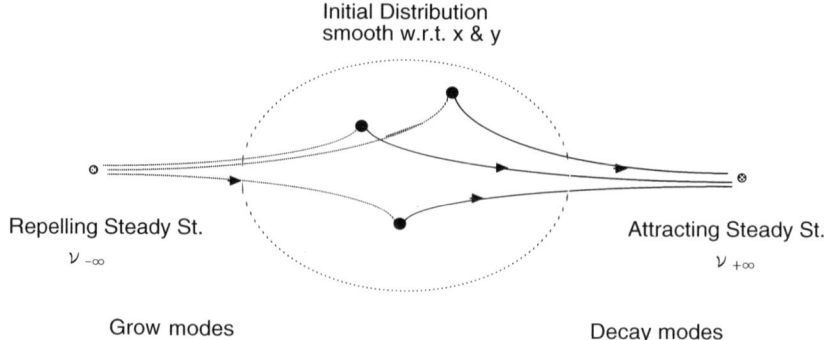

Figure 6. Schematic view of the state evolution. The steady state $v_{+\infty}$ with normal transport properties behaves as an attracting fixed point, and the steady state $v_{-\infty}$ with anti-thermodynamical properties behaves as a repelling fixed point. And any initial state unidirectionally evolves from the repeller $v_{-\infty}$ to the attractor $v_{+\infty}$. Attracting and repelling processes are described, respectively, by decay modes and grow modes.

and the state v_t at time t is given by

$$v_t = \begin{cases} \mathscr{B}_F^t v_0 & (0 \leq t < t_1) \\ \mathscr{B}_F^{t-2t_1} \mathscr{I} v_0 & (t_1 \leq t) \end{cases} \quad (38)$$

Note that the state $\mathscr{I} v_0$ is one of the initial states described before (14).

As the initial state v_0 evolves toward the steady state $v_{+\infty}$ (cf. (24)), the state just before the time reversal operation $v_{t_1-} = \mathscr{B}_F^{t_1} v_0$ is close to $v_{+\infty}$ for large t_1. On the other hand, as $-t_1 < 0$, the state $v_{t_1+} = \mathscr{B}_F^{-t_1} \mathscr{I} v_0$ just after the time reversal operation is close to the other stationary state $v_{-\infty}$. Afterwards, the state $v_t = \mathscr{B}_F^{t-2t_1} \mathscr{I} v_0$ deviates from $v_{-\infty}$ and reaches $\mathscr{I} v_0$ at time $t = 2t_1$. Then, the state v_t again approaches $v_{+\infty}$. Thus, the time reversal operation discontinuously changes a state v_{t_1-} close to $v_{+\infty}$ into a state v_{t_1+} close to $v_{-\infty}$, but does not invert the evolution. In this way, the time reversal symmetry is consistent with the unidirectional state evolution. A similar view was given by Prigogine et al. [1,51] for the behavior of entropy under time reversal experiments, where their entropy increases in dynamical evolution and discontinuously decreases by the time reversal operation.

IV. DESCRIPTION WITH BROKEN TIME REVERSAL SYMMETRY

In the previous section, forward and backward time evolutions of states are investigated and a class of initial states, which possess piecewise continuously differentiable densities with respect to x and y, are shown to approach steady states $v_{\pm\infty}$ for $t \to \pm\infty$ in a way consistent with dynamical reversibility. However, the limits $v_{\pm\infty}$ do not belong to the same class as the initial states, and the results are not fully satisfactory. To improve this point, one needs a different mathematical setting, which does not allow the asymptotic expressions of the forward and backward state evolutions ((24) and (31)) to hold simultaneously. This is also the case for the complex spectral theory. Then, we begin with a brief summary of the mathematical aspects of the complex spectral theory following the arguments of [9] (for details, see Refs. 6–14).

A. Complex Spectral Theory

Let (X, T_t, μ_0) be a classic dynamical system, where X, T_t ($-\infty < t < +\infty$) and μ_0 are, respectively, a phase space, a flow representing the dynamical evolution of each point in X, and an invariant probability measure defined on X. Then, when the system is mixing, the expectation value of an observable A at time t with respect to an initial state with density ρ_0 behaves as follows:

$$\lim_{t \to +\infty} \int_X A(z) U_t \rho_0(z) \, d\mu(z) = \int_X A(z) \, d\mu(z) \int_X \rho_0(z) \, d\mu(z)$$
$$= \int_X A(z) \, d\mu(z) \qquad (39)$$

where U_t is an evolution operator of densities defined by $U_t\rho_0(z) \equiv \rho_0(T_t^{-1}z)$ and we have used the normalization condition $\int_X \rho_0(z) \, d\mu(z) = 1$. As often discussed (see, e.g., Refs. 2 and 52), this asymptotic behavior can be identified with a relaxation process to the equilibrium state. The main objective of the complex spectral theory is to express an average $\langle A \rangle_t \equiv \int_X A(z) U_t \rho_0(z) \, d\mu(z)$ as a superposition of decaying components and to obtain them as a complete set of generalized eigenvectors of the evolution operator U_t.

We restrict ourselves to the simplest case where the expectation values at time t are superpositions of countably many exponentially decaying terms. Then, provided that the initial density ρ_0 and the real-valued observable A belong to certain classes of functions, one has

$$\langle A \rangle_t = \sum_{j=0}^{\infty} f_j(A^*) e^{-i\alpha_j t} \tilde{f}_j^*(\rho_0) \qquad (t > 0) \qquad (40)$$

where $f_j(\cdot)$ and $\tilde{f}_j(\cdot)$ are antilinear functionals, particularly

$$f_0(B) = \int_X B^*(z) \, d\mu(z), \qquad \tilde{f}_0(\rho_0) = \int_X \rho_0^*(z) \, d\mu(z)$$

with $\alpha_0 = 0$ and α_j's are complex values with $\text{Im}\alpha_j \leq 0$. Following Dirac's notation in quantum theory, the functionals are abbreviated as

$$f_j(B) = \langle B|f_j\rangle, \qquad \tilde{f}_j^*(\rho_0) = \langle \tilde{f}_j|\rho_0\rangle$$

then

$$\langle A^*|U_t\rho_0\rangle = \sum_{j=0}^{\infty} \langle A^*|f_j\rangle e^{-i\alpha_j t}\langle \tilde{f}_j|\rho_0\rangle \tag{41}$$

The right-hand side is the sesquilinear form defined by

$$\langle B|C\rangle \equiv \int_X B^*(z)C(z)\, d\mu(z)$$

As easily seen, in a Hilbert space $\mathscr{L}^2 \equiv \{A | \int_X |A(z)|^2 d\mu(z) < +\infty\}$, the evolution operator U_t is unitary: $\langle U_t A|U_t B\rangle = \langle A|B\rangle$. Therefore, in order to justify the decomposition (40), a different mathematical setting is necessary. We consider a space $\Phi_+ \subset \mathscr{L}^2$ of observables A and a space $\Phi_- \subset \mathscr{L}^2$ of initial densities ρ_0, for which the decomposition (40) is possible, and we assume that the spaces carry some topologies stronger than the Hilbert space topology, are complete with respect to them and satisfy

$$U_t^\dagger \Phi_+ \subset \Phi_+, \qquad U_t \Phi_- \subset \Phi_- \qquad (t > 0) \tag{42}$$

where $U_t^\dagger A(z) \equiv A(T_t z)$. Let $\Phi'_\pm$ be antiduals of $\Phi_\pm$, namely spaces of continuous antilinear functionals over $\Phi_\pm$ equipped with weak topologies, then the evolution operators U_t and $U_t^\dagger$ can be extended, respectively, to Φ'_+ and Φ'_- via

$$U_t f(A) \equiv f(U_t^\dagger A) \qquad (A \in \Phi_+, f \in \Phi'_+) \tag{43}$$
$$U_t^\dagger \tilde{f}(\rho) \equiv \tilde{f}(U_t \rho) \qquad (\rho \in \Phi_-, f \in \Phi'_-) \tag{44}$$

The antilinear functionals f_j and $\tilde{f}_j$ appearing in (40) are eigenvectors of the extended evolution operators:

$$U_t f_j = e^{-i\alpha_j t} f_j, \qquad U_t^\dagger \tilde{f}_j = e^{i\alpha_j^* t}\tilde{f}_j \tag{45}$$

In this sense, the decomposition (40) can be regarded as a generalized spectral decomposition of the evolution operator U_t. Equation (40) corresponds to the case where the generalized spectrum of U_t consists only of simple discrete eigenvalues. Of course, more general spectra such as continuous spectra and Jordan block structures are possible.

Because the space of continuous antilinear functionals over the Hilbert space $\mathscr{L}^2$ is naturally identified with itself, one has a couple of triples:

$$\Phi_+ \subset \mathscr{L}^2 \subset \Phi'_+, \qquad \Phi_- \subset \mathscr{L}^2 \subset \Phi'_- \tag{46}$$

which are nothing but rigged Hilbert spaces or Gelfand triples. A remarkable feature of those triples is the possibility of semigroup evolution in the extended spaces. For example, the invariance (42) implies that U_t with $t > 0$ is well-defined on Φ'_+, but U_t with $t < 0$ may not be defined on Φ'_+ because one may have $U_t^\dagger \Phi_+ \not\subset \Phi_+$ for $t < 0$. Or the evolution U_t extended to Φ'_+ may be a forward semigroup. For many examples, this is certainly the case. The noninvariance of the test function spaces for the backward time evolution implies the noninvariance of the test function spaces under the time reversal operation I: $I\Phi_\pm \not\subset \Phi_\pm$. Indeed, because of $IU_t^\dagger I = U_{-t}^\dagger$, $U_t^\dagger \Phi_+ \subset \Phi_+$ ($t > 0$) and $I\Phi_+ \subset \Phi_+$ lead to $U_{-t}^\dagger \Phi_+ \subset \Phi_+$ ($t > 0$) and, thus, $U_{-t}^\dagger \Phi_+ \not\subset \Phi_+$ ($t > 0$) implies $I\Phi_+ \not\subset \Phi_+$. Note that the possibility of the decomposition (40) as well as the invariance of the test function spaces (42) depend on systems and should be checked case by case.

Now we remark that a set of operators $\hat{\Pi}_j$ from Φ_- to Φ'_+ defined by $\hat{\Pi}_j \rho \equiv \tilde{f}_j^*(\rho) f_j$ ($\rho \in \Phi_-$) form a resolution of the natural embedding i: $\Phi_- \to \Phi'_+$ as $\sum_j \hat{\Pi}_j = i$ and commute with the evolution operator: $\hat{\Pi}_j U_t = U_t \hat{\Pi}_j$ ($t > 0$). The operators $\hat{\Pi}_j$ are nothing but subdynamics projectors involved in the early stage of the complex spectral theory [3–5].

As shown by Pollicott [48] and Ruelle [49], for expanding maps and axiom A systems, the Fourier transform of a correlation function between two dynamical variables with certain smoothness admits meromorphic extension to the complex frequency domain and may have complex poles there, which are known as the Pollicott–Ruelle resonances. Now we turn to the decomposition (40). As seen from (41), the expectation value $\langle A \rangle_t$ is a correlation function between A and ρ_0 with respect to the invariant measure μ; its Fourier transform has complex poles at $\omega = \alpha_j$, which thus correspond to the Pollicott–Ruelle resonances. For expanding maps and axiom A systems, Ruelle [49] proved existence of generalized eigenvectors and a possibility of Jordan block structures.

B. Multibaker Map Revisited

We return to the multibaker map and, for the sake of simplicity, consider the expectation values of one-particle dynamical variables, which are of the form

$$A^\#(\zeta) = \sum_{j=1}^\infty A(z_j) \tag{47}$$

where $\zeta = \{z_1, z_2, \ldots\}$ is a configuration and $A(z)$ is a function defined on the phase space Γ. In the same way as the derivation of the 1-moment measure [45], the expectation value of $A^\#$ with respect to the Poisson measure (9) can be easily calculated and is given by

$$\langle A^\# \rangle_t \equiv \int_X A^\#(\zeta) \, dP_t(\zeta) = \int_\Gamma A(z) \, dv_t(z) \tag{48}$$

or, in terms of the partially integrated distribution function $\tilde{G}_t$, by

$$\langle A^{\#}\rangle_t = \sum_{n=-\infty}^{+\infty} \int dE\, a_{E,n} \int_{[0,1]^2} A(n, \xi a_{E,n}, \eta a_{E,n}, E - nF)\, d\xi\, d_\eta\, \tilde{G}_t(n, \xi, \eta, E) \quad (49)$$

where the subscript η in d_η implies that $d_\eta \tilde{G}_t$ represents a Stieltjes integral only with respect to the argument η.

Now we consider the case of $t > 0$ and where A is supported by a finite lattice: $\{(n, x, y, E)|0 \le n \le N,\ E \in \mathbf{R}^+,\ 0 < x \le a_E, 0 < y \le a_E\}$. Let the initial density ρ_0 and the observable A be piecewise continuously differentiable with respect to both x and y; thus the decomposition (24) gives

$$\langle A^{\#}\rangle_t - \langle A^{\#}\rangle_{+\infty} = \int dE \sum_{\substack{j=1 \\ |\kappa_j|>\lambda}}^{N+1} \langle A^*|f_j(E)\rangle\, \kappa_j^t\, \langle \tilde{f}_j(E)|\delta\rho_0\rangle + \delta\mathscr{A}_t \quad (50)$$

where $\delta\mathscr{A}_t = O(t^2 \lambda^t)$, $\langle A^{\#}\rangle_{+\infty}$ is an average of $A^{\#}$ with respect to the nonequilibrium steady state $\nu_{+\infty}$:

$$\langle A^{\#}\rangle_{+\infty} = \sum_{n=0}^{N} \int dE\, a_{E,n} \int_{[0,1]^2} A(n, \xi a_{E,n}, \eta a_{E,n}, E - nF)\, d\xi\, d_\eta\, \tilde{G}_{+\infty}(n, \eta, E) \quad (51)$$

and the antilinear functional $\langle \cdot|f_j(E)\rangle$ and the linear functional $\langle \tilde{f}_j(E)|\cdot\rangle$ are defined by

$$\langle B|f_j(E)\rangle \equiv \sqrt{\frac{2}{N+2}} \sum_{n=0}^{N} a_{E,n} \int_{[0,1]^2} B^*(n, \xi a_{E,n}, \eta a_{E,n}, E - nF)\, d\xi\, d\gamma_j(n, \eta) \quad (52)$$

$$\langle \tilde{f}_j(E)|\rho\rangle \equiv \sqrt{\frac{2}{N+2}} \sum_{n=0}^{N} a_{E,n} \int_{[0,1]^2} d\bar{\gamma}_j(n, \xi)\, d\eta\, \rho(n, \xi a_{E,n}, \eta a_{E,n}, E - nF) \quad (53)$$

And $\delta\rho_0 \equiv \rho_0 - \rho_{+\infty}$ is the deviation of the initial distribution from the steady-state distribution $\rho_{+\infty}$. Equation (50) implies that $\langle A^{\#}\rangle_t$ relaxes to the steady-state average $\langle A^{\#}\rangle_{+\infty}$ with relaxation rates $|\ln \kappa_j|$.

Because of $\langle A^{\#}\rangle_t - \langle A^{\#}\rangle_{+\infty} = \langle A^*, U_t \delta\rho_0\rangle$, (50) can be rewritten as

$$\langle A^*, U_t \delta\rho_0\rangle = \int dE \sum_{\substack{j=1 \\ |\kappa_j|>\lambda}}^{N+1} \langle A^*|f_j(E)\rangle\, \kappa_j^t\, \langle \tilde{f}_j(E)|\delta\rho_0\rangle + \delta\mathscr{A}_t \quad (54)$$

where $\langle \cdot, \cdot \rangle$ stands for the sesquilinear form:

$$\langle B, \rho \rangle \equiv \int dE \sum_{n=0}^{N} \int_{[0,a_{E,n}]^2} B^*(n,x,y,E-nF) \rho(n,x,y,E-nF)\, dx\, dy \quad (55)$$

and U_t is an evolution operator of the distribution:

$$U_t \rho(n,x,y,E-nF) \equiv \rho(B_F^{-t}(n,x,y,E-nF)) \quad (56)$$

Moreover, we have

$$\langle U_t^\dagger A^* | f_j(E) \rangle = \kappa_j^t \langle A^* | f_j(E) \rangle, \quad \langle \tilde{f}_j(E) | U_t \delta\rho_0 \rangle = \kappa_j^t \langle \tilde{f}_j(E) | \delta\rho_0 \rangle \quad (t>0) \quad (57)$$

Although ρ_0 in the decomposition (54) stands for the average particle density while ρ_0 in the decomposition (41) of the complex spectral theory is the probability density, their forms are identical. Thus, the present results can be compared with the complex spectral theory.

C. Condition for Description with Broken Time Reversal Symmetry

As mentioned in the beginning of this section, the class of initial states satisfying both forward and backward state evolution formulas (24) and (31) does not contain the limits $v_{\pm\infty}$. Indeed, the initial states are piecewise continuously differentiable with respect to x and y, while the steady-state densities $\rho_{+\infty}$ and $\rho_{-\infty}$ are discontinuous on the stable and unstable manifolds of the fractal repeller (see Ref. 33), respectively, which consist of infinitely many segments. To remove this dissymmetry, it is necessary to modify the class of initial states in such a way that it contains the steady-state densities. However, inclusion of $\rho_{-\infty}$ to the initial states is not compatible with (54). Indeed, the slowest decaying mode $\tilde{f}_1$

$$\langle \tilde{f}_1 | \rho \rangle = \sqrt{\frac{2}{N+2}} \sum_{n=0}^{N} a_{E,n} \int_{[0,1]^2} d\tilde{\gamma}_1(n,\xi)\, d\eta\, \rho(n,\xi a_{E,n}, \eta a_{E,n}, E-nF) \quad (58)$$

involves an integral over a conditionally invariant measure supported by the unstable manifold of the fractal repeller, where the steady state $\rho_{-\infty}$ has discontinuities. Thus, (58) and (54) are ill-defined for $\rho_0 = \rho_{-\infty}$. In other words, let Φ_- be a space of admissible initial states satisfying (54) and including $\rho_{+\infty}$; then $\rho_{-\infty} \notin \Phi_-$. Because the state $\rho_{-\infty}$ is the time-reversed state of $\rho_{+\infty}$: $\rho_{-\infty} = I\rho_{+\infty}$, this immediately implies the noninvariance of Φ_- under time reversal operation: $I\Phi_- \not\subset \Phi_-$, or the space of admissible initial states Φ_- has broken time reversal symmetry. In short, the requirement of including steady states to a class of admissible initial states naturally leads to a description with broken time reversal symmetry.

Furthermore, if the space Φ_- carries a stronger topology than the Hilbert space topology and is invariant under the forward time evolution $U_t\Phi_- \subset \Phi_-$ ($t > 0$), the forward evolution $U_t^\dagger$ ($t > 0$) can be extended to the (anti)dual space Φ'_- and admits generalized eigenfunctions $\tilde{f}_j$ corresponding to eigenvalues κ_j. The same arguments are applied to U_t and observables A. Therefore, the difference between the time-symmetric description and that with broken time reversal symmetry lies in the different choice of admissible initial densities and observables.

V. MEASURE SELECTION AND MACROSCOPIC PROPERTIES

As mentioned in the Introduction, it is well known in the dynamical systems theory [43] that a hyperbolic system may admit uncountably many invariant measures. And one of important problems is the selection of a "physical measure." Such a problem, however, would be physically irrelevant if observable phenomena would not depend on the measures. We show that it is not the case for the multibaker map.

Let $\mu^{(\Lambda\beta)}$ be a measure defined by

$$\mu^{(\Lambda\beta)}(\Gamma_n^{\xi,\eta}(E,\Delta)) = \int_E^{E+\Delta} dE' \, a_{E'}^2 \, \Lambda^n \psi_{\Lambda\beta}(\xi) \bar{\psi}_{\Lambda\beta}(\eta) \tag{59}$$

where $\Gamma_n^{\xi,\eta}(E,\Delta)$ is a set given by

$$\Gamma_n^{\xi,\eta}(E,\Delta) = \left\{ (n,x,y,E' - nF) \Big| 0 \leq \frac{x}{a_{E',n}} < \xi, 0 \leq \frac{y}{a_{E',n}} < \eta, E < E' < E + \Delta \right\} \tag{60}$$

and $\bar{\psi}_{\Lambda\beta}(\eta) \equiv 1 - \psi_{\Lambda\beta}(1-\eta)$. The function $\psi_{\Lambda\beta}(\xi)$ is the unique solution of a functional equation:

$$\psi_{\Lambda\beta}(\xi) = \begin{cases} \frac{1-\beta}{1+\Lambda} \psi_{\Lambda\beta}\left(\frac{\xi}{l^-}\right) & (0 < \xi < l^-) \\ \beta \, \psi_{\Lambda\beta}\left(\frac{\xi - l^-}{s}\right) + \frac{1-\beta}{1+\Lambda} & (l^- < \xi < 1 - l^+) \\ \frac{\Lambda(1-\beta)}{1+\Lambda} \psi_{\Lambda\beta}\left(\frac{\xi - 1 + l^+}{l^+}\right) + \frac{1+\Lambda\beta}{1+\Lambda} & (1 - l^+ < \xi < 1) \end{cases} \tag{61}$$

where Λ and $\beta \in (0, 1)$ are positive real parameters. In (60), $a_{E,n} = a_E e^{-nF}$ ($0 \leq n \leq N$), $a_{E,n} = a_E e^F$ ($n \leq -1$), and $a_{E,n} = a_E e^{-(N+1)F}$ ($N+1 \leq n$). As is easily seen, the measure $\mu^{(\Lambda\beta)}$ is invariant under the multibaker map B_F irrespective of the parameters Λ and β.

Instead of the initial states where the particles are distributed on semiinfinite segments uniformly with respect to the Lebesgue measure, we consider those corresponding to the uniform distribution with respect to the invariant measure $\mu^{(\Lambda\beta)}$. Let the initial particle distribution $\nu_0^{(\Lambda\beta)}$ be given by

$$\nu_0^{(\Lambda\beta)}(\Gamma_n^{\xi,\eta}(E, \Delta)) = \int_E^{E+\Delta} dE' a_{E'}^2 \Lambda^n \int_0^\xi d\psi_{\Lambda\beta}(\xi') \int_0^\eta d\bar\psi_{\Lambda\beta}(\eta')$$
$$\times \bar\rho_0(n, \xi' a_{E',n}, \eta' a_{E',n}, E' - nF) \quad (62)$$

for $0 \leq n \leq N$ with density $\bar\rho_0$ and

$$\nu_0^{(\Lambda\beta)}(\Gamma_n^{\xi,\eta}(E, \Delta)) = \begin{cases} \int_E^{E+\Delta} dE' a_{E'}^2 \rho_-(E') \Lambda^{-1} \psi_{\Lambda\beta}(\xi) \bar\psi_{\Lambda\beta}(\eta) & (n \leq -1) \\ \int_E^{E+\Delta} dE' a_{E'}^2 \rho_+(E') \Lambda^{N+1} \psi_{\Lambda\beta}(\xi) \bar\psi_{\Lambda\beta}(\eta) & (n \geq N+1) \end{cases}$$
(63)

Then, if the initial density $\bar\rho_0(n, \xi a_{E,n}, \eta a_{E,n}, E - nF)$ is piecewise continuously differentiable in ξ, as in Section III, one can show that the initial measure converges to a steady-state measure $\nu_{+\infty}^{(\Lambda\beta)}$ for $t \to +\infty$. The steady-state particle distribution $\tilde\Pi_{+\infty}(n, E)$ per energy per site and the associated flow $\tilde J_{n|n+1}^{+\infty}(E)$ from the nth to the $(n+1)$th sites are given by

$$\tilde\Pi_{+\infty}(n, E) = a_E^2 \left\{ (\rho_+(E) - \rho_-(E)) \frac{\Lambda^{N+1}(1 - \Lambda^{n+1})}{1 - \Lambda^{N+2}} + \rho_-(E) \Lambda^n \right\} \quad (64)$$

$$\tilde J_{n|n+1}^{+\infty}(E) = -\frac{1-\beta}{1+\Lambda}(\rho_+(E) - \rho_-(E)) \frac{a_E^2 \Lambda^{N+1}(1 - \Lambda)}{1 - \Lambda^{N+2}} \quad (65)$$

Those results are remarkable because macroscopic properties such as the particle density per site and the associated flow depend only on the parameters of the invariant measures Λ and β, but not on the dynamical parameters l and F. As an example, suppose that a given system is known to be described by the multibaker map B_F and consider a question as to whether one can determine dynamical parameters l and F from the macroscopic properties. The answer to this question is *no* because of (64) and (65). Thus, the choice of the invariant measures does affect the macroscopic properties of the multibaker map, and a measure selection condition is certainly necessary to relate microscopic dynamics with macroscopic properties. Or the problem of the measure selection is not only mathematically important, but also physically relevant.

As explained in Ref. 43, two conditions of selecting the physical measure are proposed—that is, robustness against an external noise and describability of time averages on motions starting from randomly sampled data with respect to the Lebesgue measure. The former corresponds to the Kolmogorov measure, and the latter corresponds to the Sinai–Ruelle–Bowen measure [44]. Here we emphasize again that both criteria are not derived from internal dynamics because, in the former, a noise should be imposed from outside and, in the latter, there is no dynamical process which guarantees the distinct role of the Lebesgue measure. This observation seems to imply that the statistical behavior of a given system cannot be derived from its dynamics alone, contrary to reductionists' view.

VI. CONCLUSIONS

In this report, we have illustrated the appearance of irreversible state evolution in reversible conservative systems using a reversible area-preserving multibaker map, which is considered to be a typical hyperbolic system exhibiting deterministic diffusion.

We considered an open multibaker map where a finite chain of multibaker map is embedded between two semiinfinite free parts. Assuming, as usual (e.g., Refs. 2 and 52), that the states are described by statistical ensembles, we investigated the state evolution starting from initial states for which particles distribute uniformly on the semiinfinite segments with respect to the Lebesgue measure. The system in question is an infinitely extended system consisting of infinitely many particles with finite particle density and its state can be described by a Poisson suspension specified by a measure representing the average particle number.

When time t goes to $\pm\infty$, the average-number measure v_t (precisely speaking, its cumulative function) at t evolves toward steady states $v_{\pm\infty}$, provided that the initial density is piecewise continuously differentiable. The state $v_{+\infty}$ possesses transport properties consistent with the second law of thermodynamics, while $v_{-\infty}$ has anti-thermodynamical properties. Because the evolution to $v_{-\infty}$ for $t \to -\infty$ implies that the state v_t is repelled from $v_{-\infty}$ for $t \to +\infty$, the states v_t unidirectionally evolve from "a repeller" $v_{-\infty}$ to "an attractor" $v_{+\infty}$. This evolution is consistent with the time reversal symmetry as the time reversal operation induces jump of states near $v_{+\infty}$ to those near $v_{-\infty}$.

Reviewing the complex spectral theory and evaluating averages of one-particle dynamical variables along the present approach, their mathematical structures are compared. At least in the present multibaker map, inclusion of steady states to a class of admissible initial states naturally leads to a description with broken time-reversal symmetry.

Finally, the problem of selecting a physical measure from many invariant measures, a well-known problem in dynamical systems theory, is discussed from physical points of view, and different invariant measures are shown to correspond to different macroscopic properties such as the particle distribution per site and the associated flow. This implies that the introduction of a nondynamical measure selection condition is inevitable to bridge between microscopic dynamics and macroscopic properties and, hence, macroscopic properties cannot be derived from dynamics alone.

Acknowledgments

It was my great honor to have given a talk at the 21st Solvay Conference (November 1–5, 1998, Keihanna, Kyoto). And I would like to thank Mr. J. Solvay, Professor I. Prigogine, Professor I. Antoniou, and Professor K. Kitahara for affording me this opportunity. The author is very grateful to Professor I. Prigogine, Professor I. Antoniou, Professor Z. Suchanecki, Professor P. Gaspard, Professor J.R. Dorfman, and Dr. T. Gilbert for their collaborations, from which this work is derived. This work is a part of the project of the Institute for Fundamental Chemistry, supported by the Japan Society for the Promotion of Science—Research for the Future Program (JSPS-RFTF96 P00206) and is supported by a Grant-in-Aid for Scientific Research (B) and (C) from JSPS as well as by Waseda University Grant for Special Research Projects (Individual Research, No. 2000A-852) from Waseda University.

References

1. I. Prigogine, *From Being to Becoming*, Freeman, New York, 1980.
2. I. Prigogine, *Nonequilibrium Statistical Mechanics*, Wiley, New York, 1962; also see references therein.
3. I. Prigogine, C. George, and F. Henin, *Physica* **45**, 418 (1969).
4. I. Prigogine, C. George, F. Henin, and L. Resenfeld, *Chem. Scripta* **4**, 5 (1973).
5. C. George and F. Mayné, *J. Stat. Phys.* **48**, 1343 (1987); also see references therein.
6. T. Petrosky and I. Prigogine, *Physica A* **175**, 146 (1991).
7. H. H. Hasegawa and W. Saphir, *Phys. Rev. A* **46**, 7401 (1992); I. Antoniou and S. Tasaki, *Physica A* **190**, 303 (1992).
8. I. Antoniou and I. Prigogine, *Physica A* **192**, 443 (1993).
9. I. Antoniou and S. Tasaki, *Int. J. Quantum Chem.* **46**, 425 (1993).
10. H. H. Hasegawa and D. J. Driebe, *Phys. Lett. A* **179**, 97 (1993); *Phys. Rev. E* **50**, 1781 (1994).
11. T. Petrosky and I. Prigogine, *Adv. Chem. Phys.* **99**, 1 (1997); also see references therein.
12. I. Antoniou, M. Gadella, I. Prigogine, and G. P. Pronko, *J. Math. Phys.* **39**, 2995 (1998).
13. I. Prigogine and T. Petrosky, in *Generalized Functions, Operator Theory and Dynamical Systems*, I. Antoniou and G. Lumer, eds., Chapman and Hall/CRC, London, 1999, p. 153; also see references therein.
14. D. Driebe, *Fully Chaotic Maps and Broken Time Symmetry*, Kluwer Academic, Dordrecht, 1999; also see references therein.
15. E. C. G. Sudarshan, C. Chiu, and V. Gorini, *Phys. Rev. D* **18**, 2914 (1978); V. Gorini and G. Parravicini, in *Group Theoretical Methods in Physics*, A. Bohm et al., eds., Springer LNP, Vol. 94, Springer, Berlin, 1979, p. 219; G. Parravicini, V. Gorini, and E. C. G. Sudarshan, *J. Math. Phys.* **21**, 2208 (1980).

16. A. Bohm, *Lett. Math. Phys.* **3**, 455 (1979); A. Bohm, M. Gadella, and G. Bruce Mainland, *Am. J. Phys.* **57**, 1103 (1989); A. Bohm and M. Gadella, *Dirac Kets, Gamow Vectors and Gelfand Triplets*, Springer LNP, Vol. 348, Springer, Berlin, 1989.
17. J. Aguilar and J. M. Combes, *Commun. Math. Phys.* **22**, 269 (1971); E. Balslev and J. M. Combes, *Commun. Math. Phys.* **22**, 280 (1971); for more recent references, see, for example, *Resonances*, E. Brändas and N. Elander, eds., Springer LNP, Vol. 325, Springer, Berlin, 1989.
18. J. R. Dorfman, *An Introduction to Chaos in Non-Equilibrium Statistical Mechanics*, Cambridge University Press, Cambridge, 1999.
19. P. Gaspard, *Chaos, Scattering and Statistical Mechanics*, Cambridge University Press, Cambridge, 1998.
20. *Chaos and Irreversibility* focus issue of *Chaos* **8** (1998).
21. *Hard Ball Systems and the Lorentz Gas*, D. Szász, ed., Encyclopedia of Mathematical Science, Vol. 101, Springer, Berlin, 2000.
22. P. Gaspard and G. Nicolis, *Phys. Rev. Lett.* **65**, 1693 (1990); J. R. Dorfman and P. Gaspard, *Phys. Rev. E* **51**, 28 (1995); P. Gaspard and J. R. Dorfman, *Phys. Rev. E* **52**, 3525 (1995); P. Gaspard, *Phys. Rev. E* **53**, 4379 (1996); P. Gaspard, *Physica* **240A**, 54 (1997).
23. N. I. Chernov, G. L. Eyink, J. L. Lebowitz, and Ya. G. Sinai, *Phys. Rev. Lett.* **70**, 2209 (1993); *Commun. Math. Phys.* **154**, 569 (1993).
24. D. Ruelle, *J. Stat. Phys.* **85**, 1 (1996); *J. Stat. Phys.* **86**, 935 (1997); *J. Stat. Phys.* **95**, 393 (1999); W. Breymann, T. Tél, and J. Vollmer, *Phys. Rev. Lett.* **77**, 2945 (1996); G. Nicolis and D. Daems, *J. Phys. Chem.* **100**, 19187 (1996); *Chaos* **8**, 311 (1998); E. C. G. Cohen and L. Rondoni, *Chaos* **8**, 357 (1998).
25. D. J. Evans, E. G. D. Cohen, and G. P. Morriss, *Phys. Rev. Lett.* **71**, 2401 (1993).
26. G. Gallavotti and E. G. D. Cohen, *Phys. Rev. Lett.* **74**, 2694 (1995); *J. Stat. Phys.* **80**, 931 (1995).
27. E. G. D. Cohen, *Physica A* **213**, 293 (1995); *Physica A* **240**, 43 (1997).
28. L. Rondoni and E. C. G. Cohen, *Nonlinearity* **13**, 1905 (2000).
29. I. Claus and P. Gaspard, *J. Stat. Phys.* **101**, 161 (2000); *Phys. Rev. E* **63**, 036227 (2001).
30. P. Gaspard, I. Claus, T. Gilbert, and J. R. D. Dorfman, *Phys. Rev. Lett.* **86**, 506 (2001).
31. P. Gaspard, *J. Stat. Phys.* **68**, 673 (1992); S. Tasaki and P. Gaspard, *J. Stat. Phys.* **81**, 935 (1995).
32. T. Tél, J. Vollmer, and W. Breymann, *Europhys. Lett.* **35**, 659 (1996); T. Tél, J. Vollmer, and W. Breymann, *A multibaker model of transport in driven systems*, preprint, Eötvös University, (1997).
33. P. Gaspard, *J. Stat. Phys.* **88**, 1215 (1997).
34. J. Vollmer, T. Tél, and W. Breymann, *Phys. Rev. Lett.* **79**, 2759 (1997); W. Breymann, T. Tél, and J. Vollmer, *Chaos* **8**, 396 (1998); J. Vollmer, T. Tél, and W. Breymann, *Phys. Rev. E* **58**, 1672 (1998).
35. P. Gaspard and R. Klages, *Chaos* **8**, 409 (1998).
36. T. Gilbert and J. R. Dorfman, *J. Stat. Phys.* **96**, 225 (1999).
37. S. Tasaki and P. Gaspard, *Theor. Chem. Acc.* **102**, 385 (1999); *J. Stat. Phys.* **101**, 125 (2000).
38. L. Mátyás, T. Tél, and J. Vollmer, *Phys. Rev. E* **61**, R3295 (2000); *Phys. Rev. E* **62**, 349 (2000); J. Vollmer, T. Tél, and L. Mátyás, *J. Stat. Phys.* **101**, 79 (2000).
39. T. Tél and J. Vollmer, *Multibaker Maps and the Lorentz Gas*, in Ref. 21.
40. L. Rondoni, T. Tél, and J. Vollmer, *Phys. Rev. E* **61**, R4679 (2000).

41. T. Gilbert, J. R. Dorfman, and P. Gaspard, *Phys. Rev. Lett.* **85**, 1606 (2000); *Nonlinearity* **14**, 339 (2001).
42. T. Tél, J. Vollmer, and L. Mátyás, *Europhys. Lett.* **53**, 458 (2001).
43. J.-P. Eckmann and D. Ruelle, *Rev. Mod. Phys.* **57**, 617 (1985).
44. Ya. G. Sinai, *Russian Math. Surveys* **27**, 21 (1972); R. Bowen and D. Ruelle, *Invent. Math.* **29**, 181 (1975); D. Ruelle, *Am. J. Math.* **98**, 619 (1976).
45. I. P. Cornfeld, S. V. Fomin, and Ya. G. Sinai, *Ergodic Theory*, Springer, New York, 1982; L. A. Bunimovich et al., *Dynamical Systems, Ergodic Theory and Applications, Encyclopedia of Mathematical Sciences*, Vol. 100, Springer, Berlin, 2000.
46. S. Tasaki, T. Gilbert, and J. R. Dorfman, *Chaos* **8**, 424 (1998).
47. P. Gaspard, *Chaos* **3**, 427 (1993); S. Tasaki and P.Gaspard, *Bussei Kenkyu (Kyoto)* **66**, 21 (1996); R. F. Fox, *Chaos* **7**, 254 (1997).
48. M. Pollicott, *Invent. Math.* **81**, 413 (1985); *Invent. Math.* **85**, 147 (1986); *Ann. Math.* **131**, 331 (1990).
49. D. Ruelle, *Phys. Rev. Lett.* **56**, 405 (1986); *J. Stat. Phys.* **44**, 281 (1986); *Commun. Math. Phys.* **125**, 239 (1989); *Publ. Math. IHES* **72**, 175 (1990).
50. G. Pianigiani and J. A. Yorke, *Trans. Am. Math. Soc.* **252**, 351 (1979); T. Tél, *Phys. Rev.* **A36**, 1502 (1987); T. Tél, *Transient Chaos*, in *Directions in Chaos*, Vol. 3, Hao Bai-Lin, ed., World Scientific, Singapore, 1990.
51. B. Misra, I. Prigogine, and M. Courbage, *Physica A* **98**, 1 (1979); B. Misra and I. Prigogine, *Prog. Theor. Phys. Suppl.* **69**, 101 (1980); *Lett. Math. Phys.* **7**, 421 (1983).
52. N. N. Krylov, *Works on the Foundations of Statistical Mechanics*, Princeton University Press, Princeton, 1979.

DIFFUSION AND THE POINCARÉ–BIRKHOFF MAPPING OF CHAOTIC SYSTEMS

PIERRE GASPARD

*Center for Nonlinear Phenomena and Complex Systems,
Free University of Brussels, Brussels, Belgium*

CONTENTS

I. Introduction
II. Microscopic Chaos
III. Spatially Periodic Systems
   A. Poincaré–Birkhoff Mapping in Spatially Periodic Systems
   B. Frobenius–Perron Operator
IV. Consequences of the Eigenvalue Problem
   A. The Diffusion Coefficient and the Higher-Order Coefficients
   B. Periodic-Orbit Theory
   C. The Nonequilibrium Steady States
   D. Entropy Production
V. Conclusions
Acknowledgments
References

I. INTRODUCTION

In the present contribution, we summarize recent work on diffusion in spatially periodic chaotic systems. The hydrodynamic modes of diffusion are constructed as the eigenstates of a Frobenius–Perron operator associated with the Poincaré–Birkhoff mapping of the system. The continuous-time dynamics is included in the Frobenius–Perron operator of the mapping by using the first-return time function. We show that the transport properties of diffusion can be derived from this Frobenius–Perron operator, such as the Green–Kubo relation for the diffusion coefficient, the Burnett and higher-order coefficients, the Lebowitz–McLennan–Zubarev expression for the nonequilibrium steady states, and the entropy production of nonequilibrium thermodynamics. The hydrodynamic modes are given by conditionally invariant complex measures with a fractal-like singular character. The construction of the hydrodynamic modes is inspired by works of Prigogine [1], Balescu [2], Résibois [3], Boon [4], and others.

Recent progress in statistical mechanics has revealed the importance of chaos to understand normal transport processes such as diffusion, viscosity, or heat conductivity. During this last decade, many works have shown that microscopic systems of interacting particles are typically chaotic in the sense that their dynamics is highly sensitive to initial conditions because of positive Lyapunov exponents and that this dynamical instability leads to a dynamical randomness characterized by a positive Kolmogorov–Sinai (KS) entropy per unit time [5–12]. This dynamical randomness can induce various transport phenomena, which turn out to be *normal* in the sense that the associated fluctuations are of Gaussian character on long times and that the associated hydrodynamic modes decay exponentially.

Such results have been obtained for systems with few mutually interacting degrees of freedom. In particular, diffusion has been extensively studied in systems such as the Lorentz gases, in which a point particle undergoes elastic collisions in a periodic lattice of ions. The ions have been modeled as hard disks [13,14] or as centers of Yukawa-type potentials [15]. Normal diffusion has also been studied in simplified models such as the multibaker maps [16–19]. In the same context, Gaussian thermostated Lorentz gases have also been studied [20,21]. In these thermostated systems, the particle is submitted to an external field and to a deterministic force which is non-Hamiltonian. As a consequence, phase-space volumes are no longer preserved in thermostated systems, which constitutes a fundamental difficulty of this approach. Non-area-preserving multibaker maps have also been considered [22]. In this case, an equivalence was shown between the non-area-preserving maps and area-preserving maps with an extra variable of energy allowing the phase space to be larger as the particle gains kinetic energy [23,24]. In this contribution, we shall focus on transport in systems that preserve phase-space volumes so that the sum of all the Lyapunov exponents vanishes $\sum_i (\lambda_i^+ + \lambda_i^-) = 0$, in agreement with Liouville's theorem of Hamiltonian mechanics. For systems of statistical mechanics, Liouville's theorem is an important property because it is a consequence of the unitarity of the underlying quantum mechanics ruling the microscopic dynamics.

The purpose of the present contribution is to give a short synthesis of the studies of diffusion and other transport processes in spatially periodic chaotic systems which obeys Liouville's theorem. In these studies, a special role is played by the Poincaré–Birkhoff mapping, which provides a powerful method to reduce the continuous-time evolution of statistical ensembles of trajectories to a discrete-time Frobenius–Perron operator [25]. The concept of microscopic chaos and the escape-rate formalism will also be shortly reviewed.

The contribution is organized as follows. Section II contains a discussion about chaos in nonequilibrium statistical mechanics and about the escape-rate formalism. In Section III, we show how a flow can be reduced to a Poincaré–Birkhoff mapping and similarly how, in spatially periodic systems, Liouville's equation can be studied in terms of a special Frobenius–Perron operator of our invention. In Section IV, we show that all the known formulae of the transport theory of diffusion can be derived from our Frobenius–Perron operator and its classic Pollicott–Ruelle resonances. Applications and conclusions are given in Section V.

II. MICROSCOPIC CHAOS

In chaotic systems, the sensitivity to initial conditions generates a time horizon that is of the order of the inverse of the maximum Lyapunov exponent $\lambda_{\max}^+ = \max\{\lambda_i^+\}$, multiplied by the logarithm of the ratio of the final error $\varepsilon_{\text{final}}$ over the initial one $\varepsilon_{\text{initial}}$:

$$t_{\text{Lyapunov}} = \frac{1}{\lambda_{\max}^+} \ln \frac{\varepsilon_{\text{final}}}{\varepsilon_{\text{initial}}} \tag{1}$$

This Lyapunov time constitutes a horizon for the prediction of the future trajectory of the system [26]. Starting from initial conditions known to a precision given by $\varepsilon_{\text{initial}}$, the trajectory keeps a tolerable precision lower than $\varepsilon_{\text{final}}$ only during the time interval $0 \leq t < t_{\text{Lyapunov}}$. We notice that the Lyapunov horizon is movable in the sense that it can be extended to a longer time by decreasing the initial error $\varepsilon_{\text{initial}}$. Nevertheless, this requires to increase exponentially the precision on the initial conditions.

In a dilute gas of interacting particles, a positive Lyapunov exponent has the typical value [5,6]

$$\lambda_i^+ \sim \frac{v}{l} \ln \frac{l}{d} \tag{2}$$

where d is the particle diameter, $v \sim \sqrt{k_B T/m}$ is the mean velocity, and $l \sim (nd^2)^{-1}$ is the mean free path given in terms of the particle density n. At room temperature and pressure, a typical Lyapunov exponent takes the value $\lambda_i^+ \sim 10^{10}$ sec^{-1}, which is of the order of the inverse of the intercollisional time. There are as many positive Lyapunov exponents as there are unstable directions in phase space. Pesin's formula [27] can be used to evaluate the KS entropy per unit time of a mole of particles at equilibrium:

$$h_{KS} = \sum_i \lambda_i^+ \sim 3N_{Av} \frac{v}{l} \ln \frac{l}{d} \qquad (3)$$

where N_{Av} is Avogadro's number. The KS entropy characterizes the dynamical randomness of the motion of the particles composing the gas. The effect of this microscopic chaos has been observed in a recent experiment where a positive lower bound has been measured on the KS entropy per unit time of a fluid containing Brownian particles [28].

At equilibrium, formula (3) shows that the dynamical instability is converted into an exactly equal amount of dynamical randomness. The reason for this is that the dynamical instability cannot proceed forever in nonlinear Hamiltonian systems where the explosion is stopped by nonlinear saturations. In this way, dynamical randomness is generated by some mechanisms of nonlinear saturation during the dynamics. The exact compensation of the dynamical instability by the dynamical randomness is an interesting feature expressed by Pesin's formula (3).

In nonequilibrium situations, this exact compensation is broken leading to the generation of fractal structures in the phase space of the system of particles [11,29–32]. In the escape-rate formalism, a nonequilibrium situation is induced by absorbing boundaries in the phase space. These absorbing boundaries describe for instance an experiment in which the system is observed until a certain property reaches a certain threshold or exits a certain range of values or a domain of motion. The time of this event is recorded and the experiment is restarted. After many repetitions, a statistics of first-passage times can be performed. Very often, these first-passage times are distributed exponentially, which defines a rate of exponential decay called the *escape rate*. This rate depends on the observed property as well as on the chosen threshold—that is, on the geometry of the absorbing boundaries. If the observed property is the position of a tracer particle diffusing in a fluid, the escape rate turns out to be proportional to the diffusion coefficient. Similarly, if the observed property is the center of momentum of all the particles composing the fluid, the escape rate is proportional to the viscosity, and so on [32].

At the microscopic level, the absorbing boundaries select highly unstable trajectories along which the observed property never reaches the threshold. In chaotic systems, infinitely many such trajectories exist in spite of the fact that most trajectories reach the threshold and are absorbed at the boundary (or, equivalently, they escape out of the domain delimited by the boundaries). Accordingly, all these trajectories that evolve forever inside the boundaries without reaching them form a so-called *fractal repeller*. On this fractal repeller, the KS entropy is slightly smaller than the sum of positive Lyapunov exponents, and the difference is equal to the escape rate [27]

$$\gamma = \sum_i \lambda_i^+ - h_{KS} \qquad (4)$$

Because the escape rate γ depends on the transport coefficient, it is possible to relate the transport coefficients to the characteristic quantities of chaos, as shown elsewhere [11,30–32].

A *priori*, it is not evident that such a relationship is possible because the microscopic chaos is a property of the short time scale $(\lambda_i^+)^{-1} \sim 10^{-10}$ sec of the collisions between the particles, although the transport properties manifest themselves on the long hydrodynamic time scale. However, the nonequilibrium conditions created by the absorbing boundaries lead to a small difference between the KS entropy and the sum of positive Lyapunov exponents. This difference is of the order of the inverse of a hydrodynamic time such as the diffusion time for a tracer particle to reach the absorbing boundaries. Thanks to the escape-rate formalism [11,30–32], the connection between both types of properties thus becomes possible because a difference is taken between two properties from the short intercollisional time scale.

Subsequent work has shown the fundamental importance of the fractal repeller of the escape-rate formalism [11,19,25,33,34]. This fractal repeller is closely associated with the other fractals that appear in infinitely extended systems without absorbing boundaries. For instance, it has been possible to show that, in infinitely extended systems such as the periodic Lorentz gas, the nonequilibrium steady states of diffusion are described by singular measures (see Section IV) [19,25]. The singularities of these measures originate from the discontinuities of the invariant density in a large but finite system between two reservoirs of particles [33]. These discontinuities occur on the unstable manifolds of the fractal repeller of the trajectories evolving forever between the two reservoirs. In turn, the singular measures describing the nonequilibrium steady states are the derivatives of the diffusive hydrodynamic modes with respect to their wavenumber. As a consequence, these hydrodynamic modes are also described by singular measures with their singularities intimately related to the same fractal repeller of the escape-rate formalism. Because the singular character of these measures is at the origin of the positive entropy production of nonequilibrium thermodynamics (as shown in Section IV) [34], it turns out that the fractal repeller of the escape-rate formalism plays a basic and fundamental role in this whole context. Let us here mention that the escape-rate formalism is closely related to the scattering theory of transport by Lax and Phillips [35]. These authors have pointed out the importance of the set of trapped trajectories—today called the fractal repeller—and of its stable and unstable manifolds in order to define a classic scattering operator.

III. SPATIALLY PERIODIC SYSTEMS

A. Poincaré–Birkhoff Mapping in Spatially Periodic Systems

We consider a continuous-time system with a phase space that is periodic in some directions. This is the case for the periodic Lorentz gas in which a point particle undergoes elastic collisions in a lattice of hard disks or interacts with a lattice of Yukawa attracting potentials. In the first example [13,14], the system is defined by the Hamiltonian of kinetic energy for the free motion between the hard disks and by the condition that the position (x, y) remains outside the hard disks of diameter d forming a regular lattice $\mathscr{L}$:

$$H = \frac{p_x^2 + p_y^2}{2m} \tag{5}$$

$$\sqrt{(x - l_x)^2 + (y - l_y)^2} \geq d/2 \quad \text{for} \quad \mathbf{l} = (l_x, l_y) \in \mathscr{L} \tag{6}$$

In the second example [15], the Hamiltonian is given by

$$H = \frac{p_x^2 + p_y^2}{2m} + \sum_{\mathbf{l} \in \mathscr{L}} V(\mathbf{r} - \mathbf{l}) \quad \text{with} \quad V(r) = -\frac{\exp(-\alpha r)}{r} \tag{7}$$

Another example is a system composed of a tracer particle moving among $(N - 1)$ other particles with periodic boundary conditions, as usually assumed in numerical simulations.

Such systems are described by some differential equations

$$\frac{d\mathbf{X}}{dt} = \mathbf{F}(\mathbf{X}) \tag{8}$$

for M variables $\mathbf{X}$. If the conditions of Cauchy's theorem are satisfied, the trajectory at time t is uniquely given in terms of the initial conditions $\mathbf{X}_0$, which defines the continuous-time flow:

$$\mathbf{X}_t = \mathbf{\Phi}^t(\mathbf{X}_0) \tag{9}$$

Because of the periodicity, the phase space can be divided into a lattice of cells within each of them the vector field (8) is the same. One of these cells defines the fundamental cell of our lattice in the phase space. The flow in the full phase space can be reduced to the flow in the fundamental cell with periodic boundary conditions. The motion on the lattice of cells is followed by a vector $\mathbf{l}$ belonging to the lattice $\mathscr{L}$. A Poincaré surface of section $\mathscr{P}$ can be defined in this fundamental cell of the phase space. This surface of section is equipped with $M-1$ coordinates $\mathbf{x}$. During the time evolution, a trajectory (9) will intersect the surface of section successively at the points $\{\mathbf{x}_n\}_{n=-\infty}^{+\infty}$ and the times $\{t_n\}_{n=-\infty}^{+\infty}$. At each intersection with the surface of section, the vector locating the trajectory in the lattice is updated so that a sequence of lattice vectors $\{\mathbf{l}_n\}_{n=-\infty}^{+\infty}$ is furthermore generated by the motion.

Because the system is deterministic, each intersection $\mathbf{x}$ uniquely determines the next intersection by a nonlinear map $\boldsymbol{\varphi}(\mathbf{x})$ called the Poincaré–Birkhoff mapping, a time of first return $T(\mathbf{x})$, and a vector-valued function $\mathbf{a}(\mathbf{x}) \in \mathscr{L}$ giving the jump on the lattice [25]:

$$\begin{aligned} \mathbf{x}_{n+1} &= \boldsymbol{\varphi}(\mathbf{x}_n) \\ t_{n+1} &= t_n + T(\mathbf{x}_n) \\ \mathbf{l}_{n+1} &= \mathbf{l}_n + \mathbf{a}(\mathbf{x}_n) \end{aligned} \quad (10)$$

Reciprocally, the flow can be expressed as a suspended flow in terms of the quantities introduced in the construction of the Poincaré–Birkhoff mapping. In order to recover the M coordinates $\mathbf{X}$ of the original flow, we need to introduce an extra variable τ beside the $M-1$ coordinates $\mathbf{x}$ of the Poincaré surface of section. The extra variable τ is the time of flight since the last intersection with the surface of section. This variable ranges between zero and the time $T(\mathbf{x})$ of first return in the section: $0 \leq \tau < T(\mathbf{x})$. Thanks to these coordinates, the suspended flow takes an explicit form given by [25]

$$\tilde{\boldsymbol{\Phi}}^t(\mathbf{x}, \tau, \mathbf{l}) = \left[\boldsymbol{\varphi}^n \mathbf{x}, \tau + t - \sum_{j=0}^{n-1} T(\boldsymbol{\varphi}^j \mathbf{x}), \mathbf{l} + \sum_{j=0}^{n-1} \mathbf{a}(\boldsymbol{\varphi}^j \mathbf{x}) \right] \quad (11)$$

for $0 \leq \tau + t - \sum_{j=0}^{n-1} T(\boldsymbol{\varphi}^j \mathbf{x}) < T(\boldsymbol{\varphi}^n \mathbf{x})$. According to Cauchy's theorem and the geometric construction, there exists a function $\mathbf{G}$ that connects the variables of the suspended flow to the original ones:

$$\mathbf{X} = \mathbf{G}(\mathbf{x}, \tau, \mathbf{l}) \quad \text{with} \quad \mathbf{x} \in \mathscr{P}, \quad 0 \leq \tau < T(\mathbf{x}), \quad \mathbf{l} \in \mathscr{L} \quad (12)$$

and an isomorphism is established in this way between the suspended and the original flows:

$$\tilde{\boldsymbol{\Phi}}^t = \mathbf{G}^{-1} \circ \boldsymbol{\Phi}^t \circ \mathbf{G} \quad (13)$$

We notice that the jump vector is given by

$$\mathbf{a}(\mathbf{x}) = \mathbf{r}(\mathbf{x}) + \int_0^{T(\mathbf{x})} \mathbf{v} \circ \boldsymbol{\Phi}^\tau \circ \mathbf{G}(\mathbf{x}, 0, \mathbf{0}) \, d\tau - \mathbf{r}(\boldsymbol{\varphi} \mathbf{x}) \quad (14)$$

in terms of the velocity **v** of the tracer particle at the current position in the phase space, if $\mathbf{r}(\mathbf{x})$ denotes the position with respect to the center of the fundamental cell at the instant of the intersection **x** with the surface of section.

B. Frobenius–Perron Operator

Beyond the Lyapunov horizon, the predictability on the future trajectory is lost and statistical statements become necessary as emphasized by Nicolis [26]. Because the relaxation toward the thermodynamic equilibrium is a process taking place on asymptotically long times, its description requires the introduction of statistical ensembles of trajectories.

In a multiparticle system, the use of statistical ensembles corresponds to the repetition of the same experiment over and over again until the statistical property is established with confidence. This is the case in scattering experiments—for instance, in the study of atomic or molecular collisions. Two beams of particles are sent onto each other, and the scattering cross section of elastic or inelastic collisions can be measured. These cross sections are such statistical properties. In a scattering experiment, each beam contains an arbitrarily large number of particles arriving one after the other with statistically distributed velocities, impact parameters, and arrival times. In a beam, the particles are sufficiently separated to be considered as independent so that the experiment can be described as a succession of independent binary collisions, forming a statistical ensemble of events.

In a single-particle system such as the Lorentz gas, the statistical ensembles are introduced for the same use as aforementioned. In such systems, the statistical ensemble can also be considered for the description of a gas of independent particles bouncing in a lattice of ions. Indeed, a statistical ensemble of N trajectories of the bouncing particle is strictly equivalent to a gas of N independent particles bouncing in the system. In the limit $N \to \infty$, the phase-space distribution of the statistical ensemble is equivalent to the position–velocity distribution of the gas of independent particles. The relaxation toward the equilibrium of the former is thus equivalent to the relaxation of the latter. Our purpose is here to describe this relaxation at the microscopic level without approximation.

The statistical ensemble is described by a probability density $\rho(\mathbf{X})$ defined in the phase space. The time evolution (8)–(9) of each trajectory induces a time evolution for the whole statistical ensemble according to Liouville's equation, which takes one of the following equivalent forms [1,26]:

$$\partial_t \rho = -\text{div}(\mathbf{F}\rho) = \{H, \rho\} \equiv \hat{L}\rho \qquad (15)$$

where **F** is the vector field (8), H is the Hamiltonian, $\{\cdot, \cdot\}$ is the Poisson bracket, and $\hat{L}$ is the so-defined Liouvillian operator. As aforementioned, Liouville's equation (15) also rules the position–velocity density of a Lorentz gas of independent particles.

In the same way as the equations of motion (8) can be integrated to give the flow (9), the time integral of Liouville's equation gives the probability density ρ_t at the current time t in terms of the initial probability density ρ_0 according to

$$\rho_t(\mathbf{X}) = \exp(\hat{L}t)\rho_0(\mathbf{X}) = \rho_0(\mathbf{\Phi}^{-t}\mathbf{X}) \equiv \hat{P}^t\rho_0(\mathbf{X}) \qquad (16)$$

which defines the Frobenius–Perron operator, here given for volume-preserving systems.

In Ref. 25, we have shown how this continuous-time Frobenius–Perron operator decomposes into the discrete-time Frobenius–Perron operator associated with the Poincaré–Birkhoff mapping (10), which we assume to be area-preserving. This reduction is carried out by a Laplace transform in time which introduces the rate variable s conjugated to the time t. Furthermore, since the system is spatially periodic, we perform a spatial Fourier transform that introduces the wavenumber $\mathbf{k}$. In this way, the probability density decomposes into the so-called hydrodynamic modes of diffusion. Each of them corresponds to spatially quasiperiodic inhomogeneities of wavelength $L = 2\pi/\|\mathbf{k}\|$ in the phase-space density. Accordingly, the Frobenius–Perron operator decomposes into a continuum of Frobenius–Perron operators, one for each wavenumber $\mathbf{k}$ [25]:

$$(\hat{R}_{\mathbf{k},s} f)(\mathbf{x}) = \exp[-sT(\boldsymbol{\varphi}^{-1}\mathbf{x}) - i\mathbf{k} \cdot \mathbf{a}(\boldsymbol{\varphi}^{-1}\mathbf{x})] f(\boldsymbol{\varphi}^{-1}\mathbf{x}) \qquad (17)$$

where $f(\mathbf{x})$ is the $\mathbf{k}$-component of the density defined in the Poincaré surface of section $\mathscr{P}$. For $\mathbf{k} = 0$, the Frobenius–Perron operator (17) reduces to an operator previously derived by Pollicott [36,37].

In the Frobenius–Perron operator (17), the Laplace transform in time has introduced an exponential factor $\exp[-sT(\boldsymbol{\varphi}^{-1}\mathbf{x})]$, where $T(\boldsymbol{\varphi}^{-1}\mathbf{x})$ is the time of first return in the surface of section $\mathscr{P}$ after a previous intersection at $\boldsymbol{\varphi}^{-1}\mathbf{x}$. This factor has the following interpretation. Let us suppose that the Frobenius–Perron operator rules the time evolution of a mode with a decay rate $\gamma = -s$ such that $\operatorname{Re}\gamma > 0$. The operator (17) should describe locally in the phase space the time evolution from the previous intersection $\boldsymbol{\varphi}^{-1}\mathbf{x}$ up to the current intersection $\mathbf{x}$ with $\mathscr{P}$. During the first-return time $T(\boldsymbol{\varphi}^{-1}\mathbf{x})$, the density f decays exponentially by an amount $\exp[-\gamma T(\boldsymbol{\varphi}^{-1}\mathbf{x})]$. The first factor in Eq. (17) is there to compensate this decay in order to define the conditionally invariant density associated with the decay mode.

A similar interpretation holds for the factor $\exp[-i\mathbf{k} \cdot \mathbf{a}(\boldsymbol{\varphi}^{-1}\mathbf{x})]$. During the same segment of trajectory from $\boldsymbol{\varphi}^{-1}\mathbf{x} \in \mathscr{P}$ to $\mathbf{x} \in \mathscr{P}$, the particle has moved in the lattice by a vector $\mathbf{a}(\boldsymbol{\varphi}^{-1}\mathbf{x})$ so that the Fourier $\mathbf{k}$-component f acquires the phase $\exp[+i\mathbf{k} \cdot \mathbf{a}(\boldsymbol{\varphi}^{-1}\mathbf{x})]$ during this motion. The second factor has thus the effect to compensate this phase in order for the component f to continue to describe the density in the fundamental cell.

According to the previous discussion, we can conclude that a hydrodynamic mode of wavenumber $\mathbf{k}$ will be given by a solution of the generalized eigenvalue problem [25]:

$$\hat{R}_{\mathbf{k},s_\mathbf{k}} \psi_\mathbf{k} = \psi_\mathbf{k} \tag{18}$$

$$\hat{R}^\dagger_{\mathbf{k},s_\mathbf{k}} \tilde{\psi}_\mathbf{k} = \tilde{\psi}_\mathbf{k} \tag{19}$$

with the biorthonormality condition taken with respect to the invariant Lebesgue measure v defined in the surface of section:

$$\langle \tilde{\psi}_\mathbf{k}^* \psi_\mathbf{k} \rangle_v = 1 \tag{20}$$

In Eqs. (18)–(20), $\psi_\mathbf{k}$ is the eigenstate of the Frobenius–Perron operator (17) and $\tilde{\psi}_\mathbf{k}$ is the adjoint eigenstate. In Eq. (18), the eigenvalue of the operator (17) is taken to be equal to unity in order for $\psi_\mathbf{k}$ to become the density associated with a conditionally invariant measure, as discussed above. Indeed, it has been necessary to include already the decay rate $(-s)$ in the Frobenius–Perron operator (17) because the Poincaré–Birkhoff mapping (10) is not isochronic. Therefore, requiring that the eigenvalue of the operator (17) is equal to unity has the effect of fixing the variable s to a value which depends on the wavenumber $\mathbf{k}$ and which is intrinsic to the system. This value $s_\mathbf{k}$ is a so-called Pollicott–Ruelle resonance [36–41]. The Pollicott–Ruelle resonances have been theoretically studied in different systems such as the disk scatterers [42] or the pitchfork bifurcation [43] and, notably, in connection with relaxation and diffusion in classically chaotic quantum systems [44–49]. These resonances have also been evidenced in an experimental microwave study of disk scatterers [50,51].

Several such Pollicott–Ruelle resonances may exist. In ergodic and mixing spatially periodic systems, we may expect that there is a unique Pollicott–Ruelle resonance that vanishes with the wavenumber: $\lim_{\mathbf{k} \to 0} s_\mathbf{k} = 0$. In this limit, the corresponding eigenstate becomes the invariant state, which is the microcanonical equilibrium measure for volume-preserving Hamiltonian systems: $\lim_{\mathbf{k} \to 0} \psi_\mathbf{k} = 1$. However, for small enough nonvanishing wavenumbers $\mathbf{k} \neq 0$, the eigenstate is no longer the invariant measure and instead defines a complex conditionally invariant measure describing a hydrodynamic mode of diffusion.

In chaotic systems, these conditionally invariant measures are singular with respect to the Lebesgue measure so that their density $\psi_\mathbf{k}$ is not a function but a mathematical distribution (also called a generalized function) of a type defined by Schwartz or Gel'fand. A cumulative function can be defined as the measure of a m-dimensional rectangular domain $[\mathbf{0}, \mathbf{x}]^m$ in phase space with $0 < m \leq M - 1$:

$$F_{\mathbf{k}}(\mathbf{x}) = \int_{[\mathbf{0},\mathbf{x}]^m} \psi_{\mathbf{k}}(\mathbf{x}')d\mathbf{x}' \qquad (21)$$

which is expected to be continuous but nondifferentiable for small enough wavenumbers $\mathbf{k} \neq 0$ [17,18,25,52–54]. In Eq. (21), $d\mathbf{x}'$ defines the invariant Lebesgue measure ν in the surface of section $\mathscr{P}$. For $m = 1$, the cumulative functions (21) of the hydrodynamic modes of the multibaker maps and of the two-dimensional Lorentz gases form fractal curves in the complex plane. These fractal curves are characterized by a Hausdorff dimension that is given in terms of the diffusion coefficient and the Lyapunov exponent [55,56].

In the next section, we shall present the consequences of these results on the transport properties of diffusion.

IV. CONSEQUENCES OF THE EIGENVALUE PROBLEM

The generalized eigenvalue problem presented in the previous section defines the hydrodynamic modes of diffusion at the microscopic level of description. As summarized here below, all the relevant properties of the transport by diffusion such as the diffusion coefficient, the higher-order Burnett and super-Burnett coefficients if they exist, the nonequilibrium steady states, Fick's law, and the entropy production of nonequilibrium thermodynamics can be derived from the Frobenius–Perron operator (17).

A. The Diffusion Coefficient and the Higher-Order Coefficients

We suppose that the leading Pollicott-Ruelle resonance $s_{\mathbf{k}}$ is n-times differentiable near $\mathbf{k} = 0$: $s_{\mathbf{k}} \in \mathscr{C}^n$. Hence, the eigenvalue equations (18)–(20) can be differentiated successively with respect to the wavenumber: $\partial_{\mathbf{k}}, \partial_{\mathbf{k}}^2, \partial_{\mathbf{k}}^3, \partial_{\mathbf{k}}^4, \ldots, \partial_{\mathbf{k}}^n$ [25].

In the absence of external field, there is no mean drift and the first derivatives give

$$\partial_{\mathbf{k}} s_0 = 0 \qquad (22)$$

$$\partial_{\mathbf{k}} \psi_0(\mathbf{x}) = -i \sum_{n=1}^{\infty} \mathbf{a}(\boldsymbol{\varphi}^{-n}\mathbf{x}) \qquad (23)$$

Proceeding in a similar way up to the second derivatives gives us the matrix of diffusion coefficients (if they exist) as

$$D_{\alpha\beta} = -\frac{1}{2}\frac{\partial^2 s_0}{\partial k_\alpha \partial k_\beta} = \frac{1}{2\langle T\rangle_\nu}\sum_{n=-\infty}^{+\infty}\langle a_\alpha(a_\beta \circ \boldsymbol{\varphi}^n)\rangle_\nu \qquad (24)$$

Using Eq. (14), which relates the jump vector $\mathbf{a}(\mathbf{x})$ to the integral of the particle velocity $\mathbf{v}$, Eq. (24) implies the Green–Kubo relation:

$$D_{\alpha\beta} = \frac{1}{2} \int_{-\infty}^{+\infty} dt \langle v_\alpha (v_\beta \circ \mathbf{\Phi}^t) \rangle_\mu \qquad (25)$$

where μ is the microcanonical equilibrium measure in the original phase space.

Higher derivatives give expressions for the Burnett and super-Burnett ($B_{\alpha\beta\gamma\delta}$) coefficients if they exist [25]. These higher-order coefficients appear in the expansion of the dispersion relation of diffusion in powers of the wavenumber:

$$s_\mathbf{k} = -\sum_{\alpha\beta} D_{\alpha\beta}\, k_\alpha\, k_\beta + \sum_{\alpha\beta\gamma\delta} B_{\alpha\beta\gamma\delta}\, k_\alpha\, k_\beta\, k_\gamma\, k_\delta + \mathcal{O}(k^6) \qquad (26)$$

In a recent work [57,58], Chernov and Dettmann proved the existence of these higher-order coefficients and conjectured the convergence of the expansion (26) for the periodic Lorentz gas with a finite horizon, on the basis of the formalism described here.

B. Periodic-Orbit Theory

The periodic-orbit theory of classic systems has been extensively developed since the pioneering work by Cvitanović [59]. This theory has been applied to many dynamical systems that are either classic, stochastic, or quantum. For an overview of periodic-orbit theory, see Ref. 60.

For Axiom-A spatially periodic systems, the Fredholm determinant of the Frobenius–Perron operator (17) can be expressed as a product over all the unstable periodic orbits given by the following Zeta function [25]:

$$Z(s; \mathbf{k}) \equiv \mathrm{Det}(\hat{I} - \hat{R}_{\mathbf{k},s})$$
$$= \prod_p \prod_{m_1 \cdots m_u = 0}^{\infty} \left[1 - \frac{\exp(-s T_p - i \mathbf{k} \cdot \mathbf{a}_p)}{|\Lambda_{1p} \cdots \Lambda_{up}| \Lambda_{1p}^{m_1} \cdots \Lambda_{up}^{m_u}} \right]^{(m_1+1)\cdots(m_u+1)} \qquad (27)$$

where Λ_{ip} with $i = 1, \ldots, u$ are the instability eigenvalues of the linearized Poincaré–Birkhoff mapping (10) for the periodic orbit p. These instability eigenvalues satisfy $|\Lambda_{ip}| > 1$. T_p is the period of p and $\mathbf{a}_p$ is the vector by which the particle travels on the lattice during the period. The integer u is the number of unstable directions in the phase space.

According to the eigenvalue equation (18), the Pollicott–Ruelle resonances $s_\mathbf{k}$ are the zeroes of the Fredholm determinant: $Z(s_\mathbf{k}; \mathbf{k}) = 0$. This result can be used in order to obtain a periodic-orbit formula for the diffusion coefficient of an isotropic d-dimensional diffusive process [61–63]:

$$D = \frac{\sum_{l=0}^{\infty}(-)^l \sum_{p_1 \neq \cdots \neq p_l}(\mathbf{a}_{p_1} + \cdots + \mathbf{a}_{p_l})^2 F_{p_1} \cdots F_{p_l}}{2d \sum_{l=0}^{\infty}(-)^l \sum_{p_1 \neq \cdots \neq p_l}(T_{p_1} + \cdots + T_{p_l}) F_{p_1} \cdots F_{p_l}} \quad (28)$$

with $F_p = \prod_{i=1}^{u} |\Lambda_{ip}|^{-1}$.

We remark that the existence and analyticity of Fredholm determinants such as (27) has recently been studied for hyperbolic diffeomorphism of finite smoothness on the basis of a new theory of distributions associated with the unstable and stable leafs of the diffeomorphism [64].

C. The Nonequilibrium Steady States

At the phenomenological level, a nonequilibrium steady state can be obtained from a hydrodynamic mode in the limit where the wavelength increases indefinitely together with the amplitude because $\lim_{L \to \infty}(L/2\pi) \sin(2\pi \mathbf{g} \cdot \mathbf{r}/L) = \mathbf{g} \cdot \mathbf{r}$, where $\mathbf{g} = \nabla c$ is a gradient of concentration c. Accordingly, a nonequilibrium steady state can be obtained by the following limit [19]:

$$\Psi_{\text{nss}} \equiv -i\mathbf{g} \cdot \partial_{\mathbf{k}} \Psi_{\mathbf{k}}|_{\mathbf{k}=0} \quad (29)$$

Because the first derivatives of the hydrodynamic mode is given by Eq. (23) and because the jump vector is related to the time integral of the particle velocity by Eq. (14), it is possible to obtain the following expression for the phase-space density of a nonequilibrium steady state with a gradient of concentration $\mathbf{g} = \nabla c$:

$$\Psi_{\text{nns}}(\mathbf{X}) = \mathbf{g} \cdot \left[\mathbf{r}(\mathbf{X}) + \int_0^{\infty} \mathbf{v}(\Phi^t \mathbf{X}) \, dt \right] \quad (30)$$

where $\mathbf{r}$ and $\mathbf{v}$ are, respectively, the position and the velocity of the tracer particle that diffuses [25].

This expression can also be derived from the phase-space probability density of a nonequilibrium steady state of an open system between two reservoirs of particles at different densities [33]. These reservoirs have the effect to impose flux boundary conditions to the diffusive system. As we said in Section II, the discontinuities of this probability density occur on the unstable manifolds of the fractal repeller of the escape-rate formalism [33]. In the limit where the reservoirs are separated by an arbitrarily large distance $L \to \infty$ while keeping constant the gradient $\mathbf{g}$, expression (30) is obtained for the phase-space density with respect to the density at a point in the middle of the system. Expression (30) is known as the Lebowitz–McLennan–Zubarev nonequilibrium steady state [65–68]. This density is an invariant of motion in the sense that $\{H, \Psi_{\mathrm{nss}}\} = 0$, where $\{H, \cdot\}$ is the Poisson bracket with the Hamiltonian.

An essential property of the nonequilibrium steady state is its singular character that has appeared in the limit $L \to \infty$. Indeed, Eq. (30) defines the density of a measure which is singular with respect to the Lebesgue measure. This singular character is furthermore of fractal type because the dynamics is chaotic. The fractal-like singular character is evidenced by considering the cumulative function associated with the density (30). For the so-called multibaker model, this cumulative function is given in terms of the continuous but nondifferentiable Takagi function which is self-similar [19]. This self-similarity is a reflect of the self-similarity of the underlying fractal repeller in a finite but large open system.

A consequence of Eq. (30) is Fick's law:

$$\mathbf{j} = -D\nabla c \qquad (31)$$

obtained by computing the mean flux $\mathbf{j}$ of particles for the nonequilibrium steady state (30) [19,25].

D. Entropy Production

The fractal-like singular character of the nonequilibrium steady state in the limit of a large system has a fundamental consequence on the question of entropy production. Indeed, the usual argument leading to the constancy of Gibbs' entropy becomes questionable because this argument assumes the existence of a density *function* for the probability density of the system. Because the density function may no longer exist in the limit $L \to \infty$, we may expect a very different behavior for the time variation of the entropy [34].

As shown very recently [69], a similar problem is expected for the time-dependent relaxation toward equilibrium in a finite system. In the long-time limit $t \to \infty$, the relaxation is described in terms of the hydrodynamic modes which are singular with respect to the Lebesgue measure so that the constancy of the entropy is here again in question.

A detailed analysis of the entropy production in an elementary model of diffusion known as the multibaker map has shown that the fractal-like singular character of the nonequilibrium steady state [34]—and equivalently of the hydrodynamic modes [69]—explains why the entropy production is positive and has the following form given by nonequilibrium thermodynamics [70]:

$$\frac{d_i S}{dt} = \int D \frac{(\nabla c)^2}{c} \, d\mathbf{r} + \cdots \qquad (32)$$

where the dots denote possible corrections of higher orders in the gradient. The calculation leading to this result simply assumes a coarse-grained entropy based on a partition of phase space into arbitrarily small cells [34]. Taking an arbitrarily fine partition allows us to get rid of the arbitrariness of the partition, in analogy with the procedure used to define the Kolmogorov–Sinai entropy per unit time.

For nonequilibrium steady states, the positive entropy production (32) is obtained in the limit where an arbitrarily large system is considered before an arbitrarily fine partition [34]. Indeed, for a finite system in a nonequilibrium steady state, there is always a small scale in phase space below which the invariant measure is continuous with respect to the Lebesgue measure, so that the entropy production vanishes for fine enough partitions. The remarkable fact is that this critical scale decreases exponentially rapidly in chaotic systems so that the behavior predicted by nonequilibrium thermodynamics will predominate even in relatively small systems. This result is especially interesting because it justifies the use of nonequilibrium thermodynamics already in small parts of a biological system.

For the time-dependent relaxation toward the equilibrium in chaotic systems, the positive entropy production (32) will be obtained in the long-time limit $t \to \infty$ for any partition into arbitrarily small cells, even if the system is finite [69]. This remarkable result has its origin in the fact that the hydrodynamic modes controlling the relaxation are always singular with respect to the Lebesgue measure even in finite chaotic systems. In this regard, we may conclude that the second law of thermodynamics is a consequence of the fractal-like singular character of the hydrodynamic modes that describe the relaxation toward the equilibrium. This conclusion is natural in view of the fact that this fractal-like singular character is a direct consequence of the phase-space mixing induced by the dynamical chaos.

V. CONCLUSIONS

In this short overview, we have summarized recent results about transport by diffusion in spatially periodic chaotic systems. These results extend to reaction-diffusion systems as recently shown elsewhere [71–73].

In this context, we have shown that all the relevant properties of transport (as well as of chemical reaction) are the consequences of a Frobenius–Perron operator such as Eq. (17). The generalized eigenvalue problem based on this Frobenius–Perron operator defines the hydrodynamic modes as the slowly damped long-wavelength eigenstates corresponding to the smallest Pollicott–Ruelle resonance $s_\mathbf{k}$. These modes describe the relaxation toward equilibrium of quasiperiodic inhomogeneities in the phase-space probability density of a statistical ensemble of trajectories. The inhomogeneities are quasiperiodic in the sense that the wavelength $L = 2\pi/\|\mathbf{k}\|$ of the hydrodynamic mode adds an extra periodicity to the intrinsic one of the lattice $\mathscr{L}$.

In systems with a gas of independent particles such as the Lorentz gas, the Frobenius–Perron operator (17) directly describes the relaxation toward equilibrium of quasiperiodic inhomogeneities in the position–velocity density of particles composing the gas. Therefore, our results show explicitly how the relaxation to equilibrium can occur in a N-particle system in the limit $N \to \infty$. Indeed, in this limit, the particle density $\rho(\mathbf{X})$—or its Fourier component $f(\mathbf{x})$—becomes a smooth function which evolves according to the Frobenius–Perron operator (16)—or (17)—so that the time evolution of the gas has the nonequilibrium properties described in Section IV.

A comment is here in order that the Pollicott–Ruelle resonances $s_\mathbf{k}$ are defined for the semigroup of the forward time evolution for $t \to +\infty$. By time reversibility, there exists another semigroup for $t \to -\infty$ with time-reversed properties. The hydrodynamic modes of the forward semigroup are singular in the stable directions of the phase space, but they are smooth in the unstable directions because the dynamics is expanding in the unstable directions. In contrast, the hydrodynamic modes of the backward semigroup are singular in the unstable directions and smooth in the stable directions. This inequivalence constitutes a spontaneous breaking of the time reversal symmetry which is due to the Lyapunov dynamical instability. We notice that a dynamical instability without chaos (like in the two-disk scatterer or in the inverted harmonic potential [11]) is already enough to generate this spontaneous time reversal symmetry breaking. However, chaos is required to give a fractal character to the singular conditionally invariant measures of the hydrodynamic modes, which is essential to obtain the transport properties as well as for the entropy production (32).

We remark that the hydrodynamic modes exist only in the asymptotic expansions for $t \to \pm\infty$ of the time averages $\langle A \rangle_t$ of the physical observables $A(\mathbf{X})$. The relaxation toward equilibrium can be described in terms of exponentially decaying modes thanks to these time asymptotic expansions. Because of the dynamical instability, the forward asymptotic expansion turns out to be inequivalent to the backward asymptotic expansion. The forward asymptotic expansion for $t \to +\infty$ is obtained by the analytic continuation to complex frequencies $\exp(st) = \exp(i\omega t)$ with Im $\omega > 0$, while the backward expansion for $t \to -\infty$ is given by the analytic continuation to complex frequencies with Im $\omega < 0$. Supposing that the initial time of the experiment is taken at $t = 0$, it is a known result that the forward (resp. backward) asymptotic expansion converges only for $t > 0$ (resp. $t < 0$) [11]. Therefore, the origin $t = 0$ constitutes a horizon if we want to use the backward semigroup for future predictions on the time evolution of a statistical ensemble (or of a gas of independent particles). Somehow, this horizon is to the statistical ensemble of trajectories (or to the gas) what the Lyapunov horizon (1) is to a single trajectory. We observe that the accumulation of infinitely many trajectories in the statistical ensemble (or infinitely many particles in the gas) has created a horizon that is much stronger than the movable Lyapunov horizon (1). The semigroup horizon at $t = 0$ restricts the use of the exponentially decaying hydrodynamic modes to the future relaxation for $t > 0$. Therefore, this horizon constitutes a fundamental limit on the use of nonequilibrium thermodynamics in the description of many-particle systems. In this sense, the semigroup horizon justifies the irreversible character of nonequilibrium thermodynamics.

Acknowledgments

The author thanks Professor G. Nicolis for his support and encouragement in this research, as well as Professors J. R. Dorfman and S. Tasaki for many fruitful and inspiring discussions over the years. The "Fonds National de la Recherche Scientifique" is gratefully acknowledged for financial support. This work has also been financially supported by the InterUniversity Attraction Pole Program of the Belgian Federal Office of Scientific, Technical and Cultural Affairs.

References

1. I. Prigogine, *Nonequilibrium Statistical Mechanics*, Wiley, New York, 1962.
2. R. Balescu, *Equilibrium and Nonequilibrium Statistical Mechanics*, Wiley, New York, 1975.
3. P. Résibois and M. De Leener, *Classical Kinetic Theory of Fluids*, Wiley, New York, 1977.
4. J.-P. Boon and S. Yip, *Molecular Hydrodynamics*, Dover, New York, 1980.
5. N. N. Krylov, *Nature* **153**, 709 (1944); N. N. Krylov, *Works on the Foundations of Statistical Mechanics*, Princeton University Press, 1979; Ya. G. Sinai, *ibid.*, p. 239.
6. P. Gaspard and G. Nicolis, *Physicalia Magazine (J. Belg. Phys. Soc.)* **7**, 151 (1985).
7. R. Livi, A. Politi, and S. Ruffo, *J. Phys. A: Math. Gen.* **19**, 2033 (1986).
8. H. A. Posch and W. G. Hoover, *Phys. Rev. A* **38**, 473 (1988).
9. H. van Beijeren, J. R. Dorfman, H. A. Posch, and Ch. Dellago, *Phys. Rev. E* **56**, 5272 (1997).
10. N. I. Chernov, *J. Stat. Phys.* **88**, 1 (1997).
11. P. Gaspard, *Chaos, Scattering and Statistical Mechanics*, Cambridge University Press, Cambridge, England, 1998.
12. J. R. Dorfman, *An Introduction to Chaos in Nonequilibrium Statistical Mechanics*, Cambridge University Press, Cambridge, England, 1999.
13. L. A. Bunimovich and Ya. G. Sinai, *Commun. Math. Phys.* **78**, 247, 479 (1980).
14. N. I. Chernov, *J. Stat. Phys.* **74**, 11 (1994).
15. A. Knauf, *Commun. Math. Phys.* **110**, 89 (1987); A. Knauf, *Ann. Phys. (N.Y.)* **191**, 205 (1989).
16. P. Gaspard, *J. Stat. Phys.* **68**, 673 (1992).
17. P. Gaspard, *Chaos* **3**, 427 (1993).
18. P. Gaspard, in Y. Aizawa, S. Saito, and K. Shiraiwa, *Dynamical Systems and Chaos*, Vol. 2, World Scientific, Singapore, 1995, pp. 55–68.
19. S. Tasaki and P. Gaspard, *J. Stat. Phys.* **81**, 935 (1995).
20. A. Baranyai, D. J. Evans, and E. G. D. Cohen, *J. Stat. Phys.* **70**, 1085 (1993).
21. N. I. Chernov, G. L. Eyink, J. L. Lebowitz, and Ya. G. Sinai, *Phys. Rev. Lett.* **70**, 2209 (1993).

22. T. Tél and J. Vollmer, *Entropy Balance, Multibaker Maps, and the Dynamics of the Lorentz Gas*, in D. Szasz, ed., *Hard Ball Systems and Lorentz Gas, Encyclopedia of Mathematical Science*, Springer, Berlin, 2000, pp. 367–420.
23. S. Tasaki and P. Gaspard, *Theor. Chem. Acc.* **102**, 385 (1999).
24. S. Tasaki and P. Gaspard, *J. Stat. Phys.* **101**, 125 (2000).
25. P. Gaspard, *Phys. Rev. E* **53**, 4379 (1996).
26. G. Nicolis, *Introduction to Nonlinear Science*, Cambridge University Press, Cambridge, England, 1995.
27. J.-P. Eckmann and D. Ruelle, *Rev. Mod. Phys.* **57**, 617 (1985).
28. P. Gaspard, M. E. Briggs, M. K. Francis, J. V. Sengers, R. W. Gammon, J. R. Dorfman, and R. V. Calabrese, *Nature* **394**, 865 (1998).
29. P. Gaspard and S. A. Rice, *J. Chem. Phys.* **90**, 2225 (1989); **91**, E3279 (1989).
30. P. Gaspard and G. Nicolis, *Phys. Rev. Lett.* **65**, 1693 (1990).
31. P. Gaspard and F. Baras, *Phys. Rev. E* **51**, 5332 (1995).
32. J. R. Dorfman and P. Gaspard, *Phys. Rev. E* **51**, 28 (1995).
33. P. Gaspard, *Physica A* **240**, 54 (1997).
34. P. Gaspard, *J. Stat. Phys.* **88**, 1215 (1997).
35. P. D. Lax and R. S. Phillips, *Scattering Theory*, Academic Press, New York, 1967.
36. M. Pollicott, *Invent. Math.* **81**, 413 (1985).
37. M. Pollicott, *Invent. Math.* **85**, 147 (1986).
38. D. Ruelle, *Phys. Rev. Lett.* **56**, 405 (1986).
39. D. Ruelle, *J. Stat. Phys.* **44**, 281 (1986).
40. D. Ruelle, *J. Diff. Geom.* **25**, 99, 117 (1987).
41. D. Ruelle, *Commun. Math. Phys.* **125**, 239 (1989).
42. P. Gaspard and D. Alonso Ramirez, *Phys. Rev. A* **45**, 8383 (1992).
43. P. Gaspard, G. Nicolis, A. Provata, and S. Tasaki, *Phys. Rev. E* **51**, 74 (1995).
44. A. V. Andreev, O. Agam, B. D. Simons, and B. L. Altshuler, *Phys. Rev. Lett.* **76**, 3947 (1996).
45. O. Agam, *Phys. Rev. E* **61**, 1285 (2000); G. Blum and O. Agam, *Phys. Rev. E* **62**, 1977 (2000).
46. M. Khodas, S. Fishman, and O. Agam, *Phys. Rev. E* **62**, 4769 (2000).
47. T. Kottos and U. Smilansky, *Phys. Rev. Lett.* **85**, 968 (2000).
48. J. Weber, F. Haake, and P. Seba, *Phys. Rev. Lett.* **85**, 3620 (2000).
49. F. Haake, *Quantum Signatures of Chaos*, 2nd edition, Springer-Verlag Telos, Berlin, 2001.
50. K. Pance, W. Lu, and S. Sridhar, *Phys. Rev. Lett.* **85**, 2737 (2000).
51. W. Lu, L. Viola, K. Pance, M. Rose, and S. Sridhar, *Phys. Rev. E* **61**, 3652 (2000).
52. S. Tasaki, I. Antoniou, and Z. Suchanecki, *Phys. Lett. A* **179**, 97 (1993).
53. H. H. Hasegawa and D. J. Driebe, *Phys. Rev. E* **50**, 1781 (1994).
54. D. J. Driebe, *Fully Chaotic Maps and Broken Time Symmetry*, Kluwer, Dordrecht, 1999.
55. T. Gilbert, J. R. Dorfman, and P. Gaspard, *Nonlinearity* **14**, 339 (2001).
56. P. Gaspard, I. Claus, T. Gilbert, and J. R. Dorfman, *Phys. Rev. Lett.* **86**, 1506 (2001).
57. N. I. Chernov and C. P. Dettmann, *Physica A* **279**, 37 (2000).
58. C. P. Dettmann, *The Burnett expansion of the periodic Lorentz gas* (preprint nlin.CD/0003038, server xxx.lanl.gov).
59. P. Cvitanović, *Phys. Rev. Lett.* **61**, 2729 (1988).

60. P. Cvitanović, R. Artuso, R. Mainieri, and G. Vattay, http://www.nbi.dk/ChaosBook/
61. R. Artuso, *Phys. Lett. A* **160**, 528 (1991).
62. P. Cvitanović, P. Gaspard, and T. Schreiber, *Chaos* **2**, 85 (1992).
63. P. Cvitanović, J.-P. Eckmann, and P. Gaspard, *Chaos, Solitons & Fractals* **6**, 113 (1995).
64. A. Yu. Kitaev, *Nonlinearity* **12**, 141 (1999).
65. J. L. Lebowitz, *Phys. Rev.* **114**, 1192 (1959).
66. J. A. MacLennan Jr., *Phys. Rev.* **115**, 1405 (1959).
67. D. N. Zubarev, *Sov. Phys. Dokl.* **6**, 776 (1962).
68. D. N. Zubarev, *Nonequilibrium Statistical Thermodynamics*, Consultants, New York, 1974.
69. T. Gilbert, J. R. Dorfman, and P. Gaspard, *Phys. Rev. Lett.* **85**, 1606 (2000).
70. I. Prigogine, *Introduction to the Thermodynamics of Irreversible Processes*, Wiley, New York, 1961.
71. P. Gaspard and R. Klages, *Chaos* **8**, 409 (1998).
72. P. Gaspard, *Physica A* **263**, 315 (1999).
73. I. Claus and P. Gaspard, *J. Stat. Phys.* **101**, 161 (2000).

TRANSPORT THEORY FOR COLLECTIVE MODES AND GREEN–KUBO FORMALISM FOR MODERATELY DENSE GASES

T. PETROSKY

Center for Studies in Statistical Mechanics and Complex Systems, The University of Texas, Austin, Texas, U.S.A.; International Solvay Institutes for Physics and Chemistry, Free University of Brussels, Brussels, Belgium; and Theoretical Physics Department, University of Vrije, Brussels, Belgium

CONTENTS

I. Introduction
II. Complex Spectral Representation
III. Application to Moderately Dense Gases
IV. Velocity Autocorrelation Function
V. Ternary Correlation Subspace
VI. Renormalization of Vertices
VII. Discussions and Concluding Remarks
Acknowledgments
References

I. INTRODUCTION

In spite of many interesting results obtained in the past [1–17], the kinetic theory to moderately dense gases faces still great difficulties. The main difficulty is that the macroscopic transport equations (such as the diffusion equation) are only "Markovian" equations in which the time change is determined by the instanteneous state of the system. In contrast, the generalized master equation first derived by Prigogine and Résibois [18] (see also Ref. 19) indicates that beyond the Boltzmann type we have "memory effects" and therefore

---

Dynamical Systems and Irreversibility: A Special Volume of Advances in Chemical Physics, Volume 122, Edited by Ioannis Antoniou. Series Editors I. Prigogine and Stuart A. Rice.
ISBN 0-471-22291-7. © 2002 John Wiley & Sons, Inc.

non-Markovian behavior. Indeed, introducing the projector $P^{(0)}$ that projects the distribution function $\rho(\mathbf{r}_1, \ldots, \mathbf{r}_N, \mathbf{v}_1, \ldots, \mathbf{v}_N)$ on a function depending only on velocities, we have the well-known expression [19]

$$i\frac{\partial}{\partial t}P^{(0)}\rho(t) = \int_0^t dt'\tilde{\psi}^{(0)}(t')P^{(0)}\rho(t-t') + \tilde{D}^{(0)}(t)\rho(0) \quad (1)$$

Many attempts have been made to "markovianize" this equation (see, for example, Ref. 6). However, difficulties remain. As discussed briefly in Section III, the Markovian approximation diverges for two-dimensional systems. However, it is difficult to believe that transport theory fails in two-dimensional systems. Also, it is well known that the approach to equilibrium involves slow nonexponential processes. The memory effects retained in (1) correspond to the physical situations beyond the Boltzmann approximation valid for low concentrations.

The approach based on the complex spectral representation of the Liouville operator introduced by Prigogine and the author shows that these difficulties can be overcome [20,21]. To make this chapter more accessible, we first present a brief summary of our method (Section II). In short we consider a class of distribution functions ρ which lies outside the Hilbert space (they correspond to generalized functions or distributions). We then solve the eigenvalue problem for the Liouvillian $L = L_0 + L'$ in this extended space (L_0 corresponds to free particles and L' takes into account interactions). This permits us to introduce two complete sets of projection operators $P^{(\nu)}$ and $\Pi^{(\nu)}$:

$$\sum_\nu P^{(\nu)} = 1, \qquad \sum_\nu \Pi^{(\nu)} = 1 \quad (2)$$

where $P^{(\nu)}$ commutes with L_0 while $\Pi^{(\nu)}$ commutes with L, and ν as defined is the degree of correlations that is the number of nonvanishing wave vectors in the Fourier representation of ρ. The fundamental result of our theory is that each component $P^{(\nu)}\Pi^{(\nu)}\rho$ satisfies the closed Markovian equation [Eq. 34].

The Markovian description involves both $P^{(\nu)}$ and $\Pi^{(\nu)}$. It applies to "dressed" or "renormalized" distribution functions. For example, the one-particle velocity distribution function is given by

$$\varphi(\mathbf{v}_1, t) = \int d\mathbf{v}_2 \ldots \int d\mathbf{v}_N P^{(0)}\rho(t) = \int d\mathbf{v}_2 \ldots \int d\mathbf{v}_N \varphi_N(\mathbf{v}_1, \mathbf{v}_2, \ldots \mathbf{v}_N, t) \quad (3)$$

In contrast, $P^{(0)}\Pi^{(0)}\rho$ takes into account the interaction processes. The "dressed" velocity distribution function is

$$\varphi^{(0)}(\mathbf{v}_1, t) = \int d\mathbf{v}_2 \ldots \int d\mathbf{v}_N P^{(0)}\Pi^{(0)}\rho(t) \quad (4)$$

and this is a functional which depends on the interaction with other particles in the medium. In other words it is a "collective mode." This dressing is a *nonequilibrium effect* because for equilibrium the dressing effect disappears: $\varphi_{eq}^{(0)}(\mathbf{v}_1) = \varphi_{eq}(v_1)$, where φ_{eq} is the normalized Maxwellian.

This result is interesting, since the original distribution function $\varphi(\mathbf{v}_1, t)$ also obeys a non-Markovian equation as a result of (1). The well-known Markovian approximation for the ordinary distribution functions such as the so-called $\lambda^2 t$-approximation in weakly coupled systems is valid only for finite time interval of order of the mean free time [19]. Both for short and long time scales, the memory effects neglected in the Markovian approximation dominate in the evolution of the distribution function. Hence, the ordinary distribution functions do not satisfy Markovian transport equations. In contrast, the dressed distribution functions such as $\varphi^{(0)}(\mathbf{v}_1, t)$ satisfy Markovian equation for all times. Therefore, it is possible to define transport coefficients in terms of our dressed distribution functions. An attempt in this direction has already been made by Thedosopulu and Grecos [22]. They derived linearized hydrodynamics from kinetic theory for three-dimensional fluids. In their paper they have already shown that linearized hydrodynamics exist only for the collective modes.

In this chapter we shall mainly consider classic systems of identical particles with mass m interacting via hard-core potentials of a diameter a_0 with number density n, in d-dimensional space with $d \geq 2$. We shall apply our complex spectral representation of the Liouvillian to the system near equailibrium where we expect the *linearized* scheme of kinetic equations is valid. One of the main conclusions reached in this chapter are that Markovian equations for our dressed states exist for all dimensions. We then can define transport coefficiencies even for $d = 2$ (see Section III). Therefore the transition from the Boltzmann approximation to moderately dense fluids involves a radical change. The transport coefficients are now associated with *collective modes* and are no longer associated with ordinary reduced distribution functions.

However, since our observations of nature are established through measuring the average values of physical quantities over the *original* distribution functions, it is important to follow the time evolution of these distribution functions. The non-Markovian evolution for the distribution function in (1) can now be described as the superposition of Markovian evolution for the collective modes:

$$\varphi(\mathbf{v}_1, t) = \sum_\nu \int d\mathbf{v}_2 \ldots \int d\mathbf{v}_N P^{(0)} \Pi^{(\nu)} \rho(t) \qquad (5)$$

This is also true when we study the effect of the long-time tails on the linear response theory, the so-called Green–Kubo formalism. (For an excellent and still up-to-date review of these problems, see Ref. 6, as well as the book by Résibois and de Leener [13]. See also Refs. 14 and 15.) In this chapter we shall specially consider the evolution of the normalized dimensionless velocity autocorrelation function defined by

$$\Gamma(t) = \frac{\langle v_{1,x}(0) v_{1,x}(t) \rangle^{\text{eq}}}{\langle v_{1,x}(0)^2 \rangle^{\text{eq}}} = \int d\mathbf{v}_1 v_{1,x} \delta\varphi(\mathbf{v}_1; t) \qquad (6)$$

Here $\langle A \rangle^{\text{eq}} = \int d\mathbf{v}^N d\mathbf{r}^N A \rho_N^{\text{eq}}$ with N-particle canonical equilibrium ρ_N^{eq}, and $\delta\varphi(\mathbf{v}_1; t)$ is the one-particle reduced function defined by

$$\delta\varphi(\mathbf{v}_1; t) \equiv \beta m \int d\mathbf{v}_{(1)}^{N-1} \int d\mathbf{r}^N e^{-iLt} v_{1,x} \rho_N^{\text{eq}} \qquad (7)$$

with $\beta = 1/k_B T$. We shall show that the long-time tails contributions in (6) lead to divergences of the Green–Kubo integrals defined in (8) for $d \leq 4$.

In the Boltzmann approximation, the autocorrelation function decays in the relaxation time by the exponential law. It is well known that the memory effect in the kinetic equation for the hard-core potential leads to power law decay (the so-called long-time tail) with $t^{-d/2}$ which comes from two-mode coupling between hydrodynamic modes in the ring processes [13]. This result for $d = 3$ leads to a convergent contribution to Green–Kubo's "diffusion coefficient" defined by

$$D_{GK} = \lim_{t \to \infty} \frac{1}{m\beta} \int_0^t d\tau \Gamma(\tau) \qquad (8)$$

while leads to serious problem of the Green–Kubo formula for $d = 2$, since it leads to a divergence as $\ln t \to \infty$. Indeed, the calculation of $\Gamma(t)$, based on the assumption that a diffusive mode exists in the ordinary reduced distribution functions for $d = 2$, leads to the contradictory result that the diffusion coefficients does not exist. This problem is related to the fact that the Markovianization of the ordinary reduced distribution functions for $d = 2$ fails because of the appearance of the divergence. In contrast, as mentioned our complex spectral representation of the Liouvillian leads to a set of *well-defined* Markovian kinetic equations near equilibrium for the dressed distribution functions for all d (including $d = 2$). Using our spectral decomposition, the non-Markovian evolution can be split rigorously into independent Markovian evolutions that can be studied independently of any truncation procedure. Therefore, our complex spectral representation provides a consistent approach to estimate the long-time tail effects.

For a long time it has been accepted that the two-mode coupling in the binary correlations leads to the slowest decay process in $\Gamma(t)$ for $d > 2$, and the "critical dimension" is $d = 2$; that is, higher modes processes lead to slower decay process in the autocorrelation function for $d < 2$, while they lead to quicker decay process for $d > 2$ [23–25]. The phenomenological approach based on hydrodynamic equations has supported this result [13]. We shall show this would be true if we restrict all intermediate states *only* to hydrodynamic modes. However, we shall show that there are slower decay processes than binary correlations in higher correlation subspace, which are coming from processes where nonhydrodynamic modes are incorporated in vertices. Indeed, we shall show for moderately dense gases that the order of magninude of the long-time tail is given by

$$\bar{\Gamma}(t) \sim \sum_{\nu=2}^{\infty} \bar{\Gamma}^{(\nu)}(t) \qquad (9)$$

Here, $\bar{\Gamma}^{(\nu)}(t)$ is the slowest-decaying contribution $\Gamma^{(\nu)}(t)$ in each correlation subspace and is given by

$$\bar{\Gamma}^{(\nu)}(t) \sim g^{\nu} \left(\frac{g}{1-g}\right)^{2(\nu-2)} \left(\frac{1}{1-g}\right)^2 \tau^{2(\nu-2)} \tau^{-d(\nu-1)/2} \qquad (10)$$

where $\tau \equiv t/t_r$ is a dimensionless time measured by the unit of the relaxation time $t_r \equiv \gamma^{-1}$ with $\gamma = k_0 \langle v \rangle$, and

$$g \equiv n_0^{d-1} = k_0^d/n, \qquad \text{with } n_0 \equiv a_0^d n \qquad (11)$$

where $k_0 = l_m^{-1} = a_0^{d-1} n$ is the inverse of the mean free length and $\langle v \rangle = (m\beta)^{-1/2}$ is the thermal velocity. This shows that the actual critical dimension is $d = 4$. Moreover, this shows that Green–Kubo's diffusion coefficient D_{GK} diverges even for $d = 3$. This is in contrast to the well-defined transport coefficients associated with the Makovian kinetic equations for our dressed distribution functions.[1]

II. COMPLEX SPECTRAL REPRESENTATION

Let us present a brief summary of our method. See Ref. 20 for complete presentation. The Liouville equation for the N-particle distribution function (d.f.) $\rho(\mathbf{r}^N, \mathbf{v}^N, t)$ is given by

$$i\frac{\partial \rho}{\partial t} = L\rho \qquad (12)$$

The distribution function is normalized as $\int d\mathbf{r}^N d\mathbf{v}^N \rho = 1$.

---

[1] Many of the derivations in the present chapter have been included in detail in Ref. [30].

The Liouvillian L consists of a free Liouvillian L_0 and an interaction part L', that is,

$$L = L_0 + L' = \sum_{a=1}^{N} L_0^{(a)} + \sum_{b>a=1}^{N} L'_{ab} \qquad (13)$$

where $L_0^{(a)} = -i\mathbf{v}_a \cdot (\partial/\partial \mathbf{r}_a)$. We have the eigenstates

$$L_0|\mathbf{k}^N, \mathbf{v}^N\rangle = \sum_{a=1}^{N} \mathbf{k}_a \cdot \mathbf{v}_a |\mathbf{k}^N, \mathbf{v}^N\rangle$$

where $|\mathbf{k}^N, \mathbf{v}^N\rangle \equiv |\mathbf{k}^N\rangle \otimes |\mathbf{v}^N\rangle$. They are plane waves $\langle \mathbf{r}^N|\mathbf{k}^N\rangle = V^{-N/2} \exp[i \sum_{a=1}^{N} \mathbf{k}_a \cdot \mathbf{r}_a]$, where V is the volume of the system. The eigenstates of $L_0^{(a)}$ give a complete orthonormal basis,

$$\sum_{\mathbf{k}_a} |\mathbf{k}_a\rangle\langle\mathbf{k}_a| = 1, \qquad \int d\mathbf{v}_a |\mathbf{v}_a\rangle\langle\mathbf{v}_a| = 1 \qquad (14)$$

and

$$\langle\mathbf{k}_a|\mathbf{k}'_a\rangle = \delta^{kr}(\mathbf{k}_a - \mathbf{k}'_a), \qquad \langle\mathbf{v}_a|\mathbf{v}'_a\rangle = \delta(\mathbf{v}_a - \mathbf{v}'_a) \qquad (15)$$

where $\delta^{kr}(\mathbf{k}_a)$ is a d-dimensional Kronecker's delta. In an infinite volume limit,

$$\Omega^{-1} \sum_{\mathbf{k}_a} \to \int d\mathbf{k}_a, \qquad \Omega \delta^{kr}(\mathbf{k}_a) \to \delta(\mathbf{k}_a) \qquad (16)$$

where $\Omega \equiv V/(2\pi)^d$.

In this chapter we shall consider hard-core interactions. To describe these interactions we use the "pseudo-Liouvillian" formalism introduced by Ernst et al. [7] to take into account the singular nature of the hard-core interaction (see also Ref. 13). In this formalism the matrix element of the interaction is given by

$$\langle \mathbf{k}'^N, \mathbf{v}'^N | L'_{ab} | \mathbf{k}^N, \mathbf{v}^N \rangle = \frac{i}{V} T^{(ab)}_{\mathbf{k}'-\mathbf{k}} \delta^{kr}(\mathbf{k}'_a + \mathbf{k}'_b - \mathbf{k}_a - \mathbf{k}_b)$$
$$\times \delta^{kr}_{ab}(\mathbf{k}'^{N-2} - \mathbf{k}^{N-2}) \delta(\mathbf{v}'^N - \mathbf{v}^N) \qquad (17)$$

with the "binary collision operator" defined by

$$T^{(ab)}_\mathbf{q} = a_0^{d-1} \int_{\hat{\mathbf{s}} \cdot \mathbf{v}_{ab} > 0} d\hat{\mathbf{s}} (\hat{\mathbf{s}} \cdot \mathbf{v}_{ab})(e^{-i\mathbf{q} \cdot \hat{\mathbf{s}} a_0} \hat{b}^{(ab)}_\mathbf{v} - e^{+i\mathbf{q} \cdot \hat{\mathbf{s}} a_0}) \qquad (18)$$

where a_0 is a diameter of the hard spheres, and $\hat{\mathbf{s}}$ is the unit vector. The operator $\hat{b}_{\mathbf{v}}^{(ab)}$ replaces $\mathbf{v}_a$ and $\mathbf{v}_b$ by their pre-collisional values as $\hat{b}_{\mathbf{v}}^{(ab)} f(\mathbf{v}_a, \mathbf{v}_b) = f(\bar{\mathbf{v}}_a, \bar{\mathbf{v}}_b)$, where $\bar{\mathbf{v}}_a = \mathbf{v}_a - \hat{\mathbf{s}}(\hat{\mathbf{s}} \cdot \mathbf{v}_{ab})$, $\bar{\mathbf{v}}_b = \mathbf{v}_b + \hat{\mathbf{s}}(\hat{\mathbf{s}} \cdot \mathbf{v}_{ab})$, and $\mathbf{v}_{ab} \equiv \mathbf{v}_a - \mathbf{v}_b$. Moreover, $\delta_{ab}^{kr}(\mathbf{k}^{N-2})$ is a product of $N - 2$ Kronecker's deltas which excludes the particle a and b. We note that the total wavevector is preserved in the transition (2.6). As usual, this is a result of the translational invariance of the interaction in space.

Our method deals with the class of ensembles corresponding to the thermodynamic limit (i.e., the number of particle $N \to \infty$, and the volume $V \to \infty$, with the number density $n = N/V$ finite). This class includes the canonical distribution. It describes "persistent" interactions [20]. The distribution function ρ therefore can be decomposed into contributions from different degrees of correlations,

$$\rho = P^{(0)}\rho + \sum_a^N \sum_{\mathbf{k}_a} P^{(\mathbf{k}_a)}\rho + \sum_{b>a}^N \sum_{\mathbf{k}_a, \mathbf{k}_b} P^{(\mathbf{k}_a, \mathbf{k}_b)}\rho + \cdots \quad (19)$$

where $P^{(\nu)}$ is the projection operator which retains the νth degree of correlations [20]. For example, $P^{(0)}$ retains the "vacuum component" $\rho_0(\mathbf{v}^N)$, which is the velocity d.f. with vanishing wavevectors for all particles $\mathbf{k}^N = 0$, $P^{(\mathbf{k}_a)}\rho$ are the "inhomogeneity components" $\rho_{\mathbf{k}_a}(\mathbf{v}^N)$ with $\mathbf{k}_a \neq 0$, and $\mathbf{k}^{N-1} = 0$ for particles $a = 1$ to N, $P^{(\mathbf{k}_a, \mathbf{k}_b)}\rho$ are the "binary correlation components" $\rho_{\mathbf{k}_a, \mathbf{k}_b}$ with $\mathbf{k}_a \neq 0$ and $\mathbf{k}_b \neq 0$, and so on. To avoid heavy notations we shall abbreviate the notations as $P^{(1)} = P^{(\mathbf{k}_a)}, P^{(2)} = P^{(\mathbf{k}_a, \mathbf{k}_b)}, \ldots$ in the following expressions. We have

$$L_0 P^{(\nu)} = P^{(\nu)} L_0, \qquad P^{(\nu)} P^{(\mu)} = P^{(\nu)} \delta_{\nu,\mu}, \qquad \sum_\nu P^{(\nu)} = 1 \quad (20)$$

As we have shown (see Ref. 19), the Fourier coefficients with a smaller number of nonvanishing components in wavevectors $\mathbf{k}^N$ have a higher-order dependence on the volume factor V in the large volume limit—for example, $\rho_0/\rho_{\mathbf{k}_a} \sim V$. The appearance of these delta-function singularities leads to the usual cluster expansion of the reduced distribution functions. Then, for example, we have the reduced one-particle distribution function in the limit $\Omega \to \infty$,

$$f_1(\mathbf{r}_1, \mathbf{v}_1, t) = N \int d\mathbf{r}_2 \cdots d\mathbf{r}_N \int d\mathbf{v}_2 \cdots d\mathbf{v}_N \rho(\mathbf{r}^N, \mathbf{v}^N, t)$$

$$= n\varphi(\mathbf{v}_1, t) + \frac{n}{\Omega} \sum_{\mathbf{q}} f_{\mathbf{q}}(\mathbf{v}_1, t) e^{i\mathbf{q} \cdot \mathbf{r}_1}$$

$$\to n\varphi(\mathbf{v}_1, t) + n \int d\mathbf{q}\, f_{\mathbf{q}}(\mathbf{v}_1, t) e^{i\mathbf{q} \cdot \mathbf{r}_1} \quad (21)$$

where the volume element Ω^{-1} in the second line manifests the delta-function singularity.

Also, because of these singularities, ρ *does not* belong to the Hilbert space. As a result, the hermitian operator L acquires *complex* eigenvalues $Z_j^{(\nu)}$ breaking time symmetry. The time evolution of the system then splits into two semigroups. For Im $Z_j^{(\nu)} \leq 0$ the equilibrium is approached in our future that is for $t \to +\infty$, whereas for Im $Z_j^{(\nu)} \geq 0$ the equilibrium had been reached in our past. The domain of the two semigroups do not overlap (see Refs. 20 and 21). To be self-consistent we choose the semigroup oriented toward our future. Then we have the complex spectral representation,

$$e^{-iLt}|\rho(0)\rangle = \sum_\nu \sum_j |F_j^{(\nu)}\rangle e^{-iZ_j^{(\nu)}t} \langle \tilde{F}_j^{(\nu)}|\rho(0)\rangle \tag{22}$$

where $F_j^{(\nu)}$ and $\tilde{F}_j^{(\nu)}$ are biorthonormal sets of right- and left-eigenstates of the Liouvillian with the complex eigenvalue $Z_j^{(\nu)}$,

$$L|F_j^{(\nu)}\rangle = Z_j^{(\nu)}|F_j^{(\nu)}\rangle, \qquad \langle \tilde{F}_j^{(\nu)}|L = Z_j^{(\nu)} \langle \tilde{F}_j^{(\nu)}| \tag{23}$$

Operating the projection operator $P^{(\nu)}$ and its orthogonal operator $Q^{(\nu)} \equiv 1 - P^{(\nu)}$ to both sides of (2.12), one has a set of equations for $P^{(\nu)}$ and $Q^{(\nu)}$ components. Solving these equations, one can write the $Q^{(\nu)}$ components as a functional of the $P^{(\nu)}$ components. Then, one can find the eigenstates of L as

$$|F_j^{(\nu)}\rangle = [P^{(\nu)} + Q^{(\nu)}\mathscr{C}^{(\nu)}(Z_j^{(\nu)})]|F_j^{(\nu)}\rangle, \qquad \langle \tilde{F}_j^{(\nu)}| = \langle F_j^{(\nu)}|[P^{(\nu)} + \mathscr{D}^{(\nu)}(Z_j^{(\nu)})Q^{(\nu)}] \tag{24}$$

The "creation operator" $\mathscr{C}^{(\nu)}(z)$ and the "destruction operator" $\mathscr{D}^{(\nu)}(z)$ are defined by

$$\mathscr{C}^{(\nu)}(z) \equiv G_Q^\nu(z)LP^{(\nu)}, \qquad \mathscr{D}^{(\nu)}(z) \equiv P^{(\nu)}LG_Q^\nu(z) \tag{25}$$

with the propagator

$$G_Q^\nu(z) = Q^{(\nu)}[z - Q^{(\nu)}LQ^{(\nu)}]^{-1} \tag{26}$$

Substituting (24) into (23), we obtain the equation for the $P^{(\nu)}$ components of the eigenstates that satisfy the eigenvalue equation of the "collision operators" $\psi^{(\nu)}$, that is,

$$\psi^{(\nu)}(Z_j^{(\nu)})|u_j^{(\nu)}\rangle = Z_j^{(\nu)}|u_j^{(\nu)}\rangle, \qquad \langle \tilde{v}_j^{(\nu)}|\psi^{(\nu)}(Z_j^{(\nu)}) = \langle \tilde{v}_j^{(\nu)}|Z_j^{(\nu)} \tag{27}$$

where we have put $P^{(\nu)}|F_j^{(\nu)}\rangle = \sqrt{N_j^{(\nu)}}|u_j^{(\nu)}\rangle$ and $\langle \tilde{F}_j^{(\nu)}|P^{(\nu)} = \sqrt{N_j^{(\nu)}}\langle \tilde{v}_j^{(\nu)}|$ with a normalization constant $N_j^{(\nu)}$, and $\psi^{(\nu)}$ is given by

$$\psi^{(\nu)}(z) = P^{(\nu)}L_0 P^{(\nu)} + P^{(\nu)}L'P^{(\nu)} + P^{(\nu)}L'\mathscr{C}^{(\nu)}(z)P^{(\nu)} \qquad (28)$$

This operator consists of three components: (1) the free flow L_0, (2) an interaction linear in L', and (3) nonlinear term of the interaction where L' is incorporated in the propergagor in $\mathscr{C}^{(\nu)}(z)$.

Note that the eigenvalue problem associated to the collision operators $\psi^{(\nu)}(Z_j^{(\nu)})$ is nonlinear as the eigenvalue $Z_j^{(\nu)}$ appears inside the operators. This will play an important role. Also, the above relations show that the Liouville operator shares the same eigenvalues with the collision operators. Assuming bicompleteness in the space $P^{(\nu)}$, we may always construct a set of states $\{\langle \tilde{u}_j^{(\nu)}|\}$ biorthogonal to $\{|u_j^{(\nu)}\rangle\}$—that is, $\langle \tilde{u}_j^{(\nu)}|u_\beta^{(\mu)}\rangle = \delta_{\nu,\mu}\delta_{\alpha,\beta}$ and $\sum_j |u_j^{(\nu)}\rangle\langle \tilde{u}_j^{(\nu)}| = P^{(\nu)}$—and construct a similar relation for the set $\{\langle \tilde{v}_j^{(\nu)}|\}$ biorthogonal to $\{|v_j^{(\nu)}\rangle\}$. In general, $\langle \tilde{v}_j^{(\nu)}| \neq \langle \tilde{u}_j^{(\nu)}|$ (see Ref. 20).

In addition to the collision operator, let us introduce the "global" collision operator defined by

$$\theta^{(\nu)} \equiv \sum_j \psi^{(\nu)}(Z_j^{(\nu)})|u_j^{(\nu)}\rangle\langle \tilde{u}_j^{(\nu)}| = \sum_j |u_j^{(\nu)}\rangle Z_j^{(\nu)}\langle \tilde{u}_j^{(\nu)}| \qquad (29)$$

We also introduce the "global" creation operator $\mathbf{C}^{(\nu)}$ and destruction operator $\mathbf{D}^{(\nu)}$:

$$\mathbf{C}^{(\nu)} \equiv \sum_j \mathscr{C}^{(\nu)}(Z_j^{(\nu)})|u_j^{(\nu)}\rangle\langle \tilde{u}_j^{(\nu)}|, \qquad \mathbf{D}^{(\nu)} \equiv \sum_j |v_j^{(\nu)}\rangle\langle \tilde{v}_j^{(\nu)}|\mathscr{D}^{(\nu)}(Z_j^{(\nu)}) \qquad (30)$$

Then we have

$$|F_j^{(\nu)}\rangle = \sqrt{N_j^{(\nu)}}(P^{(\nu)} + \mathbf{C}^{(\nu)})|u_j^{(\nu)}\rangle, \qquad \langle \tilde{F}_j^{(\nu)}| = \langle \tilde{v}_j^{(\nu)}|(P^{(\nu)} + \mathbf{D}^{(\nu)})\sqrt{N_j^{(\nu)}} \qquad (31)$$

In our earlier work, we have repeatedly introduced the concept of "subdynamics" [26,27]. The relation of subdynamics to the complex spectral representation can be seen through the projection operators $\Pi^{(\nu)}$ defined by

$$\Pi^{(\nu)} \equiv \sum_j |F_j^{(\nu)}\rangle\langle \tilde{F}_j^{(\nu)}| = (P^{(\nu)} + \mathbf{C}^{(\nu)})A^{(\nu)}(P^{(\nu)} + \mathbf{D}^{(\nu)}) \qquad (32)$$

where "normalization operator" is defined by $A^{(\nu)} \equiv P^{(\nu)}(1 + \mathbf{D}^{(\nu)}\mathbf{C}^{(\nu)})^{-1}$, which gives us $N_j^{(\nu)} = \langle \tilde{u}_j^{(\nu)}|A^{(\nu)}|v_j^{(\nu)}\rangle$ for the normalization constant. They satisfy the orthogonality and completeness relations

$$L\Pi^{(\nu)} = \Pi^{(\nu)}L, \qquad \Pi^{(\nu)}\Pi^{(\mu)} = \Pi^{(\nu)}\delta_{\nu,\mu}, \qquad \sum_\nu \Pi^{(\nu)} = 1 \qquad (33)$$

$\Pi^{(\nu)}$ is an extension of $P^{(\nu)}$ to the total Liouvillian L. Because these projection operators commute with the Liouvillian, each component $\Pi^{(\nu)}|\rho\rangle$ individually satisfies the Liouville equation. For this reason, the projection operators $\Pi^{(\nu)}$ are associated with *subdynamics*. Then, (22) leads to a *Markovian equation*, which is a closed equation for the "privileged component" [i.e., $P^{(\nu)}\rho^{(\nu)}(t)$] of each of the subdynamics $\rho^{(\nu)}(t) \equiv \Pi^{(\nu)}\rho(t)$:

$$i\frac{\partial}{\partial t}P^{(\nu)}|\rho^{(\nu)}(t)\rangle = \theta^{(\nu)}P^{(\nu)}|\rho^{(\nu)}(t)\rangle \qquad (34)$$

This result is important, since as mentioned any component $P^{(\nu)}$ of the total $\rho(t) = \sum_\nu \rho^{(\nu)}(t)$ obeys a non-Markovian equation with memory effects. These effects can now be described by the superposition of Markov processes, that is,

$$\rho(t) = \sum_\nu e^{-iLt}\rho(0) = \sum_\nu [P^{(\nu)} + \mathbf{C}^{(\nu)}]e^{-i\theta^{(\nu)}t}A^{(\nu)}[P^{(\nu)} + \mathbf{D}^{(\nu)}]\rho(0) \qquad (35)$$

III. APPLICATION TO MODERATELY DENSE GASES

After this short summary we now consider the application of our approach to moderately dense systems which consist of N hard spheres ($d = 3$), or hard disks ($d = 2$) with the same mass m. We use the "pseudo-Liouvillian" formalism to take into account the singular nature of the hard-sphere interaction [13] (see also Ref. 7). We consider situations near equilibrium, then the *linearized* regime of the kinetic equations is applicable. In our argument, we follow the standard hypotheses (see Ref. 13) that (1) the classification of the processes in terms of the so-called "ring processes" is legitimate, and (2) in evaluation of the transport coefficients there is a well-defined separation of the *hydrodynamic modes* from the contribution of the *nonhydrodynamic modes*.

Let us consider the reduced one-particle d.f. for the inhomogeneous component associated to the subdynamics $\Pi^{(\mathbf{q})}$. This is defined through the N-particle d.f. by integrating over the velocities of dummy particles, that is,

$$f_1^{(\mathbf{q})}(\mathbf{v}_1, t) = \sum_j f_{j;1}^{(\mathbf{q})}(\mathbf{v}_1, t) \qquad (36)$$

where

$$f_{j;1}^{(\mathbf{q})}(\mathbf{v}_1, t) \equiv V^{N/2} \int d\mathbf{v}_{(1)}^{N-1} \langle 1_\mathbf{q}, \mathbf{v}^N | P^{(\mathbf{q})} | F_j^{(\mathbf{q})} \rangle \langle \tilde{F}_j^{(\mathbf{q})} | \rho(t) \rangle \qquad (37)$$

and we have introduced a new notation for the state $|a_\mathbf{k}, \mathbf{v}^N\rangle \equiv |\mathbf{k}_a = \mathbf{k}, \mathbf{k}^{N-1} = 0, \mathbf{v}^N\rangle$. With this reduction the kinetic equation (34) leads to Markovian equations:

$$i\frac{\partial}{\partial t} f_1^{(\mathbf{q})}(\mathbf{v}_1, t) = \Theta_1^{(\mathbf{q})} f_1^{(\mathbf{q})}(\mathbf{v}_1, t) \qquad (38)$$

or

$$i\frac{\partial}{\partial t} f_{j;1}^{(\mathbf{q})}(\mathbf{v}_1, t) = \Psi_1^{(\mathbf{q})}(Z_j^{(\mathbf{q})}) f_{j;1}^{(\mathbf{q})}(\mathbf{v}_1, t) \qquad (39)$$

for each component associated to the eigenstate of the collision operator, where the reduced one-particle collision operator for particle a is defined by [c.f. (28) and (29)]

$$\Theta_a^{(\mathbf{q})} \equiv \sum_j \Psi_a^{(\mathbf{q})}(Z_j^{(\mathbf{q})}) |r_j^{(\mathbf{q})}(a)\rangle\rangle \langle\langle \tilde{r}_j^{(\mathbf{q})}(a) | = \sum_j |r_j^{(\mathbf{q})}(a)\rangle\rangle Z_j^{(\mathbf{q})} \langle\langle \tilde{r}_j^{(\mathbf{q})}(a) | \qquad (40)$$

with

$$\Psi_a^{(\mathbf{q})}(z) \equiv \mathbf{q} \cdot \mathbf{v}_a + iK_a^{(\mathbf{q})} + \delta\Psi_a^{(\mathbf{q})}(z) \qquad (41)$$

Here $r_j^{(\mathbf{q})}(a)$ is an eigenstates of the "reduced" collision operator,

$$\Psi_a^{(\mathbf{q})}(Z_j^{(\mathbf{q})}) |r_j^{(\mathbf{q})}(a)\rangle\rangle = Z_j^{(\mathbf{q})} |r_j^{(\mathbf{q})}(a)\rangle\rangle \qquad (42)$$

and $\tilde{r}_j^{(\mathbf{q})}(a)$ is its biorthonormal state. The index a in $\Psi_a^{(\mathbf{q})}$ as well as in $r_j^{(\mathbf{q})}(a)$ denotes the label of a paricle in which we are interested. To distinguish the reduced one-particle states $|r_j^{(\mathbf{q})}(a)\rangle\rangle$ from the N-particle states $|u_j^{(v)}\rangle$, we have used the double-ket notation for the reduced states.

Corresponding to the three components mentioned in (28), the reduced collision operator consists of three components: (1) the free flow $\mathbf{q} \cdot \mathbf{v}_a$, (2) the "linearized collision operator" $K_a^{(\mathbf{q})}$ associated to L', and (3) the mode–mode coupling term $\delta\Psi_a^{(\mathbf{q})}(z)$ associated the last term in (28).

In the thermodynamic limit we have

$$K_1^{(\mathbf{q})} \Phi(\mathbf{v}_1) = \sum_\mathbf{k} \langle 1_\mathbf{q}, b_0 | \hat{K}_1 | 1_\mathbf{k}, b_0 \rangle \Phi(\mathbf{v}_1) \qquad (43)$$

where

$$\hat{K}_a = n \int d\mathbf{v}_b t^{(ab)} (1 + \hat{P}_{ab}) \varphi_{eq}(v_b) \quad (44)$$

with

$$\langle a_\mathbf{k}, b_\mathbf{l} | t^{(ab)} | a_\mathbf{p}, b_\mathbf{q} \rangle = T^{(ab)}_{\mathbf{k-p}} \delta^{kr}(\mathbf{k} + \mathbf{l} - \mathbf{p} - \mathbf{q}) \quad (45)$$

Here $|a_\mathbf{k}, b_{\mathbf{k}'}\rangle \equiv |\mathbf{k}_a = \mathbf{k}, \mathbf{k}_b = \mathbf{k}', \mathbf{k}^{N-2} = 0\rangle$, $\hat{P}_{ab}$ is the exchange operator [i.e., $\hat{P}_{ab} f(\mathbf{v}_a, \mathbf{v}_b) = f(\mathbf{v}_b, \mathbf{v}_a)$], and $\varphi^{eq}(v) = (\beta m/2\pi)^{d/2} \exp(-\beta m v^2/2)$ is the normalized Maxwellian.

For $\mathbf{q} = 0$, Eq. (43) reduces to the well-known *linearized Boltzmann collision operator* K_1^B:

$$K_1^B \Phi(\mathbf{v}_1) = n \int d\mathbf{v}_2 T_0^{(12)} (1 + \hat{P}_{12}) \varphi_{eq}(v_2) \Phi(\mathbf{v}_1) \quad (46)$$

The eigenvalue problem of this operator

$$K_1^B | \phi_j^0(1) \rangle\rangle = \lambda_j^0 | \phi_j^0(1) \rangle\rangle \quad (47)$$

has been studied in several papers (see, for example, Ref. 13). Defining the scalar product

$$\langle\langle g | f \rangle\rangle \equiv \int d\mathbf{v} \varphi_{eq}(v)^{-1} g^*(\mathbf{v}) f(\mathbf{v}) \quad (48)$$

K_1^B becomes a hermitian operator $\langle\langle g | K_1^B | f \rangle\rangle^* = \langle\langle f | K_1^B | g \rangle\rangle$. In the Hilbert space defined by this scalar product the spectrum λ_j^0 is discrete and the eigenstates $\phi_j^0(\mathbf{v}_1)$ provide a complete and orthogonal basis,

$$\sum_j |\phi_j^0(1)\rangle\rangle\langle\langle\phi_j^0(1)| = 1, \qquad \langle\langle \phi_j^0(1) | \phi_{j'}^0(1) \rangle\rangle = \delta_{j,j'} \quad (49)$$

where the first one in (49) is a formal expression of the relation $\sum_j \phi_j^0(\mathbf{v}_1) \times \phi_j^0(\mathbf{v}'_1) = \delta(\mathbf{v}_1 - \mathbf{v}'_1) \varphi_{eq}(v_1)$.

As a working example, we here present the case for $d = 2$ (see Ref. 13 for $d = 3$). The solution of the eigenvalue problem (47) leads to fourth-order degeneracy for the *collisional invariants* ϕ_α^0 that belong to the zero eigenvalue $\lambda_\alpha^0 = 0$ for $\{\alpha\} \equiv \{1, \ldots, 4\}$ (hereafter we shall use the notation $\{\alpha\}$ to indecate the eigenstates associated with the collisional invariants). They consist of

$\varphi_{eq}(v_a)$ multiplied by 1, v_{ax}, v_{ay}, or v_a^2, respectively. In order to specify the direction in space, we chose x axis toward the direction which is parallel to a unit vector $\hat{\mathbf{q}}$ of a given wavevector $\mathbf{q}$; that is, $v_{ax} = \hat{\mathbf{q}} \cdot \mathbf{v}_a$. Then the normalized collisional invariants are given by

$$\phi_\alpha^0(\mathbf{v}_a) = \chi_\alpha^0(\mathbf{v}_a)\varphi_{eq}(v_a) \qquad (50)$$

where (for $i =, 2$)

$$\chi_i^0(\mathbf{v}_a) = \frac{1}{\sqrt{2}}\left(-\frac{3}{\sqrt{17}} - (-1)^i\sqrt{m\beta}\hat{\mathbf{q}} \cdot \mathbf{v}_a + \frac{2}{\sqrt{17}}m\beta v_a^2\right) \qquad (51a)$$

$$\chi_3^0(\mathbf{v}_a) = \sqrt{m\beta}\hat{\mathbf{q}}_\perp \cdot \mathbf{v}_a \qquad (51b)$$

$$\chi_4^0(\mathbf{v}_a) = \frac{1}{\sqrt{17}}\left(5 - \frac{m\beta v_a^2}{2}\right) \qquad (51c)$$

with the unit vector $\hat{\mathbf{q}}_\perp$ which is perpendicular to $\mathbf{q}$.

For the remaining spectrum with $j \notin \{\alpha\}$ we have $\lambda_j^0 \neq 0$, and the minimum value of the nonvanishing eigenvalue is of order γ.

For inhomogeneous systems, the eigenvalue equation of the linearized Boltzmann collision operator is

$$(K_a^B - i\mathbf{q} \cdot \mathbf{v}_a)|\phi_j^{(\mathbf{q})}(a)\rangle\rangle = \lambda_j^\mathbf{q}|\phi_j^{(\mathbf{q})}(a)\rangle\rangle \qquad (52)$$

The left-eigenstate belonging to the same eigenvalue is given by $\langle\langle\phi_j^{(-\mathbf{q})}(a)|$. Together with the right-eigenstate, this satisfies bicomplete and biorthonormal relations [13].

In our calculation we use also the spectral property of the "Boltzmann–Lorentz operator" K_a^{BL} defined by [13].

$$K_a^{BL}\Phi(\mathbf{v}_a) = n\int d\mathbf{v}_b T_0^{(ab)}\varphi_{eq}(v_b)\Phi(\mathbf{v}_a) \qquad (53)$$

as well as its inhomogeneous operator $K_a^{BL} - i\mathbf{q} \cdot \mathbf{v}_a$.

Among the eigenstates $|\phi_{BL,j}^{(\mathbf{q})}(a)\rangle\rangle$ of $K_a^{BL} - i\mathbf{q} \cdot \mathbf{v}_a$, there is only one collisional invariant $\phi_{BL,1}^0(\mathbf{v}_a) = \varphi_{eq}(v_a)$ with $\mathbf{q} = 0$, which is just the Maxwellian: $K_a^{BL}\varphi_{eq}(v_a) = 0$.

We now consider the last term in (41). Retaining only two-mode processes, the operator $\delta\Psi_a^{(\mathbf{q})}$ in (41) becomes

$$\delta\Psi_1^{(\mathbf{q})}(z) \approx \delta\Psi_1^{(\mathbf{q})}(z;2)$$
$$= \frac{n\Omega}{(2\pi)^d}\int d\mathbf{k}\int d\mathbf{v}_2\langle 1_\mathbf{q},2_0|it^{(12)}g_2^{(12)}(z)it^{(ab)}(1+\hat{P}_{12})\varphi_{eq}(v_2)|1_\mathbf{k},2_0\rangle$$
$$(54)$$

where m in $\delta\Psi_1^{(\mathbf{q})}(z;m)$ denotes that this is assiciated only to the m-mode processes. The reduced two-particle propagator is defined by

$$g_2^{(ab)}(z) = \sum_{n=0}^{\infty} g_{2R}^{(ab)}(z)[i(2\pi)^{-d}t^{(ab)}g_{2R}^{(ab)}(z)]^n$$

$$= \frac{1}{z - L_0^{(ab)} - i\hat{K}_a - i\hat{K}_b - i(2\pi)^{-d}t^{(ab)}} \quad (55)$$

while the "ring" propagator is defined by

$$g_{2R}^{(ab)}(z) = \frac{1}{z - L_0^{(ab)} - i\hat{K}_a - i\hat{K}_b} \quad (56)$$

where $L_0^{(ab\ldots c)} \equiv L_0^{(a)} + L_0^{(b)} + \cdots + L_0^{(c)}$.

The propagator has to be evaluated as an analytic continued function from the upper half-plane of z. To obtain (54) we have approximated the N-particle propagator G_Q^v in (28) by retaining only the binary correlation subspace, that is,

$$G_Q^v \approx G_2 \equiv P^{(2)}[z - P^{(2)}LP^{(2)}]^{-1} \quad (57)$$

Let us separate the contribution of (54) coming from the first term $g_{2R}^{(ab)}$ in the expansion in (55) by writing $\delta\Psi_1^{(\mathbf{q})}(z;2) = \delta\Psi_1^{(\mathbf{q})}(z;2R) + \delta\Psi_1^{(\mathbf{q})}(z;2R')$, where $\delta\Psi_1^{(\mathbf{q})}(z;2R)$ is the operator obtained by replacing $g_2^{(ab)}$ by $g_{2R}^{(ab)}$ in (54), while $\delta\Psi_1^{(\mathbf{q})}(z;2R')$ corresponds to the remaining part of the expansion in (55).

The operator $\delta\Psi_1^{(\mathbf{q})}(z;2R)$ is well known [13]. For $z = \epsilon \to +i0$ it corresponds to the so-called "ring operator." This operator corresponds to a process in which an arbitrary number of collisions occur with particles in the medium during the two-mode coupling.

Before investigating the eigenvalue problem of the reduced collision operator for the inhomogeneous component $\Psi_1^{(\mathbf{q})}$, let us first discuss a simpler case of the homogeneous component with $\mathbf{q} = 0$. The derivation of the kinetic equation for the dressed one-particle velocity distribution function $\varphi_j^{(0)}(\mathbf{v}_1,t)$ associated with the vacuum-of-correlation subspace near equilibrium is quite parallel to the one presented above. The resultant equation has the same structure as (39) except that the inhomogeneous component $f_{j;1}^{(\mathbf{q})}(\mathbf{v}_1,t)$ is replaced by $\varphi_j^{(0)}(\mathbf{v}_1,t)$, and the collision operator $\Psi_1^{(\mathbf{q})}(Z_j^{(\mathbf{q})})$ is replaced by the one evaluated at $\mathbf{q} = 0$ in (41), that is,

$$\varphi_j^{(0)}(\mathbf{v}_1,t) = V^{N/2} \int d\mathbf{v}_{(1)}^{N-1} \langle 1_0, \mathbf{v}^N | P^{(0)} | F_j^{(0)} \rangle \langle \tilde{F}_j^{(0)} | \rho(t) \rangle \quad (58)$$

with

$$i\frac{\partial}{\partial t}\varphi_j^{(0)}(\mathbf{v}_1,t) = \Psi_1^{(0)}(Z_j^{(0)})\varphi_j^{(0)}(\mathbf{v}_1,t) \tag{59}$$

where

$$\Psi_1^{(0)}(Z_j^{(0)}) = iK_1^B + \delta\Psi_1^{(0)}(Z_j^{(0)}) \tag{60}$$

It is interesting to compare this equation to the traditional kinetic equation of the ordinary distribution function. Then the ring operator is considered in the literature as the asymptotic form of the *formal* kinetic equation obtained by the reduction of the generalized Master equation (1) using the so-called "Markovian approximation." In (1) the symbol "∼" denotes the "inverse Laplace transformation" of the operators defined in (28) with $v = 0$. In the Markovian approximation, one replaces $\rho(t-t')$ by $\rho(t)$ and drops the term with $\tilde{\mathscr{D}}^{(0)}(t)$ neglecting the memory effect. Moreover, taking the asymptotic limit of the upper bound of time integration to $+\infty$ in (1) with an opristimic assumption that the asymptotic limit exists, we obtain indeed a Markovian equation:

$$i\frac{\partial}{\partial t}P^{(0)}\rho(t) = \lim_{\epsilon \to 0+} \psi^{(0)}(+i\epsilon)P^{(0)}\rho(t) \tag{61}$$

From now on, ϵ denotes a positive infinitesimal $\epsilon \to 0+$, and we shall not explicitly write the "limit" notation. We reduce this Markovian equation to the one-particle velocity distribution function $\varphi(\mathbf{v}_1)$ by taking integration over dummy variables. Then, using the linear approximation near equilibrium, we obtain the formal kinetic equation [13]:

$$i\frac{\partial}{\partial t}\varphi(\mathbf{v}_1,t) = [iK_1^B + \delta\Psi_1^{(0)}(+i\epsilon)]\varphi(\mathbf{v}_1,t) \tag{62}$$

where $\varphi(\mathbf{v}_1,t)$ is the bare distribution function defined in (3).

In contrast, in our approach for the dressed state the argument of the collision operator $z = +i\epsilon$ is replaced by a finite value of $z = Z_j^{(0)} \equiv -i\xi_j^{(0)}$. This is a consequence of the nonlinearity of the eigenvalue problem [see (42)]. For the relaxing modes with $\xi_j^{(0)} \neq 0$ the order of its magnitude is given by $\xi_j^{(0)} \sim \gamma$. The physical meaning of our operator $\delta\Psi_1^{(0)}(Z_j^{(0)})$ is that the collision process takes in place in an "absorbing medium."

Approximating $\delta\Psi_1^{(0)}(z)$ by $\delta\Psi_1^{(0)}(z;2R)$ by retaining the two-mode processes, we obtain the "ring approximation" of the collision operator for moderately dense systems. The difference between our operator for the dressed distribution function and the traditional ring operator for the ordinary distribution function in (62) is essential, because our operator is finite while the ring operator as defined traditionally diverges for $d = 2$. To see this, let us decompose the ring operator into two parts: $\delta\Psi_1^{(0)}(z; 2R) = \delta\bar{\Psi}_1^{(0)}(z; 2R) + R_{nh}$, where the overbar on the operator denotes the contribution from the hydrodynamic modes with $|\mathbf{k}| < k_0$ in the integration in (54), while R_{nh} denotes the contribution from the remaining nonhydrodynamic modes. Moreover, we shall approximate $T_\mathbf{p}^{(12)}$ by $T_0^{(12)}$ in (54), so that $\hat{K}_a$ for $a = 1$ and 2 in the ring propagator (56) is approximated by K_a^B.[2]

It is well known that the diffusion modes [i.e., the share modes and heat modes associated with $\alpha = 3$ and 4 in (50)] leads to the divergence of the traditional ring operator [13]. Hence we focus here the contribution from these modes to our collision operator. The integration over the diffusion modes in (54) leads to a contribution of the order

$$\delta\bar{\Psi}_1^{(0)}(Z_j^{(0)}; 2R) \sim in\left(\frac{\gamma}{n}\right)^2 \int_0^{k_0} dk \frac{k^{d-1}}{k^2 D_\alpha^B + \xi_j^{(0)}} \tag{63}$$

where we have approximated $t^{(12)} \sim T_0^{(12)} \sim \gamma/n$ and substituted $Z_j^{(0)} = -i\xi_j^{(0)}$ into z, as well as $k^2 D_\alpha^B$ into $-i\mathbf{k} \cdot \mathbf{v}_{12} + K_1^B + K_2^B$ in the denominator of (56). The angle integration is neglected as it is irrelevant to the order of magnitude estimation (see Ref. 13, for instance). For $\xi_j^{(0)} \sim \gamma$ this integral is finite for all dimensions. For three dimensions this is well known. But (63) is finite even for $d = 2$. Indeed, by substituing the order estimate of the Boltzmann diffusion constant as $D_\alpha^B \sim D \equiv \langle v \rangle / k_0$ into (63), we obtain $\delta\bar{\Psi}_1^{(0)}(Z_j^{(0)}; 2R) \sim i\gamma n_0 \ln 2$, for $d = 2$.

It is well known that the nonhydrodynamic part of the ring diagram gives a finite correction to the linearized Boltzmann collision operator (see Refs. 1, 13, and 16). Therefore, our Markovian kinetic equation for the dressed distribution function exists for all dimensions. This is in contrast to (62) for the ordinary distribution function, because (63) is a *diverging* integral for $d = 2$ if $\xi_j^{(0)}$ is replaced by the infinitesimal ϵ. This conclusion agrees with intuition. Two-dimensional systems approach equilibrium, but non-Markovian effects play there a more important role as in three-dimensional systems.

We now consider the more complicated case of the inhomogeneous component $\Psi_1^{(\mathbf{q})}$. For this case the eigenvalue equation for the right eigenstate is given by

$$\Psi_a^{(\mathbf{q})}(Z_j^{(\mathbf{q})})|r_j^{(\mathbf{q})}(a)\rangle\rangle = Z_j^{(\mathbf{q})}|r_j^{(\mathbf{q})}(a)\rangle\rangle \tag{64}$$

---

[2] In this chapter we shall not discuss the effect of the size of hard spheres for bulk viscosity (see Refs. 13 and 28).

In the hydrodynamic case with $|\mathbf{q}| < k_0$ for $d = 3$, this equation reduces to the same eigenvalue equation as introduced by Ernst and Dorfman to discuss the nonanalytic dispersion relations in classic fluids [3]. They have shown that the eigenvalue contains the nonanalytic term of order $q^{5/2}$ term for $3d$ systems. Later these calculations have been extended to the $2d$ case [12]. Because these calculations have been already presented in detail in their papers, we shall not repeat them. The reader should consult the original papers. Here, we display only the main results. For $d = 2$ the q dependence of the eigenvalues that vanish in the limit $q \to 0$ are given by

$$\xi_\alpha^{(\mathbf{q})} = -iqc'\sigma_\alpha - q^2 D'_\alpha + q^2 \Delta_\alpha \ln\left(\frac{q}{k_0}\right) \tag{65}$$

where c', D'_α, and Δ_α are positive constants that are independent of q. Their explicit forms are not important here. The sound velocity c' and the diffusion constants D'_α include correction terms due to the ring process. These corrections are small for small n_0.

The last term in Eq. (65) is also the correction due to the ring process. This correction is nonanalytic at $q = 0$. We have the estimate $\Delta_\alpha/D'_\alpha \sim n_0$ for small n_0. For moderately dense systems the diffusion term dominates for a wide domain of q. Only for extremely small values of q satisfying $q/k_0 < \exp(-1/n_0)$, the nonanalytic term becomes larger than the diffusion term.

The results obtained in this section are interesting, because one can now define transport coefficients through the Markovian kinetic equation for the dressed distribution function for arbitrary dimensional systems. We should emphasize that the coefficients c', D'_α, and Δ_α in (65) are defined only for the distribution function of the collective modes which is a functional of the original distribution function. This is in contrast to the markovianization of the kinetic equation (62) for the ordinary distribution functions, as it leads to diverging transport coefficients for $d = 2$.

As a final part of the subject discussed in this section, let us comment on the relation between the kinetic equation (59) for the vacuum component discussed and (39) for the inhomogeneous component. Simply by substituting $\mathbf{q} = 0$ into Eq. (39) we obtain the same structure of the equation as (59). However, a care has to be taken to consider their relation. Indeed, as mentioned in Section II the essential reason why we obtain the irreversible kinetic equations in the Hamiltonian system is that the distribution functions which we consider are distributions with the delta-function singularities in the Fourier representation, and they do not belong to the Hilbert space [see the discussion after (20)]. As a result of this singularity, the inhomogeneous component $f_{j;1}^{(\mathbf{q})}(\mathbf{v}_1, t)$ with $\mathbf{q} = 0$ is *not* equal to $\varphi_j^{(0)}(\mathbf{v}_1, t)$ (cf. (21) and the volume factor in the expression).

Therefore, it should not be astonished to find the logarithmic singularity $\sim \ln q$ for $d = 2$ in the collision operator $\Psi_1^{(\mathbf{q})}$ for the inhomogeneous component at $\mathbf{q} = 0$, in spite of the fact that the collision operator $\Psi_1^{(0)}$ for the vacuum component is always well-defined. The logarithmic singularity is so weak that the coordinate representation of the kinetic equation for the inhomogeneous component is still well-defined.

IV. VELOCITY AUTOCORRELATION FUNCTION

In the previous section we studied only the evolution of the dressed distribution functions associated to $P^{(0)}\Pi^{(0)}$ or to $P^{(1)}\Pi^{(1)}$. However, the tools introduced in this chapter enable us to study more general problems such as the time evolution of original distribution functions [see (3)] as well as autocorrelation functions. This involves all subdynamics $\Pi^{(v)}$. For this situation the distinction of the original distribution functions from the dressed distribution functions is not negligible, and our nonequilibrium renormalization effects become important. As we shall show now the incorporation of the effects of all subdynamics $\Pi^{(v)}$ in the autocorrelation functions (6) introduces, as in the renormalization group, the critical dimension $d = 4$. It is for $d > 4$ that the usual Green–Kubo formalism is valid, while for $d < 4$ memory effects decay too slowly. This conclusion clashes with the traditional point of view that leads to a critical dimension $d = 2$, and it deserves a detailed discussion of the long-time tail effects, which will be presented in the subsequent sections.

Let us first illustrate the calculation of a contribution in the simplest "ring process" for binary correlation subspace $|F_j^{(2)}\rangle\langle\tilde{F}_j^{(2)}|$ which is in the $\Pi^{(2)}$ subspace in (32). In this subspace the time evolution takes place in $P^{(\mathbf{k},-\mathbf{k})}$ subspace, while the initial condition of $\delta\varphi(\mathbf{v}_1; 0)$ and the observable $v_{1,x}$ are both in the vacuum-of-correlation subspace $P^{(0)}$. Hence we have

$$\Gamma^{(2)}(t) = \int d\mathbf{v}^N v_{1,x} \langle 0, \mathbf{v}^N | P^{(0)} \mathscr{C}^{(2)}(Z_j^{(2)}) P^{(2)}$$
$$\times \exp[-i\psi^{(2)}(Z_j^{(2)})t] A^{(2)} P^{(2)} \mathscr{D}^{(2)}(Z_j^{(2)}) P^{(0)} | \rho_N^{eq} \rangle v_{1,x} \qquad (66)$$

The contribution consists of three parts: (i) the transition $P^{(0)}\mathscr{C}^{(2)}(z)P^{(2)}$, (ii) the time evolution $\exp[-i\psi^{(2)}(z)t]A^{(2)}$ in the intermediate $P^{(2)}\Pi^{(2)}P^{(2)}$ subspace, and (iii) the transition $P^{(2)}\mathscr{D}^{(2)}(z)P^{(0)}$.

To write an explicit form of the contribution in (66), it is convenient to represent the vertices $\lambda L'_{ab}$ and the propagators $(z - L_0)^{-1}$ by the Prigogine–Balescu diagram technique [19]. We first consider a process represented by the ring diagram shown in Fig. 1. In Fig. 1 the renormalized propagator is defined by the summation of the diagrams shown in Fig. 2. The left-hand part the vertex

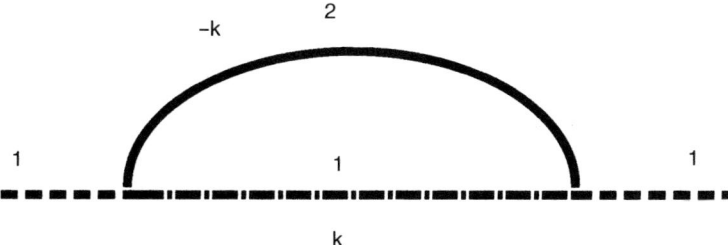

Figure 1. The ring diagram corresponding to (66) as well as to (71) in the binary correlation subspace, where the renormalized lines are defined in Fig. 2.

at left in Fig. 1 corresponds to the transition (i) mentioned above, the intermediate two lines correspond to the process (ii), and the right-hand part of the vertex at right in Fig. 1 corresponds to the transition (iii), respectively.

After the reduction by performing the integrations over the dummy variables for the particles in the medium, we obtain the contribution corresponding to $\mathscr{C}^{(2)}$ given by

$$P^{(0)}\mathscr{C}^{(2)}(Z_j^{(2)})P^{(2)} :\Rightarrow \frac{n}{-i\xi_j^{(\mathbf{k},-\mathbf{k})} - iK_1^{BL}} \langle 1_0, 2_0 | it^{(12)} | 1_\mathbf{k}, 2_{-\mathbf{k}} \rangle \quad (67)$$

where $\xi_j^{(\mathbf{k},-\mathbf{k})} = iZ_j^{(\mathbf{k},-\mathbf{k})}$ and the symbol ":⇒" denotes the correspondence between the N-particle operator and its reduced expression. Similarly for $\mathscr{D}^{(2)}$ we

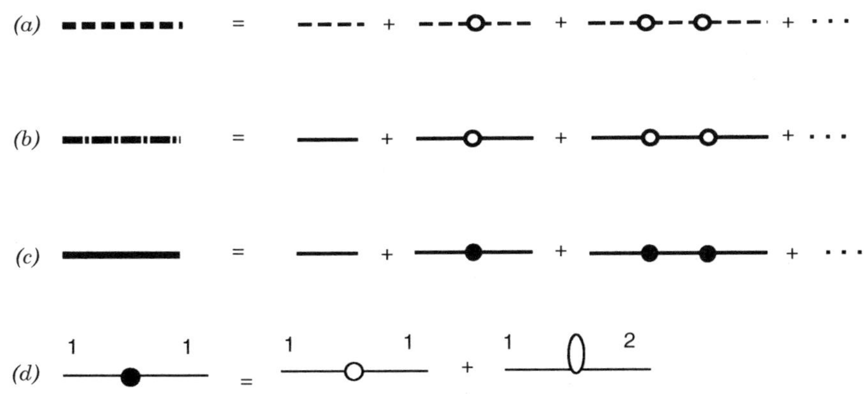

Figure 2. Renormalized lines. For the hard-core interaction we have added the new vertex [the first term in this diagram in (d)] in addition to the Prigogine–Balescu diagram.

have the correspondence

$$P^{(2)}\mathscr{D}^{(2)}(Z_j^{(2)})P^{(0)}v_{1,x}\rho_N^{eq} :\Rightarrow \frac{1}{(2\pi)^d}\langle 1_{\mathbf{k}},2_{\mathbf{k}}|it^{(12)}|1_0,2_0\rangle$$

$$\times P^{(2)}\frac{1}{-i\xi_j^{(\mathbf{k},-\mathbf{k})} - iK_1^{BL}}v_{1,x}\varphi^{eq}(v_2) \quad (68)$$

The generator of the motion in the $\Pi^{(2)}$ subspace is given by the reduced collision operator

$$\Psi^{(\mathbf{k},-\mathbf{k})}(Z_j^{(\mathbf{k},-\mathbf{k})}) = \mathbf{k}\cdot\mathbf{v}_{12} + iK_1^{BL} + iK_2^B + \delta\Psi^{(\mathbf{k},-\mathbf{k})}(Z_j^{(\mathbf{k},-\mathbf{k})}) \quad (69)$$

where $\mathbf{v}_{12} \equiv \mathbf{v}_1 - \mathbf{v}_2$. In the process corresponding to the graph in Fig. 1, the last term $\delta\Psi^{(\mathbf{k},-\mathbf{k})}(z)$ in the collision operator is neglected. This approximation corresponds to the so-called the "ring approximation" which is valid for moderately dense gases [13]. Moreover, we approximate the "renormalization operator" $A^{(2)}$ by $P^{(2)}$ in the intermediate part (ii), since the difference $P^{(2)}$ from $A^{(2)}$ leads to a higher-order correction in n_0, which is small in the moderately dense gas. Then we have the time evolution corresponding to

$$e^{-i\Psi^{(2)}(Z_j^{(2)})t}A^{(2)} :\Rightarrow P^{(2)}\exp[(-i\mathbf{k}\cdot\mathbf{v}_{12} + K_1^{BL} + K_2^B)t]P^{(2)}$$

$$= \sum_{j,j'}|\phi_{BL,j}^{(-\mathbf{k})}(1)\phi_{j'}^{(\mathbf{k})}(2)\rangle\rangle\exp[(\lambda_{BL,j}^{\mathbf{k}} + \lambda_{j'}^{-\mathbf{k}})t]\langle\langle\phi_{BL,j}^{(\mathbf{k})}(1)\phi_{j'}^{(-\mathbf{k})}(2)|$$

(70)

where $|\phi_{j}^{(\mathbf{k})}(2)\rangle\rangle$ is an eigenstate of the inhomogeneous linearized Boltzmann operator $-i\mathbf{k}\cdot\mathbf{v}_2 + K_2^B$ with an eigenvalue $\lambda_j^{\mathbf{k}}$, while $|\phi_{BL,j}^{(\mathbf{k})}(1)\rangle\rangle$ is an eigenstate of the inhomogeneous linearized Boltzmann–Lorentz operator $-i\mathbf{k}\cdot\mathbf{v}_1 + K_1^{BL}$ with an eigenvalue $\lambda_{BL,j}^{\mathbf{k}}$: see (52).

It is well known that the long-time tail effects in the autocorrelation come from the diffusion mode in the hydrodynamic modes with the small intermediate wavevector $k \ll k_0$ (see [6] for instance). For these modes we have $\xi_{5\alpha}^{(\mathbf{k},-\mathbf{k})} \sim k^2 D$ with $D = \langle v\rangle/k_0$. On the other hand, the nonvanishing eigenvalues of K_1^{BL} are of order $\gamma = k_0\langle v\rangle$. Hence, we may neglect $\xi_{5\alpha}^{(\mathbf{k},-\mathbf{k})}$ in the denominator in (67) and in (68) as compared with the eigenvalues of K_1^{BL}. Furthermore, because we are interested in the contribution from the small value of the wavevector, we may approximate $T_{\mathbf{k}}^{(12)} \approx T_0^{(12)}$ as before in Section III. Combining (67), (68), and (70) with these approximations, we have the dominant contribution of the slowest decaying contribution in this subspace as follows:

$$\begin{aligned}\hat{\Gamma}^{(2)}(t) &\approx \frac{nm\beta}{(2\pi)^d} \sum_{j\in\{\alpha\}} \int_{k<k_0} d\mathbf{k} \int d\mathbf{v}_1 d\mathbf{v}_2 v_{1,x} \frac{1}{K_1^{BL}} T_0^{(12)} \phi_{BL,1}^{(-\mathbf{k})}(\mathbf{v}_1) \phi_j^{(\mathbf{k})}(\mathbf{v}_2) \\ &\quad \times e^{(\lambda_{BL,1}^{\mathbf{k}}+\lambda_j^{-\mathbf{k}})t} \int d\mathbf{v}'_1 d\mathbf{v}'_2 \varphi_{eq}(v'_1)^{-1} \varphi_{eq}(v'_2)^{-1} \phi_{BL,1}^{(\mathbf{k})}(\mathbf{v}'_1) \phi_j^{(-\mathbf{k})}(\mathbf{v}'_2) \\ &\quad \times T_0^{(12)} \frac{1}{K_1^{BL}} v'_{1,x} \varphi_{eq}(v'_1) \varphi_{eq}(v'_2) \\ &= \frac{nm\beta}{(2\pi)^d} \sum_{j\in\{\alpha\}} \int_{k<k_0} d\mathbf{k} |\mu_\mathbf{k}^j|^2 e^{(\lambda_{BL,1}^{\mathbf{k}}+\lambda_j^{-\mathbf{k}})t} \quad \text{(for } t \gg t_r = \gamma^{-1}) \end{aligned}$$

(71)

where

$$\mu_\mathbf{k}^j = \int d\mathbf{v}_1 d\mathbf{v}_2 v_{1,x} \frac{1}{K_1^{BL}} T_0^{(12)} \phi_{BL,1}^{(-\mathbf{k})}(\mathbf{v}_1) \phi_j^{(\mathbf{k})}(\mathbf{v}_2) \tag{72}$$

and the hat on $\hat{\Gamma}^{(v)}(t)$ denotes that the contribution coming from the processes where all intermediate states are restricted only to the hydrodynamic modes. The last expression is the same one obtained by Résibois and de Leener [13]. This leads to the well-known asymptotic expression of the long-time tail effect $\sim t^{d/2}$.

In the following sections we shall estimate the contributions from higher-mode processes. However, as it is easy to expect, the complexity of the calculation rapidly increases when the degree of correlations increases. Hence we shall present only order-of-magnitude estimations of their contributions. To this end, let us first reevaluate the order estimation of the above result. As we shall see, the following estimation gives a straightforward extention to higher-order correlations.

We note that $v_{1,x}$ is not a collisional invariant of $K_{BL,1}$. This implies that

$$v_{1,x} = v_{1,x} Q_{0h}^{BL} \tag{73}$$

where $Q_{0h}^{BL} \equiv 1 - P_{0h}^{BL}$ with $P_{0h}^{BL} = |\phi_{BL,1}^{(0)}(1)\rangle\rangle\langle\langle\phi_{BL,1}^{(0)}(1)|$. Similarly, we introduce $Q_h^{BL} \equiv 1 - P_h^{BL}$ and $Q_h^B \equiv 1 - P_h^B$ for $\mathbf{p} \neq 0$ with

$$P_h^{BL}(1) = |\phi_{BL,1}^{(\mathbf{p})}(1)\rangle\rangle\langle\langle\phi_{BL,1}^{(-\mathbf{p})}(1)|, \qquad P_h^B(b) = \sum_{j\in\{\alpha\}} |\phi_j^{(\mathbf{p})}(b)\rangle\rangle\langle\langle\phi_j^{(-\mathbf{p})}(b)| \tag{74}$$

where $P_h^{BL}(1)$ and $P_h^B(b)$ are projection operators for the hydrodynamic modes.

Using these projection operators, we can write the essential part of (71) as

$$\hat{\Gamma}^{(2)}(t) \sim \frac{n}{\langle v \rangle^2} \int d\mathbf{k} \int d\mathbf{v}_1 d\mathbf{v}_2 v_{1,x} \frac{1}{\gamma} Q_{0h}^{BL} T_0^{(12)} P_h^{BL}(1) P_h^B(2) \phi_{BL,1}^{(-\mathbf{k})}(\mathbf{v}_1) \phi_\alpha^{(\mathbf{k})}(\mathbf{v}_2) e^{-k^2 Dt}$$

$$\times \int d\mathbf{v}'_1 d\mathbf{v}'_2 \phi_{BL,1}^{(\mathbf{k})}(\mathbf{v}'_1) \phi_\alpha^{(-\mathbf{k})}(\mathbf{v}'_2) P_h^{BL}(1) P_h^B(2) T_0^{(12)} \frac{1}{\gamma} Q_{0h}^{BL} v'_{1,x} \quad (75)$$

where we have replaced K_1^{BL} by γ and $\lambda_{BL,1}^{\mathbf{k}} + \lambda_j^{-\mathbf{k}}$ by $-k^2 D$.

Because $K_1^{BL} \sim nT_0^{(12)}$, the order of magnitude of the transition $T_0^{(12)}$ between hydrodynamic modes and the nonhydrodynamic mode is given by $Q_{0h}^{BL} T_0^{(12)} P_h^{BL}(1) P_h^B(2) \sim P_h^{BL}(1) P_h^B(2) T_0^{(12)} Q_{0h}^{BL} \sim \gamma/n$. With obvious notations we may abbreviate this relation by

$$Q_h T_0^{(12)} P_h \sim P_h T_0^{(12)} Q_h \sim \frac{\gamma}{n} \quad (76)$$

We substitute this estimation into (75), and then approximate the eigenstate $\phi_{BL,1}^{(\mathbf{p})}(\mathbf{v}_1)$ by its lowest order contribution $\phi_{BL,1}^0(\mathbf{v}_1) = \varphi_{eq}(v_1)$ and $\phi^{(\mathbf{q})}(\mathbf{v}_2)$ by $\phi_\alpha^0(\mathbf{v}_2)$, respectively, in the expansion in the series of $\mathbf{p}$. Moreover, we introduce the dimensionless wave vector measured by the unit of the inverse of the mean free length $\mathbf{y} = l_m \mathbf{k} = \mathbf{k}/k_0$. Using the explicit form of $\phi_\alpha^0(\mathbf{v}_2)$ presented in (50) and (51a)–(51c), we finally obtain the order of magnitude estimation of (75),

$$\hat{\Gamma}^{(2)}(t) \sim \frac{1}{n} \int d\mathbf{k} \, e^{-k^2 Dt} \sim \frac{k_0^d}{n} \int d\mathbf{y} \, e^{-y^2 \tau} \sim \frac{g}{\tau^{d/2}} \quad (77)$$

where we have used the fact that the integration over the velocities in (75) leads to $\langle v \rangle^2$ with a numerical factor of order one, and we have changed the dummy variable as $\mathbf{x} = \mathbf{y}\sqrt{\tau}$ to get the last estimation. This is consistent with the asymptotic time dependence of (71).

V. TERNARY CORRELATION SUBSPACE

Next we shall estimate the contribution from the ternary correlation subspace $|F_\alpha^{(3)}\rangle\langle \tilde{F}_\alpha^{(3)}|$ for the graph shown in Fig. 3. Renormalizing again the denominator with the linearized Boltzmann operator as in (67) and (68), we have a

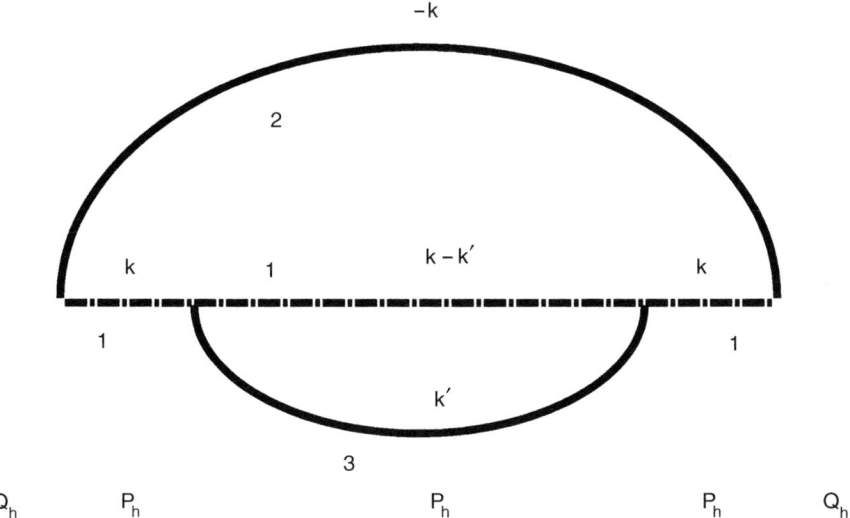

Figure 3. The ring diagram corresponding to (78) in the ternary correlation subspace. Here and in the following figures we abbreviate the renormalized lines of type (a) in Fig. 2 in the $P^{(0)}$ subspace in both ends indicated by Q_h.

contribution from the hydrodynamic modes:

$$\hat{\Gamma}^{(3)}(t) \sim n^2 \int d\mathbf{k} \int d\mathbf{k}' \frac{1}{\xi_\alpha^{(3)} + K_1^{BL}} Q_h \langle 1_0, 2_0 | t^{(12)} | 1_\mathbf{k}, 2_{-\mathbf{k}} \rangle P_h$$

$$\times \frac{1}{\xi_\alpha^{(3)} - i\mathbf{k} \cdot \mathbf{v}_{12} + K_1^{BL} + K_2^B} P_h \langle 1_\mathbf{k}, 3_0 | t^{(13)} | 1_{\mathbf{k}-\mathbf{k}'}, 3_{\mathbf{k}'} \rangle P_h$$

$$\times \exp[(-i(\mathbf{k} - \mathbf{k}') \cdot \mathbf{v}_1 + K_1^{BL} + i\mathbf{k} \cdot \mathbf{v}_2 + K_2^B - i\mathbf{k}' \cdot \mathbf{v}_3 + K_3^B)t]$$

$$\times P_h \langle 1_{\mathbf{k}-\mathbf{k}'}, 3_{\mathbf{k}'} | t^{(13)} | 1_\mathbf{k}, 3_0 \rangle P_h \frac{1}{\xi_\alpha^{(3)} - i\mathbf{k} \cdot \mathbf{v}_{12} + K_1^{BL} + K_2^B}$$

$$\times P_h \langle 1_\mathbf{k}, 2_{-\mathbf{k}} | t^{(12)} | 1_0, 2_0 \rangle Q_h \frac{1}{\xi_\alpha^{(3)} + K_1^{BL}} \tag{78}$$

where we have abbreviated the velocity integration part. We have for the diffusion modes $\xi_\alpha^{(3)} \sim (k^2 + k'^2)D$. Hence, we can again neglect $\xi_\alpha^{(3)}$ in the first and last denominators as in (66); that is, $(\xi_\alpha^{(3)} + K_1^{BL})^{-1} \approx 1/K_1^{BL} \sim n/\gamma$.

In contrast, we cannot neglect $\xi_\alpha^{(3)}$ in the denominators of the intermediate states, since eigenvalues of the linearized collision operators $-i\mathbf{k} \cdot \mathbf{v}_1 + K_1^{BL}$ and

$-i\mathbf{k} \cdot \mathbf{v}_a + K_a^B$ are of the same order $\sim k^2 D$. As a result, the intermediate denominators lead to the "infrared" singularity at $k = k' = 0$.

In (78) there appears a new type of transitions which does not exist in (75)—that is, a type of transitions between the hydrodynamic modes $P_h T_{\mathbf{k}'}^{(13)} P_h$ such as

$$P_h \langle 1_{\mathbf{k}}, 3_0 | t^{(13)} | 1_{\mathbf{k}-\mathbf{k}'}, 3_{\mathbf{k}'} \rangle P_h \sim \langle\langle \phi_{BL,1}^{(\mathbf{k})}(1) \varphi_{eq}(3) | T_{\mathbf{k}'}^{(13)} | \phi_{BL,1}^{(\mathbf{k}-\mathbf{k}')}(1) \phi_m^{(\mathbf{k}')}(3) \rangle\rangle \quad (79)$$

for the hydrodynamic modes $m \in \{\alpha\}$. These transitions vanish in the limit $k \to 0$ and $k' \to 0$. Hence, by a dimensional analysis the order of magnitude of this type of transitions is given by (e.g., for $k \sim k' < k_0$)

$$P_h T_{\mathbf{k}'}^{(13)} P_h \sim \frac{k' \gamma}{k_0 n} \quad (80)$$

This compensates the infrared singularities.

Substituting (76), (80), and the above-mentioned estimation for $\xi_\alpha^{(3)}$ into (78) and again replacing the linearized collision operators in the time-dependent part by their eigenvalues, we have an estimation as

$$\hat{\Gamma}^{(3)}(t) \sim n^2 \int d\mathbf{k} \int d\mathbf{k}' \left(\frac{1}{\gamma}\right)^2 \left(\frac{\gamma}{n}\right)^2 \left(\frac{k' \gamma}{k_0 n}\right)^2 \left(\frac{1}{(k^2 + k'^2)D}\right)^2 e^{-(k^2 + k'^2)Dt}$$

$$= \frac{1}{n^2} \int d\mathbf{k} \int d\mathbf{k}' \left(\frac{k'}{k_0}\right)^2 \left(\frac{k_0^2}{k^2 + k'^2}\right)^2 e^{-(k^2 + k'^2)Dt}$$

$$\sim g^2 \tau (\tau^{-d/2})^2 \quad (81)$$

Dividing the estimation (81) by (77), we obtain $g\tau^{(2-d)/2}$. Hence, the critical dimension for this ratio is $d = 2$. Equation (81) decays quicker than (77) for $d > 2$.

The estimation (81) is easily extended to higher-order correlations such as the contribution from the $|F_\alpha^{(v)}\rangle\langle \tilde{F}_\alpha^{(v)}|$ subspace for the graph shown in Fig. 4. Then we obtain the same critical dimension $d = 2$, as far as we restrict the intermediate states only to hydrodynamic modes.

However, this type of contribution is not the slowest decaying process. In the next section we shall show that the incorporation of nonhydrodynamic modes in intermediate vertices leads to slower decaying processes.

VI. RENORMALIZATION OF VERTICES

Let us now estimate the $|F_\alpha^{(3)}\rangle\langle \tilde{F}_\alpha^{(3)}|$ subdynamics contribution of the process shown in Fig. 5. We estimate the contibution where the intermediate states

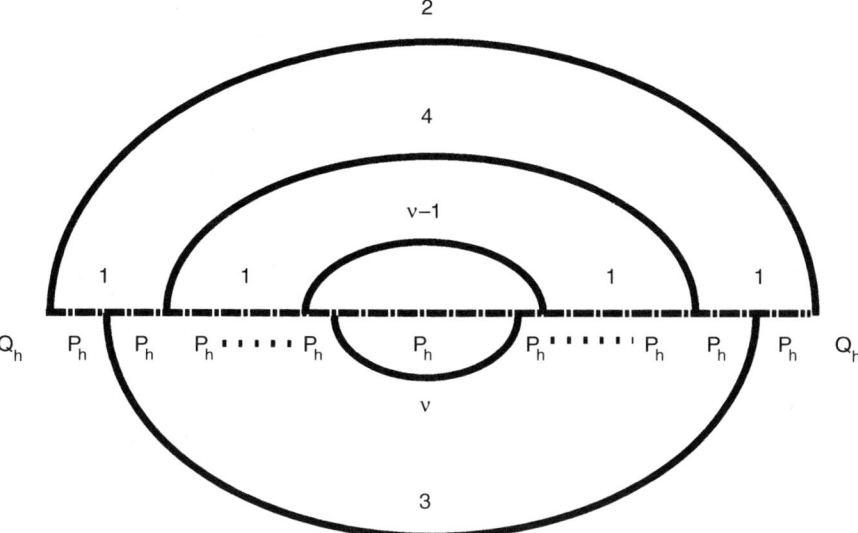

Figure 4. A ring diagram in the νth order correlation subspace with intermediate states restricted only in the hydrodynamic modes.

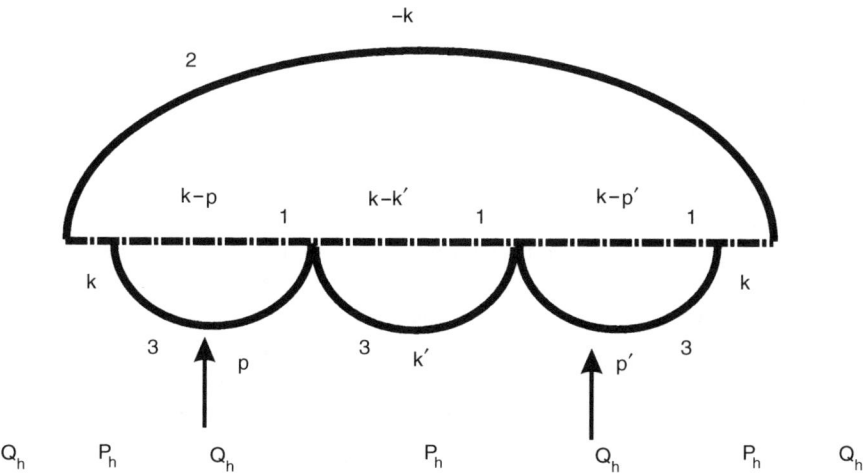

Figure 5. The ring diagram corresponding to (85) in the ternary correlation subspace. The arrows indicate the location of the intermediate rlaxing modes.

indicated by the arrows in the graph are restricted to nonhydrodynamic modes (i.e., the relaxing modes), while all other intermediate states are in hydrodynamic modes. Corresponding to (78), we have

$$\bar{\bar{\Gamma}}^{(3)}(t) \sim n^2 \int d\mathbf{k} \int d\mathbf{k}' \frac{1}{\xi_\alpha^{(3)} + K_1^{BL}} Q_h \langle 1_0, 2_0 | t^{(12)} | 1_\mathbf{k}, 2_{-\mathbf{k}} \rangle P_h$$

$$\times \frac{1}{\xi_\alpha^{(3)} - i\mathbf{k} \cdot \mathbf{v}_{12} + K_1^{BL} + K_2^B} P_h T_1^{(13)}(\mathbf{k}, \mathbf{k}') P_h$$

$$\times \exp[(-i(\mathbf{k} - \mathbf{k}') \cdot \mathbf{v}_1 + K_1^{BL} + i\mathbf{k} \cdot \mathbf{v}_2 + K_2^B - i\mathbf{k}' \cdot \mathbf{v}_3 + K_3^B)t]$$

$$\times P_h T_2^{(13)}(\mathbf{k}', \mathbf{k}) P_h \frac{1}{\xi_\alpha^{(3)} - i\mathbf{k} \cdot \mathbf{v}_{12} + K_1^{BL} + K_2^B}$$

$$\times P_h \langle 1_\mathbf{k}, 2_{-\mathbf{k}} | t^{(12)} | 1_0, 2_0 \rangle Q_h \frac{1}{\xi_\alpha^{(3)} + K_1^{BL}} \tag{82}$$

where the double overbar on $\bar{\bar{\Gamma}}^{(3)}(t)$ denotes that the contribution coming from the processes which we are looking at. Here, the transition $P_h T_1^{(13)}(\mathbf{k}, \mathbf{k}') P_h$ in (82) is corresponding to the portion indicated by the arrow on the left-hand side in Fig. 5 and is defined by

$$P_h T_1^{(13)}(\mathbf{k}, \mathbf{k}') P_h \equiv \int d\mathbf{p} P_h \langle 1_\mathbf{k}, 3_0 | t^{(13)} | 1_{\mathbf{k}-\mathbf{p}}, 3_\mathbf{p} \rangle Q_h$$

$$\times \frac{1}{-i(\mathbf{k} - \mathbf{p}) \cdot \mathbf{v}_1 + K_1^{BL} + i\mathbf{k} \cdot \mathbf{v}_2 + K_2^B - i\mathbf{p} \cdot \mathbf{v}_3 + K_3^B}$$

$$\times Q_h \langle 1_{\mathbf{k}-\mathbf{p}}, 3_\mathbf{p} | t^{(13)} | 1_{\mathbf{k}-\mathbf{k}'}, 3_{\mathbf{k}'} \rangle P_h \tag{83}$$

where the intermediate propagator is restricted in the nonhydrodynamic modes. Similarly, the transition $P_h T_2^{(13)}(\mathbf{k}', \mathbf{k}) P_h$ in (82) corresponds to the portion indicated by the arrow on the right-hand side in Fig. 5 with the nonhydrodynamic intermediate states.

The matrix elements of the transition $t^{(13)}$ between P_h and Q_h or between Q_h and P_h in (83) *do not* vanish at $k = k' = 0$ [cf. (76); this is also the case for the transition between Q_h and Q_h]. Hence the infrared singularities in the propagator are not compensated by this type of transition. This is in contrast to the transition between P_h and P_h which vanishes at the vanishing wavevectors [see (80)]. As we shall see, this difference is essential to understand the origin of the slower decay process than (78).

Moreover, we note that the eigenvalues of the linearized collision operators in the denominator in (83) is bounded from below by a positive constant of order γ for the relaxing modes in Q_h subspace. Hence, these denominators have no singularity at $k = 0$. Combining this with the property of the matrix elements of $t^{(13)}$ mentioned above, we may estimate the order of magnitude of the transitions in (83) by the aid of the dimensional analysis as [cf. (76)]

$$\lim_{k,k' \to 0} P_h T_1^{(13)}(\mathbf{k}, \mathbf{k}') P_h \sim \int d\mathbf{P} P_h T_\mathbf{p}^{(13)} Q_h \frac{1}{\gamma} Q_h T_{-\mathbf{p}}^{(13)} P_h \sim k_0^d \frac{\gamma}{n} \frac{1}{\gamma} \frac{\gamma}{n} \quad (84)$$

where the factor k_0^d comes from the integration over $\mathbf{p}$ by measuring the wavevector with its intrinsic unit k_0 of the system as $\mathbf{p} = k_0 \mathbf{y}$, as before. We have the same esitimation for $P_h T_2^{(13)}(\mathbf{k}', \mathbf{k}) P_h$ in (82).

To find the order of magnitude estimation of (82), we note a similar structure of (82) to (78). Hence, if we replace $(k'/k_0)(\gamma/n)$ in (81) by $k_0^d \gamma/n^2$, we have the estimation of (82). This gives us

$$\bar{\bar{\Gamma}}^{(3)}(t) \sim n^2 \int d\mathbf{k} \int d\mathbf{k}' \left(\frac{1}{\gamma}\right)^2 \left(\frac{\gamma}{n}\right)^2 \left(\frac{k_0^d \gamma}{n^2}\right)^2 \left(\frac{1}{(k^2 + k'^2)D}\right)^2 e^{-(k^2+k'^2)Dt}$$
$$\sim g^4 \tau^2 (\tau^{-d/2})^2 \quad (85)$$

It is remarkable that (85) is a slower decaying process than (81).

One can improve the estimate by introducing renormalized vertex $\tilde{T}^{(ab)}$ shown in Fig. 6, where all intermediate states are restricted to Q_h subspace as shown in Fig. 7. The order of magnitudes are estimated as [for $g < 1$; see (11)]

$$P_h \tilde{T}^{(ab)} Q_h \sim P_h T_\mathbf{p}^{(ab)} \left[1 + \int d\mathbf{q} \frac{1}{\gamma} Q_h T_{\mathbf{q}-\mathbf{p}}^{(ab)} + \int d\mathbf{q} d\mathbf{q}' \frac{1}{\gamma} Q_h T_{\mathbf{q}-\mathbf{p}}^{(ab)} \frac{1}{\gamma} Q_h T_{\mathbf{q}'-\mathbf{q}}^{(ab)} + \cdots \right] Q_h$$
$$\sim \frac{\gamma}{n}(1 + g + g^2 + \cdots) = \frac{\gamma}{n} \frac{1}{1-g} \quad (86)$$

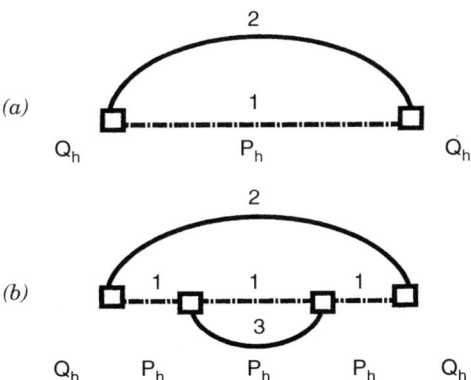

Figure 6. (a) The ring diagram with the renormalized vertices in the binary correlation subspace (c.f. Fig. 1). (b) The ring diagram with the renormalized vertices in the ternary correlation subspace (c.f. Fig. 3).

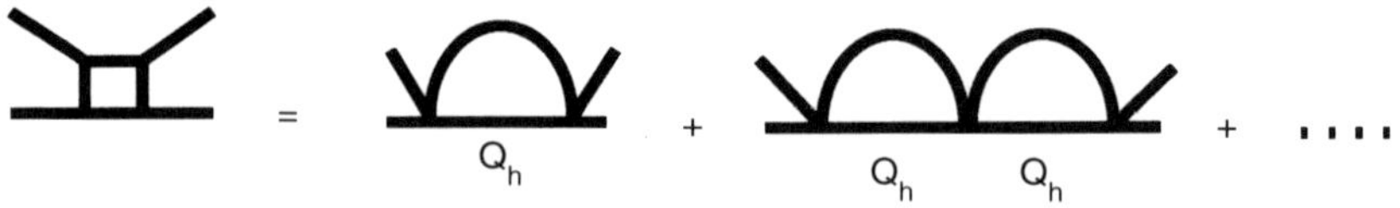

Figure 7. Renormalized vertex where all intermediate states are restricted in the relaxing modes.

where we have used an estimation $\int dp' Q_h T_{p'}^{(ab)} Q_h \sim k_0^d \gamma/n$. Similarly, we have

$$Q_h \tilde{T}^{(ab)} P_h \sim \frac{\gamma}{n} \frac{1}{1-g}, \qquad P_h \tilde{T}^{(ab)} P_h \sim \frac{\gamma}{n} \frac{g}{1-g} \tag{87}$$

With these renormalized vertices, we can estimate the processes (a) and (b) shown in Fig. 6. For the $|F_\alpha^{(2)}\rangle\langle\tilde{F}_\alpha^{(2)}|$ subdynamics contribution of (a), we have

$$\bar{\Gamma}^{(2)}(t) \sim \frac{1}{n}\left(\frac{1}{1-g}\right)^2 \int d\mathbf{k}\, e^{-k^2 Dt} \sim \frac{g\tau^{-d/2}}{(1-g)^2} \tag{88}$$

while for $|F_\alpha^{(3)}\rangle\langle\tilde{F}_\alpha^{(3)}|$ subdynamics contribution of (b)

$$\bar{\Gamma}^{(3)}(t) \sim \frac{1}{n^2}\left(\frac{1}{1-g}\right)^2\left(\frac{g}{1-g}\right)^2 \int d\mathbf{k} \int d\mathbf{k}' \left(\frac{k_0^2}{k^2+k'^2}\right)^2 e^{-(k^2+k'^2)Dt} \sim \frac{g^4 \tau^{2-d}}{(1-g)^4} \tag{89}$$

where the overbar in $\bar{\Gamma}^{(v)}(t)$ denotes the slowest decaying power law contributions with the renormalized vertices in $\Gamma^{(v)}(t)$.

Dividing the estimate (89) by (88), the ratio is given by $g\tau^{(4-d)/2}$. Hence, the critical dimension for this ratio is $d = 4$, instead of $d = 2$. Equation (89) decays quicker than (88) for $d > 4$.

This result can be also obtained by comparing the first expression in (88) to (89). Indeed, (89) contains an additional factor to (88) for $v = 3$ as

$$\frac{\bar{\Gamma}^{(v)}(t)}{\bar{\Gamma}^{(v-1)}(t)} \sim \frac{1}{n}\left(\frac{g}{1-g}\right)^2 \int d\mathbf{k}' \left(\frac{k_0^2}{k^2+k'^2}\right)^2 e^{-k^2 Dt} \sim g\left(\frac{g}{1-g}\right)^2 \tau^{(4-d)/2} \tag{90}$$

We can repeat the same order of magnitude estimation of the contribution from the $|F_\alpha^{(v)}\rangle\langle\tilde{F}_\alpha^{(v)}|$ subspace for the same type of graphs shown in Fig. 5, except that each vertex is replaced by the renormalized vertex. Then we obtain the same ratio as (90) for arbitrary $v \geq 3$.

For any v one can easily extend the above order of magunitude estimation to an arbitrary graph. The result is summarized as follows: We first count the number of lines in each intermediate state. The slowest decaying processes are given by the processes in which the number of intermediate line m increases by 1 (starting with $m = 0$) by every succesive renomarized vertex, and reaches to the maximum value $m = v$, then decreases by 1 until $m = 0$. An example of the diagram is obtained by replacing all vertices in Fig. 5 by renormalized vertices. Then, the contribution in $|F_\alpha^{(v)}\rangle\langle \tilde{F}_\alpha^{(v)}|$ subspace is obtained by iterative use of (90) starting with $v = 3$. This leads finally our main result (10).

VII. DISCUSSIONS AND CONCLUDING REMARKS

It is well known that the basic quantities in statistical mechanics are reduced distribution functions such as (3). However, it is now well understood that outside equilibrium they have a complicated time dependence including memory effects [see (1)]. It is therefore quite remarkable that there exist collective, dressed, reduced distribution functions, such as $f_1^{(\mathbf{q})}(\mathbf{v}_1, t)$ in (27), which satisfy Markovian equation in each $\Pi^{(v)}$ space.

In this chapter we studied only the evolution of the dressed distribution functions associated with $P^{(0)}\Pi^{(0)}$ or with $P^{(1)}\Pi^{(1)}$. However, our complex spectral representetion of the Liouvillian permitted us to study more general problems such as the auto-correlation function. This involved all subdynamics $\Pi^{(v)}$. For this situation the distinction of the original distribution functions from the dressed distribution functions was not negligible, and our nonequiliburium renormalization effects became important. We then derived the long-time estimation (10) for the the autocorrelation function. We can deduce several interesting conclusions from this result.

There is a critical dimension at $d = 4$, as in the renormalization group. Higher-order correlations lead to quicker decay process for $d > 4$, while they lead to slower decay process for $d < 4$. For $d < 4$, the infrared singularity at $k = 0$ is so strong that Green–Kubo's formula for the transport coefficients such as (8) diverges.

For $d < 4$, higher-order correlations lead to a diverging contribution of the summation (9) for large time, so that there is an upper limit of the time scale $\tau = \tau_0$ where (10) is applicable for times $\tau < \tau_0$. Beyond the limit the estimation based on the ring processes is likely to be invalid. The value of τ_0 can be found by putting the ratio (90) equal to one. Then we have

$$\tau_0 = \left[\frac{(1-g)^2}{g^3}\right]^{2/(4-d)} \tag{91}$$

For $n_0 = 10^{-1}$, they give us

$$\tau_0 \approx \begin{cases} 10^3 & \text{for } d = 2 \\ 10^{12} & \text{for } d = 3 \end{cases} \tag{92}$$

while for a dense case such as $n_0 = 0.5$, we obtain

$$\tau_0 \approx \begin{cases} 2 & \text{for} \quad d = 2 \\ 1.3 \times 10^3 & \text{for} \quad d = 3 \end{cases} \tag{93}$$

For $\tau \ll \tau_0$, our estimation (11) leads to $\bar{\Gamma}(t) \sim t^{-d/2}$ from the binary correlation subspace $\nu = 2$, which is consistent to the numerical simulations [2]. In order to determine whether if the correction $\bar{\Gamma}^{(3)}(t)$ due to the ternary correlation accelerates or decelerates the decay of the autocorrelation function during this time scale, we need more careful estimation of $\bar{\Gamma}^{(3)}(t)$ including the determination of the sign of this contribution by performing the integration over the angles. We intend to present such a detailed calculation elsewhere.

For $d \to \infty$, the long-time tail effects vanish, and Green–Kubo's formula is governed by a Markovian kinetic evolution of $\delta\varphi^{(0)}(\mathbf{v}_1, t)$ with the exponential decay in a single subspace $\Pi^{(0)}$.

In contrast to our conclusion of the critical dimension as $d = 4$, the phenomenological approach based on hydrodynamic equations has led to the critical dimension $d = 2$ (see Ref. 6 and the papers cited therein). Because the new type of the slower decaying processes discussed in this chapter is associated with the intermediate nonhydrodynamic processes in the vertices, it is clear why the phenomenological approach based *only* on hydrodynamic equations fails to predict these slower processes.

Let us conclude with some general remarks. The traditional approach of the kinetic theory based on the BBGKY hierarchy relays upon a *truncation* of the hierarchy at a certain order of correlations. Because the higher-order correlations become more important for asymptotic times and for $d < 4$, this truncation is incorrect.

The long-time tail effects described by (10) invalidate the Green–Kubo formalism for $d < 4$. This was well known for $d = 2$, but it is also true for $d = 3$ because there are contributions of $t^{-1}, t^{-1/2}, \ldots$ coming from multiple mode–mode couplings. Still the linear response formalism remains a valuable tool when used for times where the Markovian approximation to transport theory is valid [29]. Also, it is rigorous for $d \to \infty$. There is an amusing analogy with the mean field approach in equilibrium statistical mechanics.

In conclusion, our approach based on the spectral decomposition of the Liouville operator which avoids carefully nondynamical assumptions appears to be of special interest when going beyond the limit of dilute gases.

Acknowledgments

I want to express my deepest thanks to Professor I. Prigogine. His doubts about the validity of the truncation of the BBGKY hierarchy and the Green–Kubo formalism have inspired this work. My thanks also go to an anonymous reader who pointed out my mistake in the estimation of the transition matrix between the hydrodynamic modes, when I prepared a short note on this subject. Without his comment I could not find the new slower decaying processes presented in this chapter. I acknowledge the U.S. Department of Energy Grant No. DE-FG03-94ER14465, the Robert A. Welch Foundation Grant No. F-0365 and the European Commission Contract ESPRIT Project 21042 (CTIAC), Communauté francaise de Belgique, and the "Loterie Nationale" of Belgium for support of this work.

References

1. J. Van Leeuwen and A. Weyland, *Physica* **36**, 457 (1967).
2. B. Alder and T. Wainwright, *Phys. Rev.* **A1**, 18 (1970).
3. M. H. Ernst and J. R. Dorfman, *Physica* **61**, 157 (1972).
4. M. H. Ernst and J. R. Dorfman, *J. Stat. Phys.* **12**, 311 (1975).
5. J. R. Dorfman and E. G. D. Cohen, *Phys. Rev. A* **6**, 776 (1972); **12**, 292 (1975).
6. Y. Pomeau and P. Résibois, *Phys. Report* **19C**, 63 (1975).
7. M. H. Ernst, J. R. Dorfman, W. R. Hoegy, and J. Van Leeuwen, *Physica* **45**, 127 (1969).
8. P. Résibois and M. de Leener, *Classical Kinetic Theory of Fluids*, John Wiley & Sons, New York, 1977.
9. J. R. Dorfman and H. van Beijeren, *Statistical Mechanics, Part B*, edited by B. J. Berne, Plenum Press, New York, 1977.
10. J. R. Dorfman and T. Kirkpatrick, in *Lecture Notes in Physics: Systems Far from Equilibrium—Proceedings, Sitges*, 263 Springer-Verlag, Berlin, 1980.
11. Y. Kan and J. R. Dorfman, *Phys. Rev. A* **16**, 2447 (1977).
12. I. M. De Schepper and M. H. Ernst, *Physica* **87A**, 35 (1975).
13. P. Résibois and M. de Leener, *Classical Kinetic Theory of Fluids*, John Wiley & Sons, New York, 1977.
14. J. R. Dorfman and H. van Beijeren, *Statistical Mechanics, Part B*, edited by B. J. Berne, Plenum Press, New York, 1977.
15. J. R. Dorfman and T. Kirkpatrick, in *Lecture Notes in Physics: Systems Far From Equilibrium—Proceedings, Sitges*, 263, Springer-Verlag, Berlin, 1980.
16. Y. Kan and J. R. Dorfman, *Phys. Rev. A* **16**, 2447 (1977).
17. M. A. van der Hoef, Thesis "*Simulation Study of Diffusion in Lattice–Gas Fluids and Colloids,*" de Rijksuniversiteit te Utrecht (1992).
18. I. Prigogine and P. Résibois, *Physica* **27**, 629 (1961).
19. I. Prigogine, *Non-Equilibrium Statistical Mechanics*, Wiley Interscience, New York, 1962.
20. T. Petrosky and I. Prigogine, *Chaos, Solitons & Fractals* **7**, 441 (1996).
21. T. Petrosky and I. Prigogine, *Advances in Chemical Physics* Vol. 99, John Wiley & Sons, New York, 1997, p. 1.
22. M. Thedosopulu and A. P. Grecos, *Physica* **95A**, 35 (1979).
23. P. Résibois, *Physica* **70**, 431 (1973).
24. P. Résibois and Y. Pomeau, *Physica* **72**, 493 (1974).
25. M. Theodrosopulu and P. Résibois, *Physica* **82A**, 47 (1976).
26. I. Prigogine, C. George, and F. Henin, *Physica* **45**, 418 (1969).
27. T. Petrosky and H. Hasegawa, *Physica* **160A**, 351 (1989).
28. J. Piasecki, in *Fundamental Problems in Statistical Mechanics*, Vol. IV, 1977.
29. P. Résibois, *J. Chem. Phys.* **41**, 2979 (1964).
30. T. Petrosky, *Foundation of Physics*, **29**, 1417 (1999); **29**, 1581 (1999).

NEW KINETIC LAWS OF CLUSTER FORMATION IN *N*-BODY HAMILTONIAN SYSTEMS

Y. AIZAWA

Department of Applied Physics, Faculty of Science and Engineering, Waseda University, Tokyo, Japan

CONTENTS

I. Two Kinetic Phases Embedded in the Cluster
II. Universality of the Log-Weibull Distribution Characterizing Arnold Diffusion
III. Quasi-Structure Surviving as a Whole Body
Acknowledgments
References

I. TWO KINETIC PHASES EMBEDDED IN THE CLUSTER

An equilibrium cluster is described by the ergodic measure $\mu(x)$ defined in the one-particle phase space x,

$$\mu(x) = \langle r \rangle \mu_c(x) + (1 - \langle r \rangle) \mu_g(x) \tag{1}$$

where $\langle r \rangle$ stands for the mean fraction of the phase space corresponding to the clustering motions; $\mu_c(x)$ and $\mu_g(x)$ are the normalized characteristic measures to describe the cluster phase and the gaseous phase, respectively. The probability density $P_g(T_g)$ for the residence time T_g in gaseous phase is usually approximated by the Poissonian, $P_g(T_g) = \langle T_g \rangle^{-1} \exp[-T_g/\langle T_g \rangle]$, but on the other hand the probability density $P_c(T_c)$ for the residence time T_c in cluster phase is quite different from the Poissonian:

$$P_c(T_c) = p \cdot \frac{dQ_W}{dT_c} + (1-p) \cdot \frac{dQ_L}{dT_c} = \frac{dQ_c}{dT_c} \tag{2}$$

Dynamical Systems and Irreversibility: A Special Volume of Advances in Chemical Physics, Volume 122, Edited by Ioannis Antoniou. Series Editors I. Prigogine and Stuart A. Rice. ISBN 0-471-22291-7. © 2002 John Wiley & Sons, Inc.

where $Q_W = \exp[-AT_c^{-\alpha}]$ (negative Weibull distribution), $Q_L = \exp[-B \times (\log T_c)^{-\beta}]$ (log-Weibull distribution), and p is the fraction of the negative Weibull component [1–3]. Figure 1 demonstrates the accumulated probability $Q_c(x)(x = T_c)$, where the scaling regimes corresponding to two components Q_W(or P_W) and Q_L(or P_L) are clearly observed, and particularly the intrinsic long-time tails of $Q_L(x)$ are systematically prolonged when the cluster size becomes large; when the total energy E decreases, the size of the cluster increases. Two kinetic phases generally coexist in big clusters.

The shape of a cluster depends on the strength of interaction between the cluster and the environment. When the member particles are violently exchanged in the case of a small cluster, the shape is quite irregular, but in a large cluster the shape is almost globular and the variation of the shape is very slow and majestic. This is the reason why the $1/f$ fluctuation is often observed in clustering motions. Figure 2 is a typical example of cluster formation in N-body hamiltonian systems which we have reported in a previous paper [1], where the stability of member particles is demonstrated by the gray scale which represents the Gauss–Riemannian curvature; the inside of a cluster has positive curvature, and the outside has negative one. The Riemannian geometrization, which was used in the analysis of the Mixmaster universe model [4], is successfully applied for the rigorous definition of the cluster.

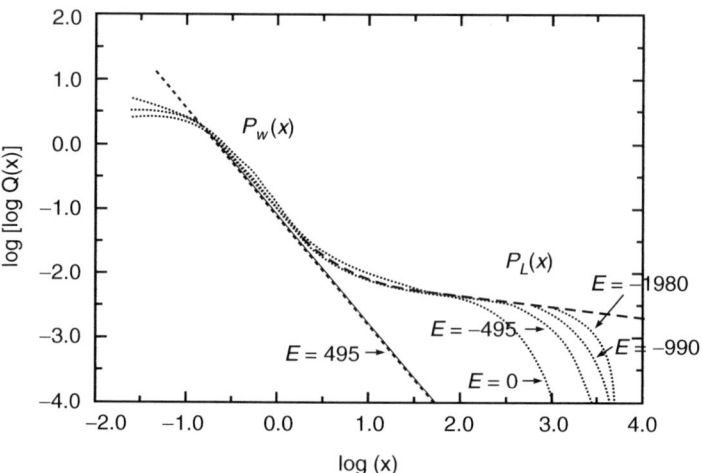

Figure 1. Distribution function $Q(x)$ for the trapping time x. $P_W(x)$ is the negative-Weibull and $P_L(x)$ is the log-Weibull. The parameter E is the total energy that controls the size of the cluster.

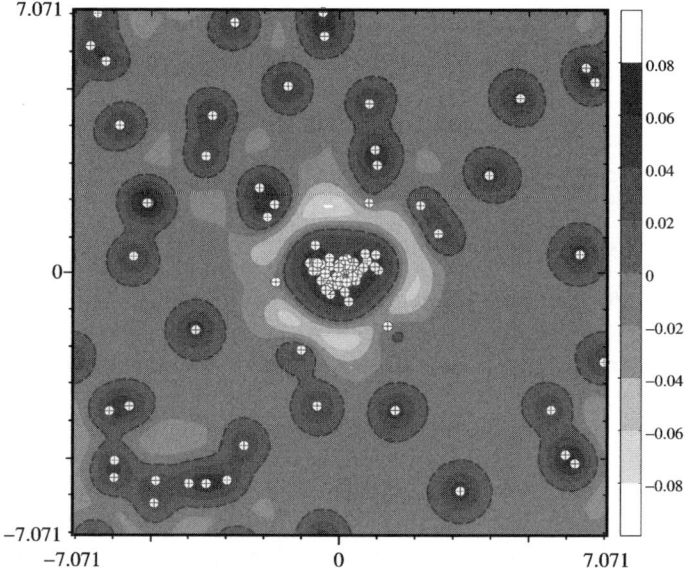

Figure 2. A snapshot of the cluster formation at $E = 0$, where the number of member particles is almost 50% of the total particles.

II. UNIVERSALITY OF THE LOG-WEIBULL DISTRIBUTION CHARACTERIZING ARNOLD DIFFUSION

The log-Weibull component becomes larger and larger when the cluster size increases and the cluster shape approaches to a globular one. This implies that the one-particle motion can be approximated by an integrable hamiltonian $H(p,q)$ if we consider the clustering motions in a large cluster;

$$H(p,q) = H_0(p) + \epsilon H_1(p,q,t) \qquad (3)$$

where H_0 is the effective integrable Hamiltonian and the $\epsilon H_1(p,q,t)$ is the perturbation due to the small derivation from the globular cluster. Equation (3) is the standard form of nearly integrable hamiltonian systems, where we can use the Nekhoroshev theorem [7] for slowly drifting motions such as the Arnold diffusion in the Fermi–Pasta–Ulam models for quartz oscillators [5]. The distribution function $P(T)$ of the characteristic time T for the diffusion was derived many years ago [6] in the following form: $P(T) \propto \frac{1}{T(\log T)^c} (T \gg 1)$. It is easily obtained that the distribution function $P(T)$ for the Arnold diffusion is nothing but the log-Weibull distribution function demonstrated in Fig. 1; if we put $c = 2$, the intrinsic long-time tails in cluster formation (in two-dimensional simulations) are completely understood in terms of the Arnold diffusion [3].

III. QUASI-STRUCTURE SURVIVING AS A WHOLE BODY

Clusters appear almost always in the transitional regime between two different thermodynamical phases as a stable kinetic phase. The only difference from the ordinary phase in thermodynamic limit is that the cluster is an extremely small open system with finite scales in time as well as in space, where microscopic fluctuations influence prominently on the whole processes extending from the birth to the death of the quasi-structure. The cluster discussed here behaves like a giant particle composed of many microscopic particles. The internal structure of the cluster has been explored for a long time, but no one succeeded in finding out any rigid structures on the inside of the cluster. However, our simulations elucidate that the internal structures are clearly understood in terms of the coexistence of two different kinetic laws embedded in a cluster. The self-organization of these two different types of kinetic phases is essential in order that the cluster can survive for long period, and the stability seems to be protected by its own internal mechanisms, which are inherent to the cluster itself. We can say that a cluster should be understood as an entity with active nature, and it is never a passive entity only adapting to the environment.

Acknowledgments

The author thanks Prof. I. Prigogine, Dr. I. Antoniou, and Dr. T. Petrosky for their valuable comments and encouragement.

References

1. Y. Aizawa, K. Sato, and K. Ito, *Prog. Theor. Phys.* **103**(3), 519–540 (2000).
2. M. Nakato and Y. Aizawa, *Chaos, Solitons & Fractals.* **11**, 171–185 (2000).
3. Y. Aizawa, *Prog. Theor. Phys. Suppl.* **139**, 1–11 (2000).
4. Y. Aizawa, N. Koguro, and I. Antoniou, *Prog. Theor. Phys.* **98**, 1225–1250 (1997).
5. Y. Aizawa, *J. Korean Phys. Soc.* **28**, 310–314 (1995).
6. Y. Aizawa, *Prog. Theor. Phys.* **81**(2), 249–253 (1989); Y. Aizawa et al., *Prog. Theor. Phys. Suppl.* **98**, 36–82 (1989).
7. N. N. Nekhoroshev, *Russ. Math. Surveys* **32**(6), 1–65 (1977).

PART THREE

QUANTUM THEORY, MEASUREMENT, AND DECOHERENCE

QUANTUM PHENOMENA OF SINGLE ATOMS

H. WALTHER

Sektion Physik der Universität München and Max Planck Institut für Quantenoptik, Garching, Federal Republic of Germany

CONTENTS

I. Introduction
II. Experiments with the One-Atom Maser
   A. Quantum Jumps and Atomic Interferences in the Micromaser
   B. Entanglement in the Micromaser
   C. Trapping States
   D. Generation of GHZ States
   E. The One-Atom Maser and Ultracold Atoms
III. Ion Trap Experiments
   A. Resonance Fluorescence of a Single Atom
   B. The Ion-Trap Laser
IV. Conclusions
References

I. INTRODUCTION

Using the modern techniques of laser spectroscopy, it has become possible to observe the radiation interaction of single atoms. The techniques thus made it possible to investigate the radiation–atom interaction on the basis of single atoms. The most promising systems in this connection seem to be single atoms in cavities and also single atoms in traps. The studies in cavities allow us to select one interacting mode and thus represent the ideal system with respect to a quantum treatment. In high-Q cavities a steady-state field of photons can be generated displaying nonclassic photon statistics. It thus becomes possible to study the interaction also in the limit of nonclassic or sub-Poissonian fields.

Dynamical Systems and Irreversibility: A Special Volume of Advances in Chemical Physics,
Volume *122*, Edited by Ioannis Antoniou. Series Editors I. Prigogine and Stuart A. Rice.
ISBN 0-471-22291-7. © 2002 John Wiley & Sons, Inc.

Single trapped ions allow us to observe among other phenomena quantum jumps and antibunching in fluorescence radiation. The fluorescent channel represents the interaction with many modes; however, it is also possible to combine single mode cavities with trapped atoms, as, for example, in the case of the proposed ion-trap laser.

A new and interesting twist in radiation–atom interaction can be added when ultracold atoms are used in both cavities and traps. In this case the distribution of the matter wave plays an important role besides the standing electromagnetic wave in the cavity, and their interaction is determined by their respective overlap leading to new effects.

In the following we will review experiments of single atoms in cavities and traps performed in our laboratory. Furthermore, new proposals for experiments with ultracold atoms will be discussed. We start with the discussion of the one-atom maser.

II. EXPERIMENTS WITH THE ONE-ATOM MASER

The one-atom maser or micromaser uses a single mode of a superconducting niobium cavity [1–4]. In the experiments, values of the quality factor as high as 3×10^{10} have been achieved for the resonant mode, corresponding to an average lifetime of a photon in the cavity of 0.2 s. The photon lifetime is thus much longer than the interaction time of an atom with the maser field; during the atom passes through the cavity, the only change of the cavity field that occurs is due to the atom-field interaction. Contrary to other strong coupling experiments in cavities (optical or microwave; see, e.g., Ref. 5 for a comparison between the different setups), it is possible with our micromaser to generate a steady-state field in the cavity which has nonclassic properties so that the interaction of single atoms in those fields can be investigated. Furthermore, the generation process of those fields has been studied and is well understood. The experiment is quite unique in this respect; this also holds in comparison with the one-atom laser [6] which has been omitted in the survey given in Table 1 of Ref. 5.

The atoms used in our micromaser experiments are rubidium Rydberg atoms pumped by laser excitation into the upper level of the maser transition, which is usually induced between neighboring Rydberg states. In the experiments the atom–field interaction is probed by observing the population in the upper and lower maser levels after the atoms have left the cavity. The field in the cavity consists only of single or a few photons depending on the atomic flux. Nevertheless, it is possible to study the interaction in considerable detail. The dynamics of the atom–field interaction treated with the Jaynes–Cummings model was investigated by selecting and varying the velocity of the pump atoms [2]. The counting statistics of the pump atoms emerging from the cavity allowed us to measure the nonclassic character of the cavity field [3,4] predicted by the micromaser theory. The maser field can be investigated in this way since there is entanglement between the maser field and the state in which the atom leaves the cavity [7,8]. It also has been observed that under suitable experimental conditions the maser field exhibits metastability and hysteresis [9]. The first of the maser experiments have been performed at cavity temperatures of 2 or 0.5 K. In the more recent experiments the temperature was reduced to roughly 0.1 K by using an improved setup in a dilution refrigerator [9]. For a review of the previous work see Raithel et al. [10].

In the following we give a brief review of recent experiments which deal with the observation of quantum jumps of the micromaser field [9] and with the observation of atomic interferences in the cavity [11]. New experiments on the correlation of atoms after the interaction with the cavity field will be briefly mentioned. Furthermore, we will discuss the generation of number or Fock states, and we will also describe new possibilities opening up when ultracold atoms are used for the experiments.

A. Quantum Jumps and Atomic Interferences in the Micromaser

Under steady-state conditions, the photon statistics $P(n)$ of the field of the micromaser is essentially determined by the pump parameter, $\Theta = N_{ex}^{1/2} \Omega\, t_{int}/2$ [10,13]. Here, N_{ex} is the average number of atoms that enter the cavity during the decay time of the cavity field τ_{cav}, Ω is the vacuum Rabi floppy frequency, and t_{int} is the atom–cavity interaction time. The quantity $\langle v \rangle = \langle n \rangle / N_{ex}$ shows the generic behavior shown in Fig. 1.

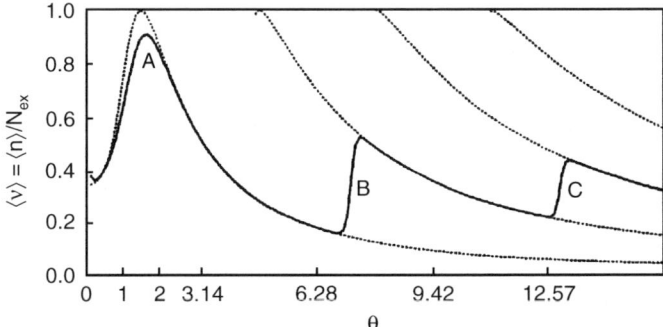

Figure 1. Mean value of $v = n/N_{ex}$ versus the pump parameter $\Theta = \Omega t_{int} \sqrt{N_{ex}}/2$, where the value of Θ is changed via N_{ex}. The solid line represents the micromaser solution for $\Omega = 36$ KHz, $t_{int} = 35\mu s$, and temperature $T = 0.15$ K. The dotted lines are semiclassic steady-state solutions corresponding to fixed stable gain = loss equilibrium photon numbers [14]. The crossing points between a line $\Theta =$ const and the dotted lines correspond to the values where minima in the Fokker–Planck potential $V(v)$ occur (see details in the text).

It suddenly increases at the maser threshold value $\Theta = 1$ and reaches a maximum for $\Theta \approx 2$ (denoted by A in Fig. 1). At threshold the characteristics of a continuous phase transition [12,13] are displayed. As Θ further increases, $\langle v \rangle$ decreases and reaches a minimum at $\Theta \approx 2\pi$ and then abruptly increases to a second maximum (B in Fig. 1). This general type of behavior recurs roughly at integer multiples of 2π, but becomes less pronounced with increasing Θ. The reason for the periodic maxima of $\langle v \rangle$ is that for integer multiples of $\Theta = 2\pi$ the pump atoms perform an almost integer number of full Rabi flopping cycles and start to flip over at a slightly larger value of Θ, thus leading to enhanced photon emission. The periodic maxima in $\langle v \rangle$ for $\Theta = 2\pi, 4\pi$, and so on, can be interpreted as first-order phase transitions [12,13]. The field strongly fluctuates for all phase transitions being caused by the presence of two maxima in the photon number distribution $P(n)$ at photon numbers n_l and $n_h (n_l < n_h)$.

The phenomenon of the two coexisting maxima in $P(n)$ was also studied in a semiheuristic Fokker–Planck (FP) approach [12]. There, the photon number distribution $P(n)$ is replaced by a probability function $P(v, \tau)$ with continuous variables $\tau = t/\tau_{\text{cav}}$ and $v(n) = n/N_{\text{ex}}$, the latter replacing the photon number n. The steady-state solution obtained for $P(v, \tau)$, $\tau \gg 1$, can be constructed by means of an effective potential $V(v)$, showing minima at positions where maxima of $P(v, \tau)$, $\tau \gg 1$, are found. Close to $\Theta = 2\pi$ and multiples thereof, the effective potential $V(v)$ exhibits two equally attractive minima located at stable gain–loss equilibrium points of maser operation [14] (see Fig. 1). The mechanism at the phase transitions mentioned is always the same: A minimum of $V(v)$ loses its global character when Θ is increased, and it is replaced in this role by the next one. This reasoning is a variation of the Landau theory of first-order phase transitions, with $\sqrt{v}$ being the order parameter. This analogy actually leads to the notion that in the limit $N_{\text{ex}} \to \infty$ the change of micromaser field around integer multiples $\Theta = 2\pi$ can be interpreted as first-order phase transitions.

Close to first-order phase transitions, long field evolution time constants are expected [12,13]. This phenomenon was experimentally demonstrated in Ref. 9, along with related phenomena, such as spontaneous quantum jumps between equally attractive minima of $V(v)$, bistability, and hysteresis. Some of those phenomena are also predicted in the two-photon micromaser [15], for which qualitative evidence of first-order phase transitions and hysteresis is reported.

The experimental setup used is shown in Fig. 2. It is similar to that described by Rempe and Walther [4] and by Benson et al. [9]. As before, ^{85}Rb atoms were used to pump the maser. They are excited from the $5S_{1/2}$, $F = 3$ ground state to $63P_{3/2}$, $m_J = \pm 1/2$ states by linearly polarized light of a frequency-doubled c.w. ring dye laser. The polarization of the laser light is linear and parallel to the likewise linearly polarized maser field, and therefore only $\Delta m_J = 0$ transitions are excited. Superconducting niobium cavities resonant with the transition to the

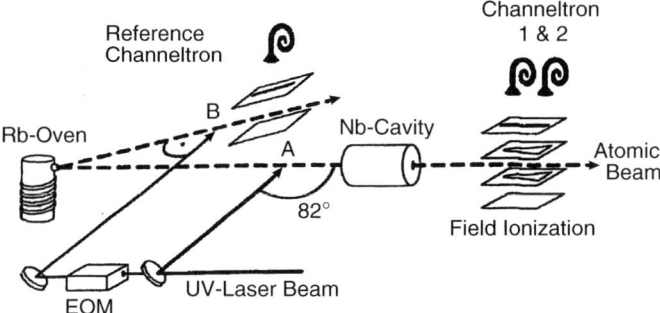

Figure 2. Sketch of the experimental setup. The rubidium atoms emerge from an atomic beam oven and are excited at an angle of 82° at location A. After interaction with the cavity field, they enter a state-selective field ionization region, where channeltrons 1 and 2 detect atoms in the upper and lower maser levels, respectively. A small fraction of the UV radiation passes through an electro-optic modulator (EOM), which generates sidebands of the UV radiation. The blueshifted sideband is used to stabilize the frequency of the laser onto the Doppler-free resonance monitored with a secondary atomic beam produced by the same oven (location B).

$61D_{3/2}, m_J = \pm 1/2$ states were used; the corresponding resonance frequency is 21.506 GHz. The experiments were performed in a ^{3}He/^{4}He dilution refrigerator with cavity temperatures $T \approx 0.15$ K. The cavity Q values ranged from 4×10^9 to 8×10^9. The velocity of the Rydberg atoms and thus their interaction time t_{int} with the cavity field were preselected by exciting a particular velocity subgroup with the laser. For this purpose, the laser beam irradiated the atomic beam at an angle of approximately 82°. As a consequence, the UV laser light (linewidth $\approx$ 2 MHz) is blue-shifted by 50–200 MHz by the Doppler effect, depending on the velocity of the atoms.

Information on the maser field and interaction of the atoms in the cavity can be obtained solely by state-selective field ionization of the atoms in the upper or lower maser level after they have passed through the cavity. For different t_{int} the atomic inversion has been measured as a function of the pump rate by comparing the results with micromaser theory [12,13], and the coupling constant Ω is found to be $\Omega = (40 \pm 10)$ krad/s.

Depending on the parameter range, essentially three regimes of the field evolution time constant τ_{field} can be distinguished. Here we only discuss the results for intermediate time constants. The maser was operated under steady-state conditions close to the second first-order phase transition (C in Fig. 1). The interaction time was $t_{\text{int}} = 47\mu s$ and the cavity decay time was $\tau_{\text{cav}} = 60$ ms. The value of N_{ex} necessary to reach the second first-order phase transition was $N_{\text{ex}} \approx 200$. For these parameters, the two maxima in $P(n)$ are manifested in spontaneous jumps of the maser field between the two maxima with a time constant of $\approx$5 s. This fact and the relatively large pump rate led to the clearly

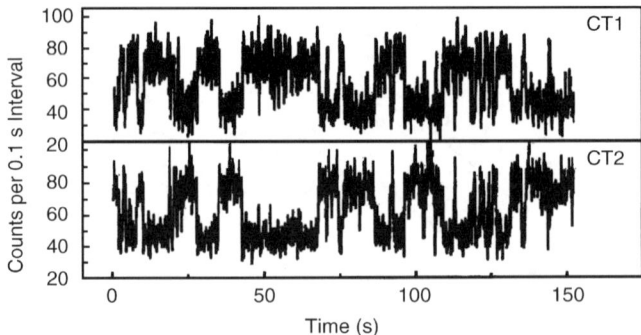

Figure 3. Quantum jumps between two equally stable operation points of the maser field. The chaneltron counts are plotted versus time (CT1 = upper state and CT2 = lower state signals). The signals of the two different detectors show a counterphase behavior; this makes it easy to discriminate between signal and noise.

observable field jumps shown in Fig. 3. Because of the large cavity field decay time, the average number of atoms in the cavity was still as low as 0.17. The two discrete values for the counting rates correspond to the metastable operating points of the maser, which correspond to ≈70 and ≈140 photons. In the FP description, the two values correspond to two equally attractive minima in the FP potential $V(v)$. If one considers, for instance, the counting rate of lower-state atoms (CT2 in Fig. 3), the lower (higher) plateaus correspond to time intervals in the low (high) field metastable operating point. If the actual photon number distribution is averaged over a time interval containing many spontaneous field jumps, the steady-state result $P(n)$ of the micromaser theory is recovered.

In the parameter ranges where switching occurs much faster than in the case shown in Fig. 3, the individual jumps cannot be resolved; therefore, different methods have to be used for the measurement. Furthermore, hysteresis is observed at the maser parameters for which the field jumps occur. Owing to lack of space, these results cannot be discussed here. For a complete survey on the performed experiments, see Ref. 9.

The next topic we would like to discuss is the observation of atomic interferences in the micromaser [11]. Because a nonclassic field is generated in the maser cavity, we were able for the first time to investigate atomic interference phenomena under the influence of nonclassic radiation. Owing to the bistable behavior of the maser field, the interferences display quantum jumps; thus the quantum nature of the field becomes directly visible in the interference fringes. Interferences occur because a coherent superposition of dressed states is produced by mixing the states at the entrance and exit holes of the cavity. Inside the cavity the dressed states develop differently in time, giving rise to Ramsey-type interferences [16] when the maser cavity is tuned through resonance.

The setup used in the experiment is identical to the one described previously [9]. However, the flux of atoms through the cavity is by a factor of 5–10 higher than in the previous experiments, where the $63P_{3/2}$–$61D_{5/2}$ transition was used. For the experiments the Q value of the cavity was 6×10^9, corresponding to a photon decay time of 42 ms.

Figure 4 shows the standard maser resonance in the uppermost plot which is obtained when the resonator frequency is tuned. At large values of N_{ex} ($N_{ex} > 89$) sharp, periodic structures appear. These typically consist of (a) a smooth wing on the low-frequency side and (b) a vertical step on the high-frequency side. The clarity of the pattern rapidly decreases when N_{ex} increases to 190 or beyond. We will see later that these structures have to be interpreted as interferences. It can be seen that the atom-field resonance frequency is red-shifted with increasing N_{ex}, and the shift reaches 200 kHz for $N_{ex} = 190$. Under these conditions there are roughly 100 photons on the average in the cavity. The

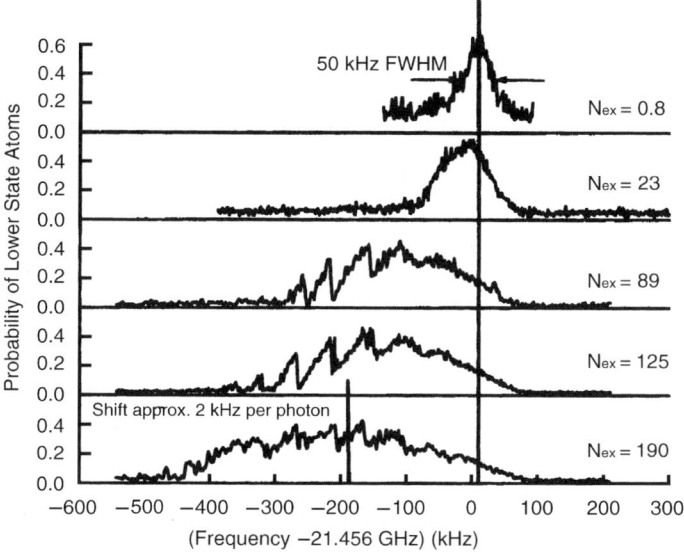

Figure 4. Shift of the maser resonance $63P_{3/2}$–$61D_{5/2}$ for fast atoms ($t_{int} = 35\,\mu s$). The upper plot shows the maser line for low pump rate ($N_{ex} < 1$). The FWHM linewidth (50 kHz) sets an upper limit of ≈ 5 mV/cm for the residual electric stray fields in the center of the cavity. The lower resonance lines are taken for the indicated large values of N_{ex}. The plots show that the center of the maser line shifts by about 2 kHz per photon. In addition, there is considerable field-induced line broadening that is approximately proportional to $\sqrt{N_{ex}}$. For $N_{ex} \geq 89$ the lines display periodic structures, which are discussed in the text.

large red-shift cannot be explained by AC Stark effect, which for 100 photons would amount to about 1 kHz for the transition used. Therefore it is obvious that other reasons must be responsible for the observed shift.

It is known from previous maser experiments that there are small static electric fields in the entrance and exit holes of the cavity. It is supposed that this field is generated by patch effects at the surface of the niobium metal caused by rubidium deposits caused by the atomic beam or by microcrystallites formed when the cavities are tempered after machining. The tempering process is necessary to achieve high-quality factors. The influence of those stray fields is only observable in the cavity holes; in the center of the cavity they are negligible owing to the large atom-wall distances.

When the interaction time t_{int} between the atoms and the cavity field is increased, the interference structure disappears for $t_{int} > 47$ μs [11]. This is due to the fact that there is no nonadiabatic mixing any more between the substates when the atoms get too slow.

In order to understand the observed structures, the Jaynes–Cummings dynamics of the atoms in the cavity has to be analyzed. This treatment is more involved than that in connection with previous experiments, since the higher maser field requires detailed consideration of the field in the periphery of the cavity, where the additional influence of stray electric fields is more important.

The usual formalism for the description of the coupling of an atom to the radiation field is the dressed atom approach [17], leading to splitting of the coupled atom-field states, depending on the vacuum Rabi-flopping frequency Ω, the photon number n, and the atom-field detuning δ. We face a special situation at the entrance and exit holes of the cavity. There we have a position-dependent variation of the cavity field, as a consequence of which Ω is position-dependent. An additional variation results from the stray electric fields in the entrance and exit holes. Owing to the Stark effect, these fields lead to a position-dependent atom-field detuning δ.

The Jaynes–Cummings Hamiltonian only couples pairs of dressed states. Therefore, it is sufficient to consider the dynamics within such a pair. In our case, prior to the atom-field interaction the system is in one of the two dressed states. For parameters corresponding to the periodic substructures in Fig. 4, the dressed states are mixed only at the beginning of the atom–field interaction and at the end. The mixing at the beginning creates a coherent superposition of the dressed states. Afterwards the system develops adiabatically, whereby the two dressed states accumulate a differential dynamic phase Φ that strongly depends on the cavity frequency. The mixing of the dressed states at the entrance and exit holes of the cavity, in combination with the intermediate adiabatic evolution, generates a situation similar to a Ramsey two-field interaction.

The maximum differential dynamic phase Φ solely resulting from dressed-state coupling by the maser field is roughly 4π under the experimental conditions used here. This is not sufficient to explain the interference pattern of Fig. 4, where we have at least six maxima corresponding to a differential phase of 12π. This means that an additional energy shift differently affecting upper and lower maser states is present. Such a phenomenon can be caused by the above-mentioned small static electric fields present in the holes of the cavity. The static field causes a position-dependent detuning δ of the atomic transition from the cavity resonance; as a consequence we get an additional differential dynamic phase Φ. In order to interpret the periodic substructures as a result of the variation of Φ with the cavity frequency, the phase Φ has to be calculated from the atomic dynamics in the maser field.

The quantitative calculation can be performed on the basis of the micromaser theory. The calculations reproduce the experimental finding that the maser line shifts to lower frequencies when N_{ex} is increased [11]. The mechanism for that can be explained as follows: The high-frequency edge of the maser line does not shift with N_{ex} at all, since this part of the resonance is produced in the central region of the cavity, where practically no static electric fields are present. The low-frequency cutoff of the structure is determined by the location where the mixing of the dressed states occurs. With decreasing cavity frequency, those points shift closer to the entrance and exit holes, with the difference between the particular cavity frequency and unperturbed atomic resonance frequency giving a measure of the static electric field at the mixing locations. Closer to the holes the passage behavior of the atoms through the mixing locations becomes nonadiabatic for the following reasons: First, the maser field strength reduces toward the holes. This leads to reduced repulsion of the dressed states. Second, the stray electric field strongly increases toward the holes. This implies a larger differential slope of the dressed state energies at the mixing locations, and therefore it leads to a stronger nonadiabatic passage. At the same time the observed signal extends further to the low-frequency spectral region. Because the photon emission probabilities are decreasing toward lower frequencies, their behavior finally defines the low-frequency boundary of the maser resonance line. With increasing N_{ex} the photon number n increases. As for larger values of n, the photon emission probabilities get larger; also, increasing N_{ex} leads to an extension of the range of the signal to lower frequencies. This theoretical expectation is in agreement with the experimental observation.

In the experiment it is also found that the maser line shifts toward lower frequencies with increasing t_{int}. This result also follows from the developed model: The red shift increases with t_{int} because a longer interaction time leads to a more adiabatic behavior in the same way as a larger N_{ex} does.

The calculations reveal that on the vertical steps displayed in the signal the photon number distribution has two distinctly separate maxima similar to those observed at the phase transition points discussed above. Therefore, the maser field should exhibit hysteresis and metastability under the present conditions as well. The hysteresis indeed shows up when the cavity frequency is linearly scanned up and down with a modest scan rate [11]. When the maser is operated in steady-state and the cavity frequency is fixed to the steep side of one of the fringes, we also observe spontaneous jumps of the maser field between two metastable field states.

The calculations also show that on the smooth wings of the more pronounced interference fringes the photon number distribution $P(n)$ of the maser field is strongly sub-Poissonian. This leads us to the conclusion that we observe Ramsey-type interferences induced by a nonclassic radiation field. The sub-Poissonian character of $P(n)$ results from the fact that on the smooth wings of the fringes the photon gain reduces when the photon number is increased. This feedback mechanism stabilizes the photon number resulting in a sub-Poissonian photon distribution.

B. Entanglement in the Micromaser

Owing to the interaction of the Rydberg atom with the maser field there is an entanglement between field and the state in which a particular atom is leaving the cavity.

This entanglement was studied in several papers, see, for example, Refs. 8 and [18]. Furthermore, there is a correlation between the states of the atoms leaving the cavity subsequently. If, for example, atoms in the lower maser level are studied [19], an anticorrelation is observed in a region for the pump parameter Θ where sub-Poissonian photon statistics is present in the maser field. Recently, measurements [20] of these pair correlations have been performed giving a rather good agreement with the theoretical predictions by Briegel et al. [21]. Because the cavity field plays an important role in this entanglement, the pair correlations disappear when the time interval between subsequent atoms get larger than the storage time of a photon in the cavity.

C. Trapping States

The trapping states are a steady-state feature of the micromaser field peaked in a single photon number, and they occur in the micromaser as a direct consequence of field quantization. At low cavity temperatures the number of blackbody photons in the cavity mode is reduced and trapping states begin to appear [22,23]. They occur when the atom field coupling, Ω, and the interaction time, t_{int}, are chosen such that in a cavity field with n_q photons each atom undergoes an integer number, k, of Rabi cycles. This is summarized by the condition

$$\Omega t_{\text{int}} \sqrt{n_q + 1} = k\pi \tag{1}$$

When (1) is fulfilled, the cavity photon number is left unchanged after the interaction of an atom and hence the photon number is "trapped." This will

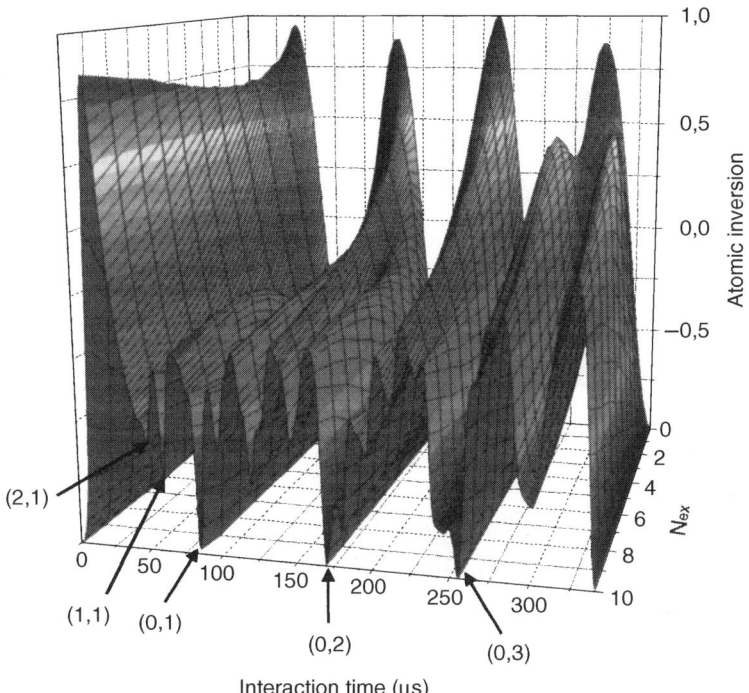

Figure 5. Inversion of the micromaser field under conditions that the trapping states occur. The thermal photon number in 10^{-4}. As the pump rate increases, the formation of the trapping states from the vaccum show up. See also Fig. 6. The trapping states are characterized by (n_q, k) whereby n_q is the photon number and k is the number of Rabi cycles.

occur over a large range of the atomic pump rate N_{ex}. The trapping state is therefore characterized by the photon number n_q and the number of integer multiples of full Rabi cycles k.

The buildup of the cavity field can be seen in Fig. 5, where the emerging atom inversion is plotted against interaction time and pump rate. The inversion is defined as $I(t_{int}) = P_g(t_{int}) - P_e(t_{int})$, where $P_{e(g)}(t_{int})$ is the probability of finding an excited state (ground state) atom for a particular interaction time t_{int}. At low atomic pump rates (low N_{ex}) the maser field cannot build up and the maser exhibits Rabi oscillations due to the interaction with the vacuum field. At the positions of the trapping states, the field builds up until it reaches the trapping state condition. This manifests itself as a reduced emission probability and hence as a dip in the atomic inversion. Once in a trapping state the maser will remain there regardless of the pump rate. Therefore the trapping states show up as valleys in the N_{ex} direction. Figure 6 shows the photon number

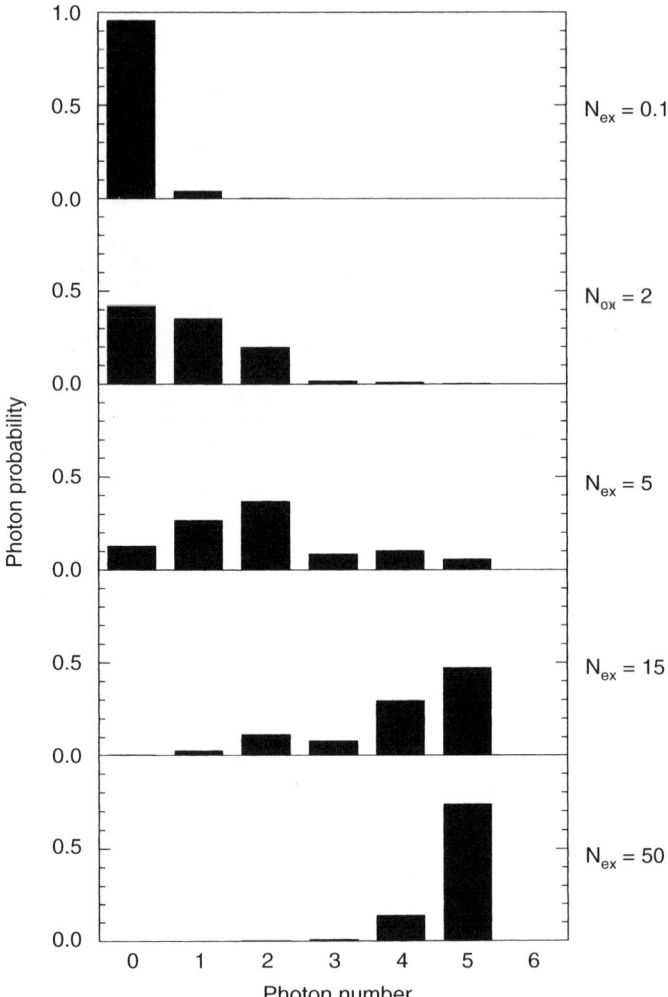

Figure 6. Photon number distribution of the maser field as a function of the atomic flux N_{ex}. Shown is a numerical simulation. At $N_{ex} = 50$ a Fock state with rather high purity is achieved.

distribution as the pump rate is increased for the special condition of the five-photon trapping state. The photon distribution develops from a thermal distribution toward higher photon numbers until the pump rate is high enough for the atomic emission to be affected by the trapping state condition. As the pump rate is further increased, and in the limit of a low thermal photon number, the field continues to build up to a single trapped photon number and the steady-state distribution approaches a Fock state. There is in general a slight deviation from a pure Fock state resulting from photon losses. Therefore the state with the next lower photon number has a small probability. A lost photon will be replaced by the next incoming atom; however, there is a small time interval until this actually happens, depending on the availability of an atom.

Owing to blackbody radiation at finite temperatures, there is a probability of having thermal photons enter the mode. The presence of a thermal photon in the cavity disturbs the trapping state condition and an atom can emit a photon. This causes the photon number of the field to jump to a value above the trapping condition n_q and a cascade of emission events will follow resulting in a build up of a new photon distribution with an average photon number $n > n_q$ (Fig. 7). The steady-state behavior of the maser field thus reacts very sensitively on the presence of thermal photons and the number of lower state atoms increases.

Note that under readily achievable experimental conditions, it is possible for the steady-state field in the cavity to approach Fock states with a high fidelity. Figure 8 summarizes results of simulations of the micromaser field corresponding to Fock states from $n = 0$ to $n = 5$. The experimental realization

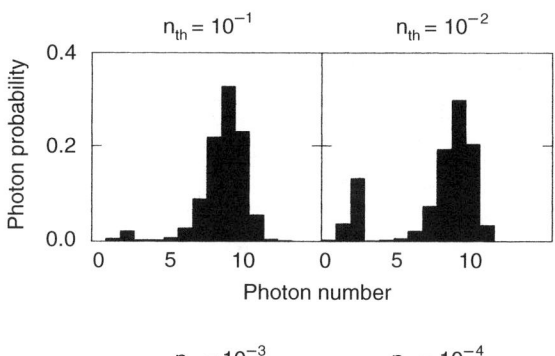
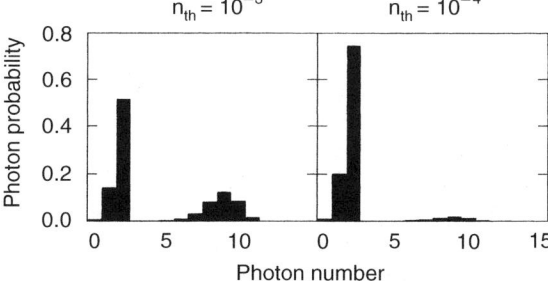

Figure 7. Photon number distribution of the maser field. This simulation shows the strong influence of thermal photons on the steady-state photon number distribution ($N_{ex} = 25$).

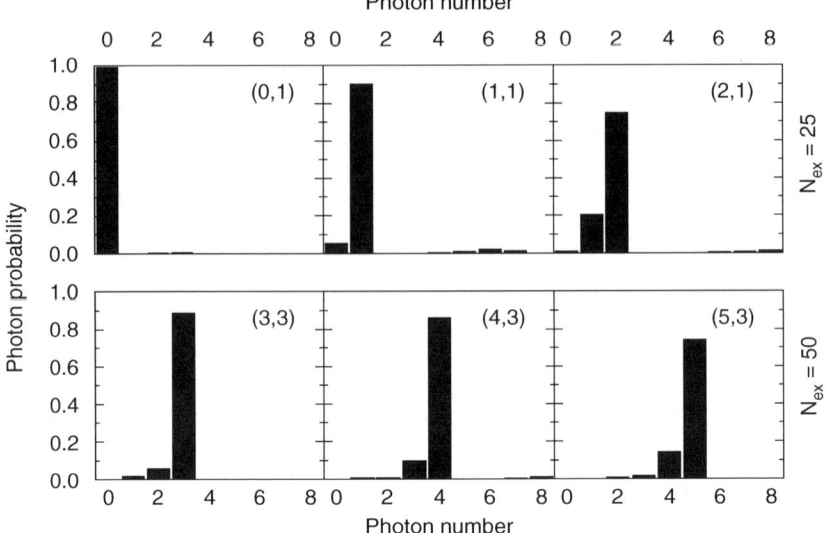

Figure 8. Purity of Fock states generated in the micromaser under trapping state conditions. Shown are computer simulations for Fock states between $n = 0$ and $n = 5$. The thermal photon number in $n_{th} = 10^{-4}$.

requires pump rates of $N_{ex} = 25$ to $N_{ex} = 50$, a temperature of about 100 mK, and a high selectivity of atomic velocity [23].

The results displayed in Fig. 8 show again the phenomenon that is also evident in Fig. 6: The deviation from the pure Fock state is mainly caused by losses; therefore a contribution of the state with the next lower photon number is also present in the distribution. This is caused by the fact that a finite time is needed in the maser cavity to replace a lost photon.

Experimental evidence that trapping states can be achieved is shown in Fig. 9a and 9b (see Ref. 23 for details). The measurement was taken at a cavity temperature of 0.3 K corresponding to a thermal photon number of 0.054. The plot shows a rescaled inversion that takes the increasing loss of Rydberg atoms at longer interaction times into account. The shown result gives a good qualitative agreement with the Monte Carlo simulations performed in Ref. 10.

In Fig. 9a, which displays the results for smaller N_{ex}, the trapping states corresponding to the vacuum $(n_q, k) = (0, 1)$, one photon $(1, 1)$, and two photons $(2, 1)$ and $(2, 2)$ are clearly visible while at higher N_{ex} (Fig. 9b) the vacuum and one-photon trapping states become less visible. With increasing N_{ex} the width of the trapping states decreases [22] and the influence of detuning and velocity averaging becomes larger. At higher pump rates it is also important that

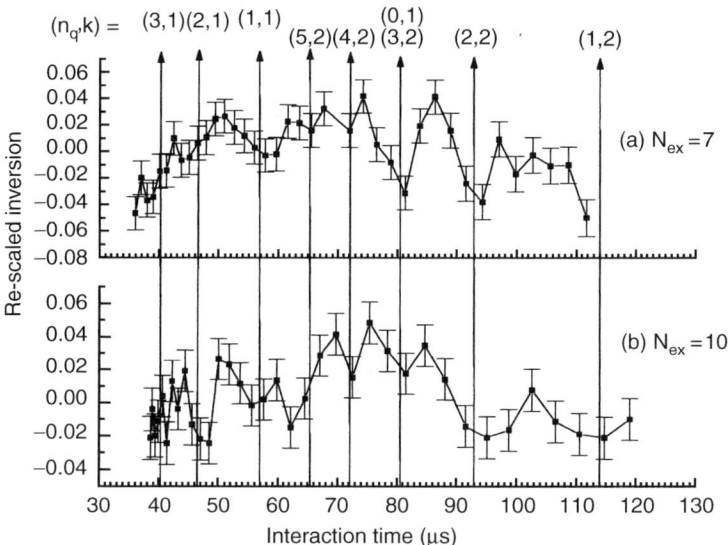

Figure 9. Experimental measurement of trapping states in the maser field (Weidinger et al. [23]). Trapping states appear in a reduced probability of finding ground-state atoms. The rescaling considers the decay of the Rydberg states on their way from preparation to detection.

the proportion of two atom events grows, reducing the visibility of trapping states.

In the following we would like to mention another method that can be used to generate Fock states of the field of the one-atom maser. When the atoms leave the cavity of the micromaser they are in an entangled state with the cavity field [20]. A method of state reduction was suggested by Krause et al. [24] to observe the buildup of the cavity field to a known Fock state. By state reduction of the outgoing atom, also the field part of the entangled atom-field state is projected out and the photon number in the field either increases or decreases depending on the state of the observed outgoing atom. If the field is initially in a state $|n\rangle$, then an interaction of an atom with the cavity leaves the cavity field in a superposition of the states $|n\rangle$ and $|n+1\rangle$ and leaves the atom in a superposition of the internal atomic states $|e\rangle$ and $|g\rangle$.

$$\Psi = \cos(\phi)|e\rangle|n\rangle - i\sin(\phi)|g\rangle|n+1\rangle \qquad (2)$$

where ϕ is an arbitrary phase. The state selective field ionization measurement of the internal atomic state reduces also the field to one of the states $|n\rangle$ or $|n+1\rangle$. State reduction is independent of interaction time; hence a ground-state atom always projects the field onto the $|n+1\rangle$ state independent of the time spent in the cavity. This results in an *a priori* probability of the maser field being in a specific but unknown number state [24]. If the initial state is the vacuum, $|0\rangle$, then a number state created is equal to the number of ground-state atoms that were collected within a suitably small fraction of the cavity decay time.

In a system governed by the Jaynes–Cummings Hamiltonian, spontaneous emission is reversible and an atom in the presence of a resonant quantum field undergoes Rabi oscillations. That is, the relative populations of the excited and ground states of the atom oscillate at a frequency $\Omega\sqrt{n+1}$, where Ω is the atom field coupling constant. Experimentally, we measure again the atomic inversion, $I = P_g - P_e$. In the presence of dissipation a fixed photon number n in a particular mode is not observed and the field always evolves into a mixture of such states. Therefore the inversion is generally given by

$$I(n, t_{\text{int}}) = -\sum_n P_n \cos\left(2\Omega\sqrt{n+1}\, t_{\text{int}}\right) \qquad (3)$$

where P_n is the probability of finding n photons in the mode, and t_{int} is the interaction time of the atoms with the cavity field.

The experimental verification of the presence of Fock states in the cavity takes the form of a pump–probe experiment in which a pump atom prepares a quantum state in the cavity while the Rabi phase of the emerging probe atom measures the quantum state. The signature that the quantum state of interest has been prepared is simply the detection of a defined number of ground-state atoms. To verify that the correct quantum state has been projected onto the cavity field a probe atom is sent into the cavity with a variable, but well-defined, interaction time. Because the formation of the quantum state is independent of interaction time, we need not change the relative velocity of the pump and probe atoms, thus reducing the complexity of the experiment. In this sense we are performing a reconstruction of a quantum state in the cavity using a similar method to that described by Bardoff et al. [25]. This experiment reveals the maximum amount of information that can be found relating to the cavity photon number. We have recently used this method to demonstrate the existence of Fock states up to $n = 2$ in the cavity [26].

D. Generation of GHZ States

The following proposal for the creation of states of the Greenberger–Horne–Zeilinger (GHZ) type [28,29] is an application of the vacuum trapping state. By the use of the vacuum trapping state a field determination during the measurement is not necessary, which simplifies strongly the preparation of the

GHZ states. We assume that the cavity is initially empty, and two excited atoms traverse it consecutively. The velocity of the first atom, and its consequent interaction time, is such that it emits a photon with probability $(\sin\varphi_1)^2 = 51.8\%$, where $\varphi_1 = 0.744\,\pi$ is the corresponding Rabi angle ϕ of (2) for $n = 0$. The second atom arrives with the velocity dictated by the vacuum trapping condition; for $n = 0$ it has $\phi = \pi$ in (2), so that $\phi = \sqrt{2}\,\pi$ for $n = 1$. Assuming that the duration of the whole process is short on the scale set by the lifetime of the photon in the cavity, we thus have

$$|0, e, e\rangle \xrightarrow[\text{atom}]{\text{first}} |0, e, e\rangle \cos(\varphi_1) - i|1, g, e\rangle \sin(\varphi_1)$$

$$\xrightarrow[\text{atom}]{\text{second}} -|0, e, e\rangle \cos(\varphi_1) - |2, g, g\rangle \sin(\varphi_1) \sin(\sqrt{2}\,\pi)$$

$$- i|1, g, e\rangle \sin(\varphi_1) \cos(\sqrt{2}\,\pi) \quad (4)$$

where, for example, $|1, g, e\rangle$ stands for "one photon in the cavity and first atom in the ground state and second atom excited." With the above choice of $\sin(\varphi_1) = 0.720$, we have $\cos(\varphi_1) = \sin(\varphi_1)\sin(\sqrt{2}\,\pi) = -0.694$, so that the two components with even photon number ($n = 0$ or $n = 2$) carry equal weight and occur with a joint probability of 96.3%. The small 3.7% admixture of the $n = 1$ component can be removed by measuring the parity of the photon state [30] and conditioning the experiment to even parity. The two atoms and the cavity field are then prepared in the entangled state:

$$\Psi_{GHZ} = \frac{1}{\sqrt{2}}(|0, e, e\rangle + |2, g, g\rangle) \quad (5)$$

which is a GHZ state of the Mermin kind [31] in all respects.

E. The One-Atom Maser and Ultracold Atoms

In this section we discuss the case where the micromaser is pumped by ultracold atoms; in this limit the center-of-mass motion has to be treated quantum mechanically, especially when the kinetic energy $(\hbar k)^2/2M$ of the atoms is of the same order or smaller than the atom–field [32] interaction energy $\hbar\Omega$.

For simplicity, we consider here the situation where an atom in the excited state $|e\rangle$ is incident upon a cavity that contains n photons so that the combined atom–field system is described by the state $|e, n\rangle = (|\gamma^+_{n+1}\rangle + |\gamma^-_{n+1}\rangle)/\sqrt{2}$. The dressed-state components $|\gamma^+_{n+1}\rangle$ and $|\gamma^-_{n+1}\rangle$, which are the eigenstates of the atom–field interaction Hamiltonian, encounter different potentials giving rise to different reflection and transmission of the atom. Appropriate relative phase shifts between the dressed-state components during the atom–field interaction may result in the state $(|\gamma^+_{n+1}\rangle - |\gamma^-_{n+1}\rangle)/\sqrt{2} = |g, n+1\rangle$, which corresponds to the emission of a photon and a transition to the lower atomic level $|g\rangle$. Likewise, changes in the relative reflection and transmission amplitudes may lead to a de-excitation of the atom.

For thermal atoms, the emission probability shown in Fig. 10 displays the usual Rabi oscillations as a function of the interaction time τ. For very slow atoms, however, the emission probability is a function of the interaction length L and shows resonances such as the ones observed in the intensity transmitted by a

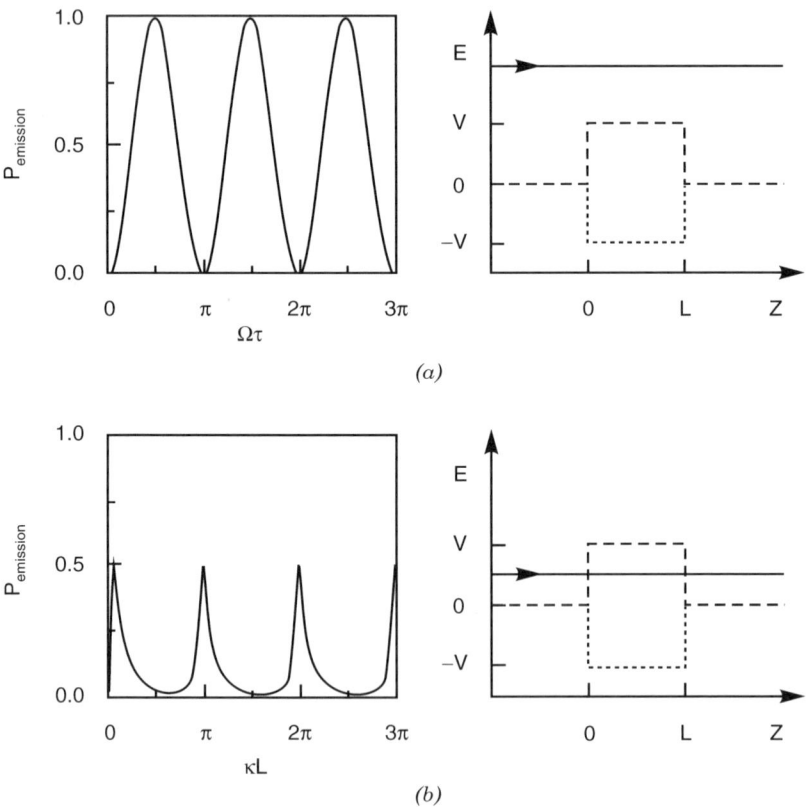

Figure 10. Emission probability for (a) thermal atoms with $k/\kappa = 10$ versus the interaction time $\Omega\tau$ and (b) ultracold atoms with $k/\kappa = 0.1$ versus the interaction length κL, and the corresponding repulsive (*dashed lines*) and attractive (*dotted lines*) atom–field potential. The constant κ is defined by $(\hbar\kappa)^2/2m = \hbar\Omega$.

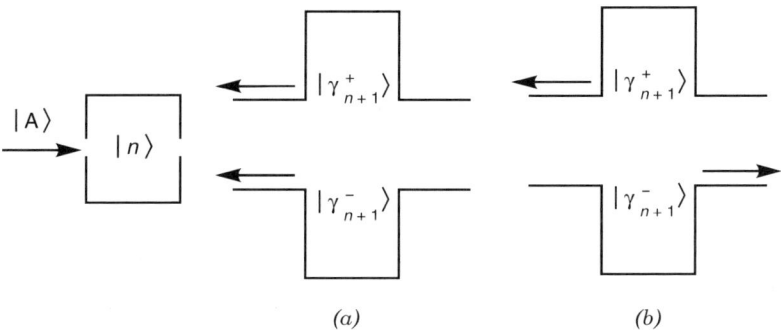

Figure 11. Reflection and transmission of the atoms at the potential barrier for the $|\gamma_{n+1}^{+}\rangle$ and at the potential well for the $|\gamma_{n+1}^{-}\rangle$ component (a) out of the mazer resonance and (b) on resonance.

Fabry–Perot resonator. The resonances occur when the cavity length is an integer multiple of half the de Broglie wavelength of the atom inside the potential well.

Figure 11 illustrates the reflection and transmission of the atom for a cavity whose mode function is a mesa function, which approximates the lowest TM mode of a cylindrical cavity. For very cold atoms, the dressed-state component that encounters the potential barrier is always reflected. In general, the other dressed-state component is also reflected at the well. The situation changes dramatically if the cavity length is an integer multiple of half the de Broglie wavelength. In this case, the $|\gamma_{n+1}^{-}\rangle$ is completely transmitted, which implies a 50% transmission probability for the atom. A detailed calculation [32] shows that in such a situation the emission probability for a photon is 1/2 for each of the two dressed-state components, yielding an overall emission probability $P_{\text{emission}} = 1/2$.

So far, we have discussed the motion and atom–field interaction of a single-atom incident upon the cavity. Due to the unusual emission probability, a beam of ultracold atoms can produce unusual photon distributions such as a shifted thermal distribution. For details about this microwave amplification by z-motion-induced emission of radiation (mazer), the reader is referred to the trilogy [33–35].

In order to see the mazer resonances for atoms with a certain velocity spread, the interaction length L has to be small. Whereas in the usual cylindrical micromaser cavities the smallest cavity length is given by half the wavelength of the microwaves, cavities of the reentrant type, as depicted in Fig. 12, allow for an interaction length much smaller than the wavelength. With such a device, an experiment with realistic parameters seems possible [34].

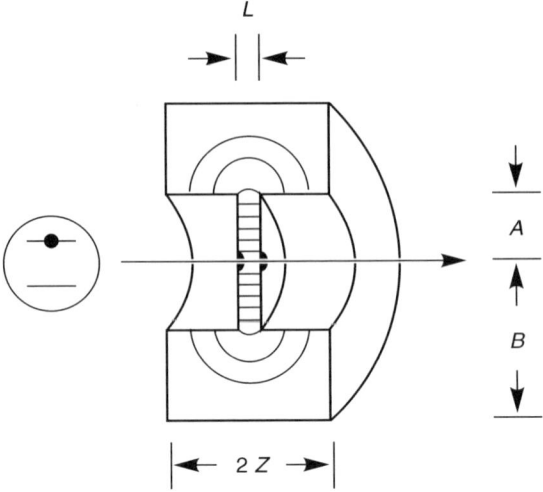

Figure 12. Possible experimental setup with a reentrant cavity.

III. ION TRAP EXPERIMENTS

Besides the experiments performed with atoms in a cavity the trapped ion techniques provide another way to investigate quantum phenomena in radiation atom interaction. In the following recent experiments and new proposals for experiments will be reviewed.

A. Resonance Fluorescence of a Single Atom

Resonance fluorescence of an atom is a basic process in radiation–atom interactions and has therefore always generated considerable interest. The methods of experimental investigation have changed continuously due to the availability of new experimental tools. A considerable step forward occurred when tunable and narrow band dye laser radiation became available. These laser sources are sufficiently intense to easily saturate an atomic transition. In addition, the lasers provide highly monochromatic light with coherence times much longer than typical natural lifetimes of excited atomic states. Excitation spectra with laser light using well-collimated atomic beam lead to a width being practically the natural width of the resonance transition; therefore it became possible to investigate the frequency spectrum of the fluorescence radiation with high resolution. However, the spectrograph used to analyze the reemitted radiation was a Fabry–Perot interferometer, the resolution of which did reach the natural width of the atoms, but was insufficient to reach the laser linewidth (see, e.g., Hartig et al. [36] and Cresser et al. [37]). A considerable progress in this direction was achieved by investigating the fluorescence spectrum of ultracold atoms in an optical lattice in a heterodyne experiment [38]. In these measurements a linewidth of 1 kHz was achieved; however, the quantum aspects of the resonance fluorescence such as antibunched photon statistics cannot be investigated under these conditions becasue they wash out when more than one atom is involved.

Thus the ideal experiment requires a single atom to be investigated. For some time it has been known that ion traps enable us to study the fluorescence from a single laser-cooled particle practically at rest, thus providing the ideal case for the spectroscopic investigation of the resonance fluorescence. The other essential ingredient for achieving a high resolution is the measurement of the frequency spectrum by heterodyning the scattered radiation with laser light as demonstrated with many cold atoms [38]. Such an optimal experiment with a single trapped Mg^+ ion is reviewed in the following. The measurement of the spectrum of the fluorescent radiation at low excitation intensities is presented. Furthermore, the photon correlation of the fluorescent light has been investigated under practically identical excitation conditions. The comparison of the two results shows a very interesting aspect of complementarity because the heterodyne measurement corresponds to a "wave" detection of the radiation, whereas the measurement of the photon correlation is a "particle" detection scheme. It will be shown that under the same excitation conditions the wave detection provides the properties of a classic atom (i.e., a driven oscillator), whereas the particle or photon detection displays the quantum properties of the atom. Whether the atom displays classic or quantum properties thus depends on the method of observation.

The spectrum of the fluorescence radiation is given by the Fourier transform of the first-order correlation function of the field operators, whereas the photon statistics and photon correlation is obtained from the second-order correlation function. The corresponding operators do not commute, and thus the respective observations are complementary. The present theory on the spectra of fluorescent radiation following monochromatic laser excitation can be summarized as follows: Fluorescence radiation obtained with low incident intensity is also monochromatic owing to energy conservation. In this case, elastic scattering dominates the spectrum, and thus one should measure a monochromatic line at the same frequency as the driving laser field. The atom stays in the ground state most of the time, and absorption and emission must be considered as one process with the atom in principle behaving as a classic oscillator. This case was treated on the basis of a quantized field many years ago by Heitler [39]. With increasing intensity upper and lower states become more strongly coupled leading to an inelastic component, which increases with the square of the intensity. At low intensities, the elastic part dominates since it depends linearly on the intensity. As the intensity of the exciting light increases, the atom spends more time in the upper state and the effect of the vacuum fluctuations comes into play through spontaneous emission. The inelastic component is added to the spectrum, and the elastic component goes through a maximum where the Rabi flopping frequency $\Omega = \Gamma/\sqrt{2}$ (Γ is the natural linewidth) and then disappears with growing Ω. The inelastic part of the spectrum gradually broadens as Ω increases and for $\Omega > \Gamma/2$ sidebands begin to appear [37,40].

The experimental study of the problem requires, as mentioned above, a Doppler-free observation. In order to measure the frequency distribution, the fluorescent light has to be investigated by means of a high-resolution spectrometer. The first experiments of this type were performed by Schuda et al. [41] and later by Walther [42], Hartig et al. [36], and Ezekiel and co-workers [43]. In all these experiments, the excitation was performed by single-mode dye laser radiation, with the scattered radiation from a well-collimated atomic beam observed and analyzed by a Fabry–Perot interferometer.

Experiments to investigate the elastic part of the resonance fluorescence giving a resolution better than the natural linewidth have been performed by Gibbs and Venkatesan [44] and Cresser et al. [37].

The first experiments that investigated antibunching in resonance fluorescence were also performed by means of laser-excited collimated atomic beams. The initial results obtained by Kimble, Dagenais, and Mandel [45] showed that the second-order correlation function $g^{(2)}(t)$ had a positive slope that is characteristic of photon antibunching. However, $g^{(2)}(0)$ was larger than $g^{(2)}(t)$ for $t \to \infty$ due to number fluctuations in the atomic beam and to the finite interaction time of the atoms [46,47]. Further refinement of the analysis of the experiment was provided by Dagenais and Mandel [47]. Rateike et al. [48] used a longer interaction time for an experiment in which they measured the photon correlation at very low laser intensities (see Cresser et al. [37] for a review). Later, photon antibunching was measured using a single trapped ion in an experiment that avoids the disadvantages of atom number statistics and finite interaction time between atom and laser field [49].

As pointed out in many papers, photon antibunching is a purely quantum phenomenon (see, e.g., Cresser et al. [37] and Walls [50]). The fluorescence of a single ion displays the additional nonclassic property that the variance of the photon number is smaller than its mean value (i.e., it is sub-Poissonian) [49,51].

The trap used for the present experiment was a modified Paul trap, called an endcap trap [52]. The trap consists of two solid copper–beryllium cylinders (diameter 0.5 mm) arranged collinearly with a separation of 0.56 mm. These correspond to the cap electrodes of a traditional Paul trap, whereas the ring electrode is replaced by two hollow cylinders, one of which is concentric with each of the cylindrical endcaps. Their inner and outer diameters are 1 and 2 mm, respectively, and they are electrically isolated from the cap electrodes. The fractional anharmonicity of this trap configuration, determined by the deviation of the real potential from the ideal quadrupole field, is below 0.1% (see Schrama et al. [52]). The trap is driven at a frequency of 24 MHz with typical secular frequencies in the xy-plane of approximately 4 MHz. This required a radio-frequency voltage with an amplitude on the order of 300 V to be applied between the cylinders and the endcaps.

The measurements were performed using the $3^2 S_{1/2} - 3^2 P_{3/2}$ transition of a $24\,\mathrm{Mg}^+$ ion at a wavelength of 280 nm. The heterodyne measurement is performed as follows. The dye laser excites the trapped ion while the fluorescence is observed in a direction of about 54° to the exciting laser beam. However, both the observation direction and the laser beam are in a plane perpendicular to the symmetry axis of the trap. A fraction of the laser radiation is removed with a beamsplitter and then frequency-shifted [by 137 MHz with an acousto-optic modulator (AOM)] to serve as the local oscillator. An example of a heterodyne signal is displayed in Fig. 13. The signal is the narrowest optical heterodyne spectrum of resonance fluorescence reported to date. Thus our experiment provides the most compelling confirmation of Weisskopf's prediction of a coherent component in resonance fluorescence. The linewidth observed implies that exciting laser and fluorescent light are coherent over a length of 400,000 km. Further details on the experiment are given in Refs. 53 and 54.

Investigation of photon correlations employed the ordinary Hanbury–Brown and Twiss setup. The setup was essentially the same as described by Diedrich and Walther [49]. The results are shown and discussed in Ref. 53 also.

The presented experiment describes the first high-resolution heterodyne measurement of the elastic peak in resonance fluorescence of a single ion. At identical experimental parameters we also measured antibunching in the photon correlation of the scattered field. Together, both measurements show that, in the limit of weak excitation, the fluorescence light differs from the excitation radiation in the second-order correlation but not in the first-order correlation. However, the elastic component of resonance fluorescence combines an extremely narrow frequency spectrum with antibunched photon statistics, which means that the fluorescence radiation is not second-order coherent as expected from a classic point of view [55]. The heterodyne and the photon correlation measurement are complementary because they emphasize either the classical wave properties or the quantum properties of resonance fluorescence, respectively.

B. The Ion-Trap Laser

There have been several theoretical papers on one-atom lasers in the past [56–60]. This system provides a testing ground for new theoretical concepts and results in the quantum theory of the laser. Examples are atomic coherence effects [61] and dynamic (i.e., self-generated) quantum-noise reduction

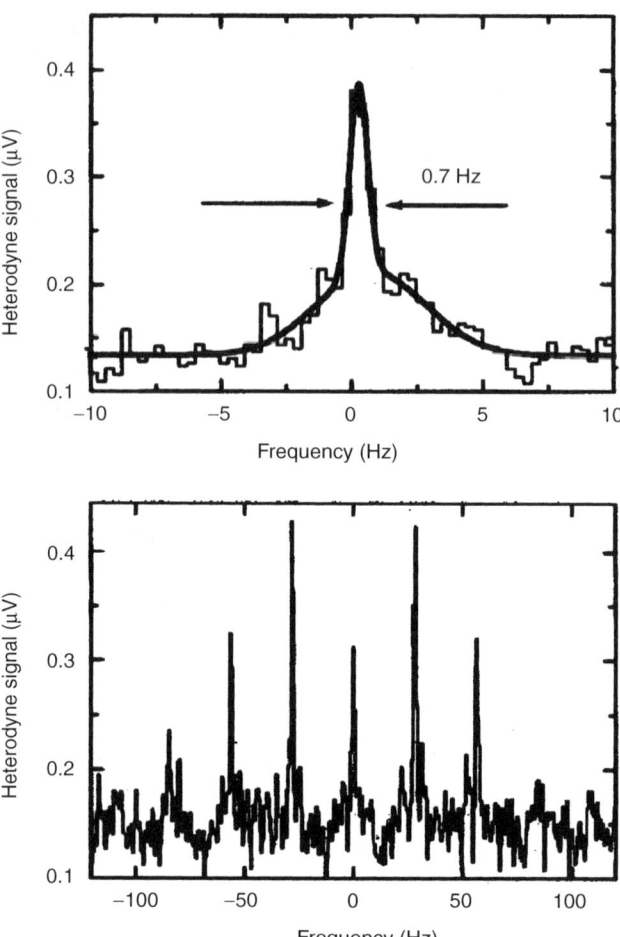

Figure 13. Heterodyne spectrum of a single trapped 24 Mg$^+$ ion. *Left*: Resolution bandwidth 0.5 Hz. The solid line is a Lorentzian fit to the experimental data; the peak appears on top of a small pedestal being 4 Hz wide. The latter signal is due to random phase fluctuations in the spatially separated sections of the light paths of local oscillator and fluorescent light; they are generated by variable air currents in the laboratory. *Right*: Heterodyne spectrum of the coherent peak with sidebands generated by mechanical vibrations of the mount holding the trap. The vibrations are due to the operation of a rotary pump in the laboratory. For details see Ref. 54.

[59,62,63]. All these aspects are a consequence of a pump process whose complex nature is not accounted for in the standard treatment of the laser. So far there is one experiment where laser action could be demonstrated with one atom at a time in the optical resonator [6]. A weak beam of excited atoms was used to pump this one-atom laser.

A formidable challenge for an experiment is to perform a similar experiment with a trapped ion in the cavity. Mirrors with an ultrahigh finesse are required, and a strong atom–field coupling is needed. After the emission of a photon, the ion has to be pumped before the next stimulated emission can occur. Similar to what occurs in the resonance fluorescence experiments that show antibunching [45,49], there is a certain time gap during which the ion is unable to add another photon to the laser field. It has been shown [59] that this time gap plays a significant role in the production of a field with sub-Poissonian photon statistics.

We have investigated the theoretical basis for an experimental realization of the ion-trap laser. Our analysis takes into account details such as the multilevel structure, the coupling strengths, and the parameters of the resonator. It has been a problem to find an ion with an appropriate level scheme. We could show that it is possible to produce a laser field with the parameters of a single Ca^+ ion. This one-atom laser displays several features, which are not found in conventional lasers: the development of two thresholds, sub-Poissonian statistics, lasing without inversion, and self-quenching. The details of this work are reported in Refs. 64 and 65. In a subsequent paper [66], also the center-of-mass motion of the trapped ion was quantized. This leads to additional features of the ion trap laser, especially a multiple vacuum Rabi-splitting is observed.

The Ca^+ scheme is sketched in Fig. 14a. It contains a Λ-type subsystem: The ion is pumped coherently from the ground state to the upper laser level $4P_{1/2}$, and stimulated emission into the resonator mode takes place on the transition to $3D_{3/2}$ at a wavelength of 866 nm. Further pump fields are needed to close the pump cycle and to depopulate the metastable levels.

Although spontaneous relaxation from the upper laser level to the ground state takes place at a relatively large rate of 140 MHz and suppresses the atomic polarization on the laser transition, laser light is generated for realistic experimental parameters due to atomic coherence effects within the Λ subsystem. The occurrence of laser action is demonstrated in Fig. 14b for a resonator with a photon damping rate $A = 1$ MHz and a vacuum Rabi frequency $g = 14.8$ MHz on the laser transition. For the numerical calculation of the realistic scheme, the Zeeman substructure and the polarizations of the fields have to be taken into account. With increasing coherent pump Ω, the mean photon number inside the resonator first increases and then decreases. Both the increase and decease of the intensity are accompanied by maxima in the intensity fluctuations, which can be interpreted as thresholds. Laser action takes place in between these two thresholds. This is confirmed by the Poissonian-like photon distribution given in the inset of Fig. 14b. In addition, the linewidth of the output spectrum is in the laser region up to 10 times smaller than below the first and beyond the second threshold [65]. Note that for a thermal distribution the solid and dashed curves in Fig. 14b for the intensity and the intensity fluctuations would coincide.

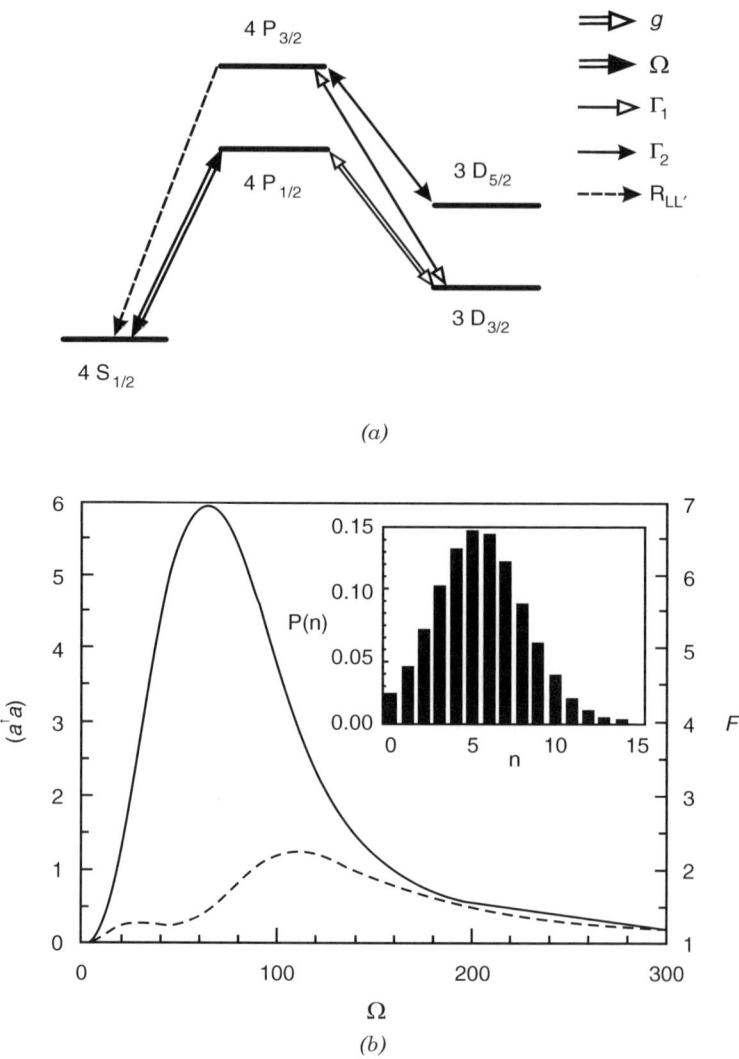

Figure 14. (a) Schematic representation of the Ca$^+$ scheme for the ion-trap laser. (b) Mean photon number $\langle a^\dagger a \rangle$ (*solid curve*) and Fano factor F (*dashed curve*) versus the coherent pump strength Ω. The parameters are $A = 1$, $g = 14.8$, $\Gamma_1 = 40$, and $\Gamma_2 = 100$. The inset shows the photon distribution for $\Omega = 50$. All rates are in MHz.

For a nonvanishing Lamb–Dicke parameter η, higher vibrational states will be excited during the pump and relaxation processes; the amplitude of the atomic motion will increase. Therefore, the ion will in general not remain at an antinode of the resonator mode, and the strength of the atom–field coupling will decrease. However, the atom can be prevented from heating up by detuning a coherent pump field. The coupling strength is given by the product of a constant g_0 depending on the transition probability and a motion-dependent function that is determined by an overlap integral involving the motional wavefunction of the atom and the mode function of the field [66].

In a simple two-level laser model with decay rate R_{AB} and pump rate R_{BA}, the cooling process may be incorporated by coupling the atomic motion to a thermal reservoir with cooling rate B and thermal vibron number μ. In such a simple model, the discrete nature of the quantized motion shows up below threshold in a multiple vacuum Rabi splitting of the output spectrum [66]. This is illustrated in Fig. 15. The pairs of peaks correspond to different vibrational states with different atom–field coupling.

The cooling mechanism is most transparent in the special case of resolved-sideband cooling. The coherent pump may be detuned to the first lower vibrational sideband so that with each excitation from $4S_{1/2}$ to $4P_{1/2}$, one vibron is annihilated and the CM motion is cooled. Eventually, all the population will collect in the motional ground state of the atomic ground state $4S_{1/2}$ and cannot participate in the lasing process. The coherent pump strength is now given by Ω_0 times a motion-dependent function. In order to maintain laser action in the presence of the cooling, an additional broadband pump field Γ may be applied to the cooling transition. Figure 16 indicates that a field with a mean photon

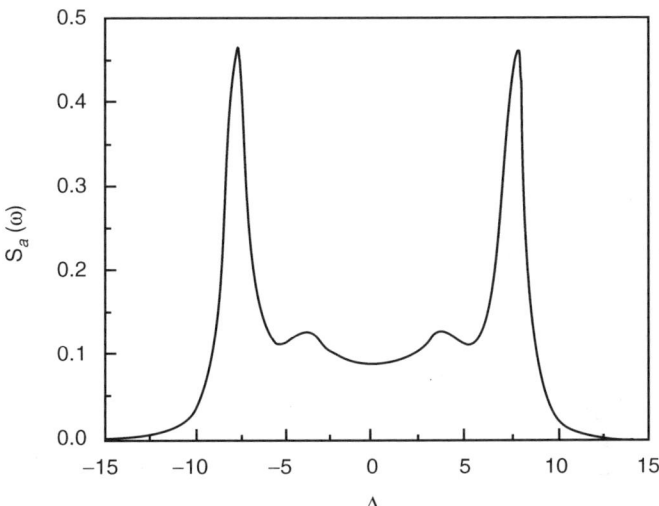

Figure 15. Multiple vacuum Rabi splitting in the output spectrum $S_a(\omega)$ for the two-level atom with quantized CM motion. The parameters are $A = 0.1$, $B = 0.05$, $\mu = 0.5$, $R_{AB} = 0.1$, $R_{BA} = 0.001$, and $\eta = 0.7$. All rates are in units of g_0.

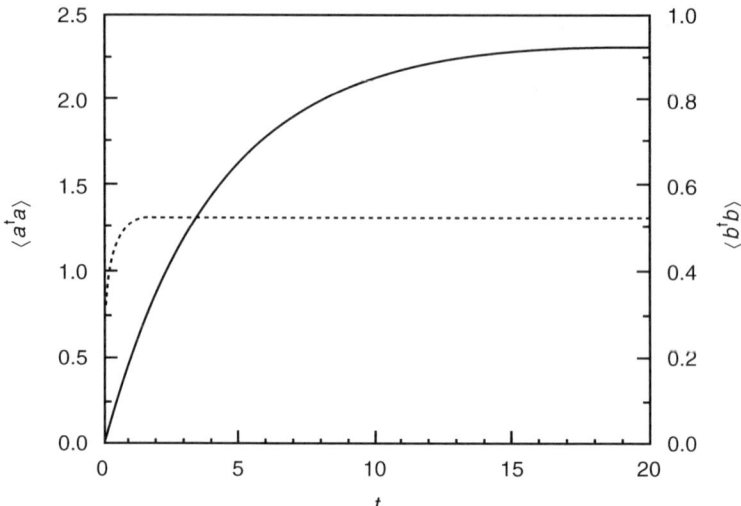

Figure 16. Time evolution of the mean photon number (*solid curve*) and the mean vibron number (*dashed curve*) in the Ca$^+$ ion-trap laser with sideband cooling. The parameters are $A = 0.5$, $g_0 = 14.8$, $\Omega_0 = 100$, $\Gamma = \Gamma_1 = 40$, $\Gamma_2 = 100$, and $\eta = 0.1$ on the laser transition. Initially, the atom is in the ground state and the vibronic distribution is thermal with $\langle b^\dagger b \rangle = 0.1$. All rates are in MHz.

number $\langle a^\dagger a \rangle = 2.3$ is generated while the mean vibron number is restricted to a value of $\langle b^\dagger b \rangle = 0.5$. If a larger mean vibron number is acceptable, the pump rate Γ can be increased and more population takes part in the laser action. This leads to considerably larger mean photon numbers. The calculation shows that it is possible to incorporate a cooling mechanism in a multilevel one-atom laser scheme and to obtain significant lasing also for nonperfect localization of the atom. Although it is difficult to reach the resolved-sideband limit in an experiment, cooling may still be achieved in the weak-binding regime by detuning a coherent pump field.

IV. CONCLUSIONS

In this chapter, recent experiments with single atoms in cavities and traps are reviewed. It is especially pointed out that using ultracold atoms will lead to new interesting aspects in atom–matter interaction. The possibility that now ultracold atoms are available bring such experiments into reach in the near future.

The quantum-mechanical CM motion of the atoms incident upon a micromaser cavity is equivalent to a scattering problem that involves both a repulsive and an attractive potential. The emission probability for an initially excited ultracold atom exhibits sharp resonances when the de Broglie wavelength fits resonantly into the cavity. These resonances may be observed experimentally with the help of a reentrant cavity. Whereas the eigenstates of the atomic motion are continuously distributed for the maser, the motion is confined to a trapping potential in the one-atom laser. The discrete nature of the CM motion in the trap is reflected below threshold by multiple vacuum Rabi splitting. In order to prevent the atom from being continuously heated by the pump and relaxation processes, sideband cooling has been incorporated into the model. The recently proposed Ca^+ ion-trap laser is used to illustrate the possibility of one-atom lasing in the presence of a cooling mechanism.

There is one very interesting application of the "mazer" which should be briefly mentioned here: The device can act as a filter for matter waves and can thus be used to increase the coherence length of an atomic beam, in the same way that a Fabry–Perot can be used to increase the coherence length of a light wave. This application is discussed in Ref. 67.

References

1. D. Meschede, H. Walther, and G. Müller, *Phys. Rev. Lett.* **54**, 551 (1985).
2. G. Rempe, H. Walther, and N. Klein, *Phys. Rev. Lett.* **58**, 353 (1987).
3. G. Rempe, F. Schmidt-Kaler, and H. Walther, *Phys. Rev. Lett.* **64**, 2783 (1990).
4. G. Rempe and H. Walther, *Phys. Rev. A* **42**, 1650 (1990).
5. H. J. Kimble, O. Carnal, N. Georgiades, H. Mabuchi, E. S. Polzik, R. J. Thompson, and Q. A. Turchette, *Atomic Physics*, Vol. 14, D. J. Wineland, C. E. Wieman, and S. J. Smith, eds., American Insitute of Physics, New York, 1995, pp. 314–335.
6. K. An, J. J. Childs, R. R. Dasari, and M. S. Feld, *Phys. Rev. Lett.* **73**, 3375 (1994).
7. C. Wagner, R. J. Brecha, A. Schenzle, and H. Walther, *Phys. Rev. A* **47**, 5068 (1993).
8. M. Löffler, B.-G. Englert, and H. Walther, *Appl. Phys. B* **63**, 511 (1996).
9. O. Benson, G. Raithel, and H. Walther, *Phys. Rev. Lett.* **72**, 3506 (1994).
10. G. Raithel, C. Wagner, H. Walther, L. M. Narducci, and M. O. Scully, *Advances in Atomic, Molecular, and Optical Physics*, Supplement 2, P. Berman, ed., Academic Press, New York, 1994, pp. 57–121.
11. G. Raithel, O. Benson, and H. Walther, *Phys. Rev. Lett.* **75**, 3446 (1995).
12. P. Filipowicz, J. Javanainen, and P. Meystre, *Phys. Rev. A* **34**, 3077 (1986).
13. L. A. Lugiato, M. O. Scully, and H. Walther, *Phys. Rev. A* **36**, 740 (1987).
14. P. Meystre, *Progress in Optics*, Vol. 30, E. Wolf, ed., Elsevier Science Publishers, New York, 1992, pp. 261–355.
15. J. M. Raimond, M. Brune, L. Davidovich, P. Goy, and S. Haroche, *Atomic Physics* **11**, 441 (1989).
16. N. F. Ramsey, *Molecular Beams*, Clarendon Press, Oxford, 1956, pp. 124–134.
17. C. Cohen-Tannoudji, J. Dupont-Roc, and G. Grynberg, *Atom–Photon Interactions*, John Wiley & Sons, New York, 1992, pp. 407–514.
18. C. Wagner, A. Schenzle, and H. Walther, *Opt. Commun.* **107**, 318 (1994).
19. H. Walther, *Phys. Reports* **219**, 263 (1992).

20. B.-G. Englert, M. Löffler, O. Benson, B. Varcoe, M. Weidinger, and H. Walther, *Fortschr. Phys.* **46**, 897 (1998).
21. H. J. Briegel, B.-G. Englert, N. Sterpi, and H. Walther, *Phys. Rev. A* **49**, 2962 (1994).
22. P. Meystre, G. Rempe, and H. Walther, *Opt. Lett.* **13**, 1078 (1988).
23. M. Weidinger, B. T. H. Varcoe, R. Heerlein, and H. Walther, *Phys. Rev. Lett.* **82**, 3795 (1999).
24. J. Krause, M. O. Scully, and H. Walther, *Phys. Rev. A* **36**, 4547 (1987).
25. P. J. Bardoff, E. Mayr, and W. P. Schleich, *Phys. Rev. A* **51**, 4963 (1995).
26. B. T. H. Varcoe, S. Brattke, M. Weidinger, and H. Walther, *Nature* **403**, 743 (2000).
27. M. Brune, F. Schmidt-Kaler, A. Maali, J. Dreyer, E. Hagley, J. M. Raimond, and S. Haroche, *Phys. Rev. Lett.* **76**, 1800 (1996).
28. D. M. Greenberger, M. Horne, and A. Zcilinger, *Bell's Theorem, Quantum Theory, and Conceptions of the Universe*, M. Kafatos, ed., Kluwer, Dordrecht, 1989.
29. D. M. Greenberger, M. Horne, M. Shimony, and A. Zeilinger, *Am. J. Phys.* **58**, 1131 (1990).
30. B.-G. Englert, N. Sterpi, and H. Walther, *Opt. Commun.* **100**, 526 (1993).
31. N. D. Mermin, *Physics Today* **43**(6), 9 (1990).
32. M. O. Scully, G. M. Meyer, and H. Walther, *Phys. Rev. Lett.* **76**, 4144 (1996).
33. G. M. Meyer, M. O. Scully, and H. Walther, *Phys. Rev. A* **56**, 4142 (1997).
34. M. Löffler, G. M. Meyer, M. Schröder, M. O Scully, and H. Walther, *Phys. Rev. A* **56**, 4153 (1997).
35. M. Schröder, K. Vogel, W. P. Schleich, M. O. Scully, and H. Walther, *Phys. Rev. A* **57**, 4164 (1997).
36. W. Hartig, W. Rasmussen, R. Schieder, and H. Walther, *Z. Phys. A* **278**, 205 (1976).
37. J. D. Cresser, J. Häger, G. Leuchs, F. M. Rateike, and H. Walther, *Top. Curr. Phys.* **27**, 21 (1982).
38. P. S. Jessen, C. Gerz, P. D. Lett, W. D. Philipps, S. L. Rolston, R. J. C. Spreuuw, and C. I. Westbrook, *Phys. Rev. Lett.* **69**, 49 (1992).
39. W. Heitler, *The Quantum Theory of Radiation*, 3rd edition, Oxford University Press, Oxford, England, 1954, pp. 196–204.
40. B. R. Mollow, *Phys. Rev.* **188**, 1969 (1969).
41. F. Schuda, C. Stroud, Jr., and M. Hercher, *J. Phys. B* **1**, L198 (1974).
42. H. Walther, *Lecture Notes in Physics* **43**, 358 (1975).
43. F. Y. Wu, R. E. Grove, and S. Ezekiel, *Phys. Rev. Lett.* **35**, 1426 (1975); R. E. Grove, F. Y. Wu, and S. Ezekiel, *Phys. Rev. Lett. A* **15**, 227 (1977).
44. H. M. Gibbs and T. N. C. Venkatesan, *Opt. Commun.* **17**, 87 (1976).
45. H. J. Kimble, M. Dagenais, and L. Mandel, *Phys. Rev. Lett.* **39**, 691 (1977).
46. E. Jakeman, E. R. Pike, P. N. Pusey, and J. M. Vaugham, *J. Phys. A* **10**, L257 (1977).
47. H. J. Kimble, M. Dagenais, and L. Mandel, *Phys. Rev. A* **18**, 201 (1978); M. Dagenais and L. Mandel, *Phys. Rev. A* **18**, 2217 (1978).
48. F. M. Rateike, G. Leuchs, and H. Walther, results cited in Ref. 27.
49. F. Diedrich and H. Walther, *Phys. Rev. Lett.* **58**, 203 (1987).
50. D. F. Walls, *Nature* **280**, 451 (1979).
51. R. Short and L. Mandel, *Phys. Rev. Lett.* **51**, 384 (1983).
52. C. A. Schrama, E. Peik, W. W. Smith, and H. Walther, *Opt. Commun.* **101**, 32 (1993).
53. J. T. Höffges, H. W. Baldauf, T. Eichler, S. R. Helmfrid, and H. Walther, *Opt. Commun.* **133**, 170 (1997).

54. J. T. Höffges, H. W. Baldauf, W. Lange, and H. Walther, *J. Mod. Opt.* **55**, 1999 (1997).
55. R. Loudon, *Rep. Prog. Phys.* **43**, 913 (1980).
56. Y. Mu and C. M. Savage, *Phys. Rev. A* **46**, 5944 (1992).
57. C. Ginzel, H. J. Briegel, U. Martini, B.-G. Englert, and A. Schenzle, *Phys. Rev. A* **48**, 732 (1993).
58. T. Pellizzari and H. J. Ritsch, *Mod. Opt.* **41**, 609 (1994); *Phys. Rev. Lett.* **72**, 3973 (1994); P. Horak, K. M. Gheri, and H. Ritsch, *Phys. Rev. A* **51**, 3257 (1995).
59. H.-J. Briegel, G. M. Meyer, and B.-G. Englert, *Phys. Rev. A* **53**, 1143 (1996); *Europhys. Lett.* **33**, 515 (1996).
60. M. Löffler, G. M. Meyer, and H. Walther. *Phys. Rev. A* **55**, 3923 (1997).
61. For a recent review see E. Arimondo, *Progress in Optics*, Vol. XXXV, E. Wolf, ed., Elsevier, Amsterdam, 1996, pp. 257–354.
62. A. M. Khazanov, G. A. Koganov, and E. P. Gordov, *Phys. Rev. A* **42**, 3065 (1990); T. C. Ralph and C. M. Savage, *Phys. Rev. A* **44**, 7809 (1991); H. Ritsch, P. Zoller, C. W. Gardiner, and D. F. Walls, *Phys. Rev. A* **44**, 3361 (1991).
63. K. M. Gheri and D. F. Walls, *Phys. Rev. A* **45**, 6675 (1992); H. Ritsch and M. A. M. Marte, *Phys. Rev. A* **47**, 2354 (1993).
64. G. M. Meyer, H.-J. Briegel, and H. Walther, *Europhys. Lett.* **37**, 317 (1997).
65. G. M. Meyer, M. Löffler, and H. Walther, *Phys. Rev. A* **56**, R1099 (1997).
66. M. Löffler, G. M. Meyer, and H. Walther, *Europhys. Lett.* **40**, 263 (1997).
67. M. Löffler and H. Walther, *Europhys. Lett.* **41**, 593 (1998).

QUANTUM SUPERPOSITIONS AND DECOHERENCE: HOW TO DETECT INTERFERENCE OF MACROSCOPICALLY DISTINCT OPTICAL STATES

F. T. ARECCHI

Department of Physics, University of Florence, Florence, Italy; and National Institute of Applied Optics (INOA), Florence, Italy

A. MONTINA

Department of Physics, University of Florence, Florence, Italy

CONTENTS

I. Quantum Superposition and Decoherence
II. Optical Implementation of Mesoscopic Quantum Interference
III. "Which Path" Experiment with a Large Photon Number
Acknowledgment
References

I. QUANTUM SUPERPOSITION AND DECOHERENCE

There is a basic difference between the predictions of quantum theory for quantum systems that are closed (isolated) and open (interacting with their environments.) In the case of a closed system, the Schrödinger equation and the superposition principle apply literally. In contrast, the superposition principle is not valid for open quantum systems. Here the relevant physics is quite different, as has been shown by many examples in the context of condensed matter physics,

---

Dynamical Systems and Irreversibility: A Special Volume of Advances in Chemical Physics, Volume 122, Edited by Ioannis Antoniou. Series Editors I. Prigogine and Stuart A. Rice. ISBN 0-471-22291-7. © 2002 John Wiley & Sons, Inc.

quantum chemistry, and so on. The evolution of open quantum systems has to be described in a way violating the assumption that each state in the Hilbert space of a closed system is equally significant. Decoherence is a negative selection process that dynamically eliminates nonclassic states.

The distinguishing feature of classic systems, the essence of "classic reality," is the persistence of their properties—the ability of systems to exist in predictably evolving states, to follow a trajectory which may be chaotic, but is deterministic. This suggests the relative stability—or, more generally, predictability—of the evolution of quantum states as a criterion that decides whether they will be repeatedly encountered by an observer and can be used as ingredients of a "classic reality." The characteristic feature of the decoherence process is that a generic initial state will be dramatically altered on a characteristic decoherence time scale: Only certain stable states will be left on the scene.

Quantum measurement is a classic example of a situation in which a coupling of a macroscopic quantum apparatus A and a microscopic measured system S forces the composite system into a correlated, but usually exceedingly unstable, state. In a notation where $|A_0\rangle$ is the initial state of the apparatus and $|\psi\rangle$ the initial state of the system, the evolution establishing an A–S correlation is described by

$$|\psi\rangle|A_0\rangle = \sum_k \alpha_k |\sigma_k\rangle|A_0\rangle \rightarrow \sum_k \alpha_k |\sigma_k\rangle|A_k\rangle = |\Phi\rangle \qquad (1)$$

An example is the Stern–Gerlach apparatus. There the states $|\sigma_k\rangle$ describe orientations of the spin, and the states $|A_k\rangle$ are the spatial wavefunctions centered on the trajectories corresponding to different eigenstates of the spin. When the separation of the beams is large, the overlap between them tends to zero ($\langle A_k|A'_k\rangle \sim \delta_{kk'}$). This is a precondition for a good measurement. Moreover, when the apparatus is not consulted, A–S correlations would lead to a mixed density matrix for the system S:

$$\rho_s = \sum_k |\alpha_k|^2 |\sigma_k\rangle\langle\sigma_k| = Tr|\Phi\rangle\langle\Phi| \qquad (2)$$

However, this premeasurement quantum correlation does not provide a sufficient foundation to build a correspondence between the quantum formalism and the familiar classic reality. It only allows for Einstein–Podolsky–Rosen quantum correlations between A and S, which imply the entanglement of an arbitrary state—including nonlocal, nonclassic superpositions of the localized status of the apparatus (observer)—with the corresponding relative state of the other system. This is a prescription for a Schrödinger cat, not a resolution of the measurement problem. What is needed, therefore, is an effective superselection rule that "outlaws" superpositions of these preferred "pointer states." This rule cannot be absolute: There must be a time scale sufficiently short, or an interaction strong enough, to render it invalid, because otherwise measurements could not be performed at all. Superselection should become more effective when the size of the system increases. It should apply, in general, to all objects and allow us to reduce elements of our familiar reality—including the spatial localization of macroscopic system—from Hamiltonians.

Environment-induced decoherence has been proposed to fit these requirements [1]. The transition from a pure state $|\Phi\rangle\langle\Phi|$ to the effectively mixed ρ_{AS} can be accomplished by coupling the apparatus A to the environment ϵ. The requirement to get rid of unwanted, excessive, EPR-like correlations (1) is equivalent to the demand that the correlations between the pointer states of the apparatus and the measured system ought to be preserved in spite of an incessant measurement-like interaction between the apparatus pointer and the environment. In simple models of the apparatus, this can be assured by postulating the existence of a pointer observable with eigenstates (or, more precisely, eigenspaces) that remain unperturbed during the evolution of the open system. This "nondemolition" requirement will be exactly satisfied when the pointer observable O commutes with the total Hamiltonian generating the evolution of the system:

$$[(H + H_{\text{int}}), O] = 0 \tag{3}$$

For an idealized quantum apparatus, this condition can be assumed to be satisfied and—provided that the apparatus is in one of the eigenstates of O—leads to an uneventful evolution:

$$|A_k\rangle|\epsilon_0\rangle \rightarrow |A_k\rangle|\epsilon_k(t)\rangle \tag{4}$$

However, when the initial state is a superposition corresponding to different eigenstates of O, the environment will evolve into an $|A_k\rangle$-dependent state:

$$\left(\sum_k \alpha_k |A_k\rangle\right)|\epsilon_0\rangle \rightarrow \sum_k \alpha_k |A_k\rangle|\epsilon_k(t)\rangle \tag{5}$$

The decay of the interference terms is inevitable. The environment causes decoherence only when the apparatus is forced into a superposition of states, which are distinguished by their effect on the environment. The resulting continuous destruction of the interference between the eigenstates of O leads to an effective environment-induced superselection. Only states which are stable in spite of decoherence can exist long enough to be accessed by an observer so that they can count as elements of our familiar, reliably existing reality.

Effective reduction of the state vector follows immediately. When the environment becomes correlated with the apparatus,

$$|\Phi\rangle|\epsilon_0\rangle \to \sum_k \alpha_k |A_k\rangle |\sigma_k\rangle |\epsilon_k(t)\rangle = |\Psi\rangle \qquad (6)$$

but the apparatus is not consulted (so that it must be traced out), we have

$$\rho_{AS} = \text{Tr}|\Psi\rangle\langle\Psi| = \sum_k |\alpha_k|^2 |A_k\rangle\langle A_k| |\sigma_k\rangle\langle\sigma_k| \qquad (7)$$

Only correlations between the pointer states and the corresponding relative states of the system retain their predictive validity. This form of ρ_{AS} follows, provided that the environment becomes correlated with the set of states $\{|A_k\rangle\}$ (it could be any other set) and that it has acted as a good measuring apparatus, so that $\langle\epsilon_k(t)|\epsilon_{k'}(t)\rangle = \delta_{kk'}$ (the states of the environment and the different outcomes are orthogonal).

Let us consider a system S ruled by a Hamiltonian H_0 and coupled to the environment through the term

$$H' = \nu x E \qquad (8)$$

where ν is the coupling strength, x is a coordinate of the system, and E is an environment operator. As we trace the overall density operator over an ensemble of environments with temperature T, the system's density matrix in the coordinate representation, $\rho(x, x')$, evolves according to the following master equation [2]:

$$\frac{d\rho}{dt} = \frac{1}{i\hbar}[H_0, \rho] - \gamma(x - x')(\partial_x - \partial_{x'})\rho - \eta \frac{kT}{\hbar^2}(x - x')^2 \rho \qquad (9)$$

where $\eta := \nu^2/2$, and $\gamma := \eta/2m$ is the drift coefficient that rules the evolution of the first moments. "Negative selection" consists of the rapid decay of the off-diagonal elements of $\rho(x, x')$. Indeed, for $\hbar \to 0$, the last term on the right-hand side of (9) prevails, providing the solution

$$\rho(x, x', t) = \rho(x, x', 0) \exp\left(-\eta \frac{kT}{\hbar^2}(x - x')^2\right) \qquad (10)$$

With $\Delta x = x - x'$, we see that an initial offset $\rho(x, x', 0)$ decays after a decoherence time

$$\tau_D = \frac{1}{\gamma}\left(\frac{\lambda_{DB}}{\Delta x}\right)^2 \qquad (11)$$

where

$$\lambda_{DB} = \frac{\hbar}{p} = \frac{\hbar}{\sqrt{2mkT}} \qquad (12)$$

is the thermal de Broglie length. At length scales $\Delta x \gg \lambda_{DB}$, we have $\tau_D \ll 1/\tau$, such that the system decoheres rapidly and then continues with the standard Brownian decay on the time scale $1/\gamma$.

II. OPTICAL IMPLEMENTATION OF MESOSCOPIC QUANTUM INTERFERENCE

The possibility of interference between macroscopically distinct states (the so-called Schrödinger cats [3]) has been suggested by Leggett [4–6] for the case of two opposite magnetic flux states associated with a SQUID.

Recently, two experiments on Schrödinger cats have been demonstrated. In the first one [7] the two different states $|\pm \alpha\rangle$ are coherent states of the vibrational motion of a ^{9}Be$^+$ion within a one-dimensional ion trap. The maximum separation reported between the two states corresponds to about $2|\alpha| = 6$. In the second one [8] the two different states are coherent states of a microwave field, with a maximum separation up to about 3.3.

An optical experiment would consist of generating the superposition of two coherent states of an optical field and detecting their interference. Generating a superposition of coherent states requires some nonlinear optical operations, and different proposals have been formulated, based respectively on $\chi^{(3)}$ and $\chi^{(2)}$ nonlinearities. In the first one [9] a coherent state, injected onto a $\chi^{(3)}$ medium, evolves toward the superposition of two coherent states 180° out of phase with each other. However, for all practically available $\chi^{(3)}$ values, the time necessary to generate the superposition state, which scales as $1/\chi^{(3)}$, is always much longer than the decoherence time. We recall that for a superposition $(|\alpha\rangle + |-\alpha\rangle)/\sqrt{2}$ of two coherent states, the decoherence time is given by the damping time of the field, divided by the square distance $(4|\alpha|^2)$ [9].

The second proposal, by Song, Caves, and Yurke (SCY) [10], consists of an optical parametric amplifier (OPA) pumped by a coherent field, generating an entangled state of signal (S) and readout (R) modes. Passing the S mode through a further OPA, and measuring its output field conditioned upon the photon number on the R mode, should yield interference fringes, associated with the coherent superposition of two separate states. However, the fringe visibility is extremely sensitive to the R detector efficiency, and as a result the SCY interference has not been observed so far.

We have recently introduced a modified version of SCY, whereby fringes can still be observed at the efficiencies of currently available detectors [11]. The price to be paid is a very low count rate, which is, however, compensated for by the use of a high-frequency pulsed laser source. Our setup is shown in Fig. 1.

Choosing the back-evasion condition of Ref. 12 it can be shown that the state of the two field modes at the output of the first OPA apparatus is

$$|\psi\rangle = e^{-iT\hat{X}_S \hat{Y}_R}|0,0\rangle \qquad (13)$$

where $T = 2\sinh(r)$, r being proportional to the product of the pump laser amplitude and the nonlinear susceptibility $\chi^{(2)}$ of the parametric amplifier [13], and $\hat{X}_S = (\hat{a}_S + \hat{a}_S^+)/\sqrt{2}$, $\hat{Y}_R = (\hat{a}_R - \hat{a}_R^+)/(i\sqrt{2})$.

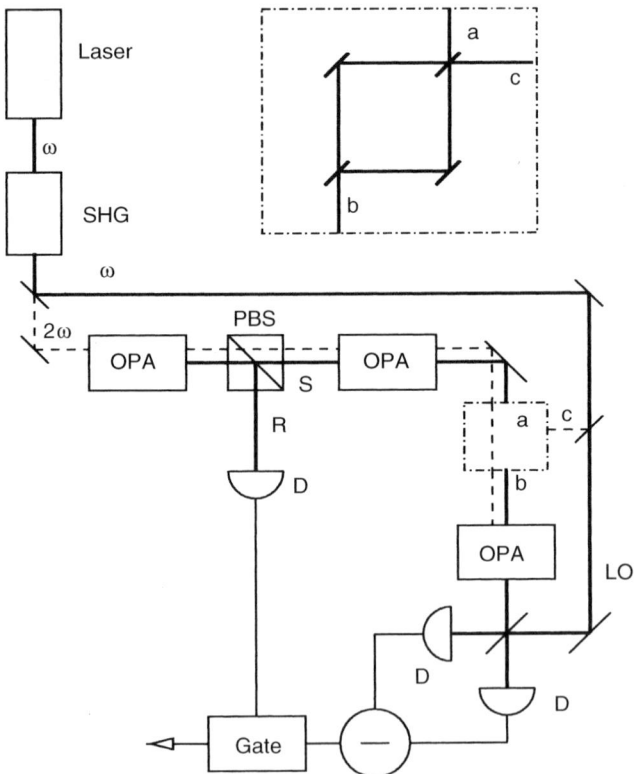

Figure 1. Layout of the proposed experiment: SHG, second harmonic generation; OPA, optical parametric amplifiers (including polarization rotators); PBS, polarizing beam splitter; R, readout channel; S, signal channel; D, detectors; LO, local oscillator for homodyne. The homodyne detection is performed via a balanced scheme. The dashed–dotted box on the S channel (magnified in the inset) denotes the optional insertion of a Mach–Zehnder interferometer with two inputs, a and c, and one output, b. Branch c include a phase adjustment in order to build the superposition state given by Eq. (25). When no interferometer is inserted, a coincides with b.

In the representation of the eigenstates $|x_S, y_R\rangle$ of $\hat{X}_S$ e $\hat{Y}_R$, $|\psi\rangle$ is written as

$$\psi(x_S, y_R) = \langle x_S, y_R|\psi\rangle = e^{-iTx_Sy_R}\psi_o(x_S, y_R) \qquad (14)$$

where $\psi_o(x_S, y_R)$ is the wavefunction of the vacuum state.

By a photodetection measurement on mode R, we obtain the (not normalized) state ψ_n^S of mode S conditioned upon the photon number n on R, that is,

$$\psi_n^S(x_S) = \int_{-\infty}^{\infty} dy_R \psi(x_S, y_R)\psi_n^*(y_R) \qquad (15)$$

where n is the photon number detected on R and $\psi_n(y_R)$ is the number state $|n\rangle$ in the y_R representation of the R mode.

The integral

$$P(n) = \int_{-\infty}^{\infty} |\psi_n^S(x)|^2 \, dx \qquad (16)$$

gives the probability that n photons are in mode R.

The probability distribution of x_s, conditioned on the photon number in mode R, is [13]

$$P(x_S|n) = \frac{|\psi_n^S(x_S)|^2}{P(n)}$$
$$= \frac{(2n)!!(1 + T^2/2)^{(2n+1)/2}}{\pi^{1/2}n!(2n-1)!!} x_S^{2n} e^{-(1+T^2/2)x_S^2} \qquad (17)$$

Both the dependence of $P(n)$ upon n [Eq. (16)] and the dependence of $P(x|n)$ upon x [Eq. (17)] for n between 0 and 10 have been visualized in Figs. 1 and 2 of Ref. 11.

For $n > 0$ the conditional probability (17) is approximated by the sum of two Gaussians whose distance increases with n. The width of each of the two peaks is smaller than that corresponding to a coherent state. SCY suggested to increase the peak separation by passing the S signal through a degenerate OPA, described by the evolution operator

$$U_1(r_1) = e^{-r_1(\hat{a}_S\hat{a}_S - \hat{a}_S^+\hat{a}_S^+)} \qquad (18)$$

The output of this second OPA consists of the superposition of two near-coherent states.

When measuring the quadrature Y_S at the output of the second OPA for a fixed photon number n detected on the R channel, interference fringes should appear as a result of the superposition.

The probability distributions $P(y_S|n)$ of Y_S for n that goes from 0 to 10 and for $T = 3$ are reported in Fig. 5 of Ref. 11. Of course, if we sum up several of them with their weights $P(n)$, the interference fringes cancel out. From this fact, it is easily understood how critical the quantum efficiency of the R photodetector is.

Let us suppose that the R photodetector has an efficiency $\eta_R < 1$. For the time being, we refer to a single photomultiplier detector. Selecting the laser wavelength, the quantum efficiency of the photocatode can be $\eta_R = 0.05$. If n photons impinge on it, the probability of detecting m photons is given by the binomial distribution

$$P(m|n) = \binom{n}{m} \eta_R^m (1 - \eta_R)^{n-m} \qquad (19)$$

Thus, the probability of y_S conditioned by the detection of m photons on R is given by

$$P_\eta(y_S|m) = \sum_{n \geq m} P(y_S|n) P(n|m)$$

$$= \sum_{n \geq m} \frac{P(y_S|n) P(m|n) P(n)}{N(m)} \qquad (20)$$

where $P(n)$ is given by Eq. (16) and the normalization factor in the denominator is $N(m) = \sum_n P(m|n) P(n)$.

The P_η are reported in Fig. 2, using the parameters chosen in [13] ($T = 3$), for some values of the efficiency and for m that goes from 1 to 5. With $\eta_R = 0.7$ the fringes practically disappear, and therefore no superposition is observed.

The last term of Eq. (20), based on Bayes theorem, says that in order to get the distribution of y_S, conditioned by the detection of m photons, we must consider all distributions $P(y_S|n)$ for $n \geq m$, each one weighted by the probability $P(n|m)$ of n photons when m of them have been counted. With the parameters considered in Ref. 13, $P(n|4)$ has the behavior reported in Fig. 3a (we have set $m = 4$). The uncertainty on n implies a reduction of the fringe visibility on y_S.

We aim at reducing the width of the distribution $P(n|m)$, based on the available efficiency of commercial detectors. The only parameter that we can change is the gain T of the first OPA. Reducing the value of gain T, the distribution $P(n)$ decays faster for increasing n. In Fig. 3b we have reported

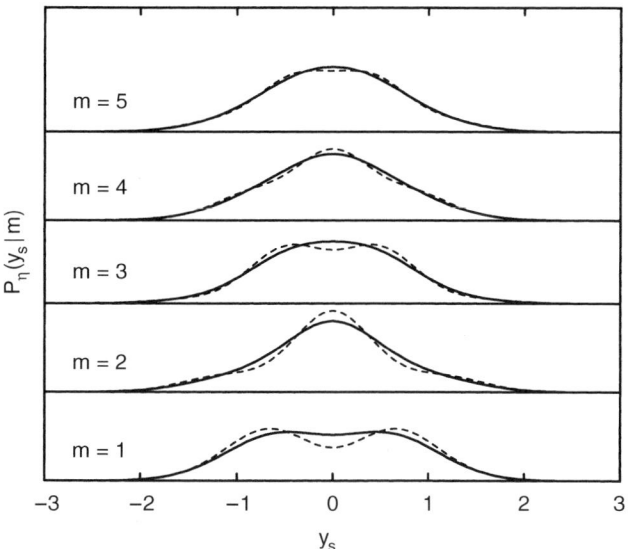

Figure 2. Probability distribution $P_\eta(y_S|m)$ of y_S for $T = 3$ and different efficiencies of the readout detector: $\eta_R = 0.7$ (dashed line), $\eta_R = 0.5$ (solid line).

$P(n|4)$ for $\eta_R = 0.3$ and for $T = 2$, 1, and 0.4. In this last case we note a sharp reduction of the probability for $n > m$; therefore if the detector counts m, there is a very small probability of having $n > m$ photons.

To confirm such a guess, we report in Fig. 4 the distributions $P_\eta(y_S|m)$ for some values of T and for $\eta_R = 0.3$.

The very remarkable fact is that for $T = 0.4$, the fringe visibility is not practically affected by lowering the quantum efficiency. An alternative detection scheme replaces the single photomultiplier with an array of single-photon detectors [14,15]. In such a case the binomial distribution (8) no longer holds, and one should instead recur to Eq. (11) of Ref. 12. This change does not affect the fringe visibility.

Lowering T has no practical influence on the separation of the two near-coherent states at the exit of the second OPA for the same photon number m in mode R.

However, there is a price to pay, indeed: A small T lowers the probability of photon detection on mode R. In Fig. 5 of Ref. 9 we have reported the distribution $N(m)$, for $\eta_R = 0.05$ and $T = 0.4$. $N(4)$ is less than 10^{-10}; thus even if we utilize a pulsed laser with frequency 80 MHz and select $m = 4$, we have less than one favorable event every 100 seconds.

Thus, we must compromise between the fringe visibility and the counting rate.

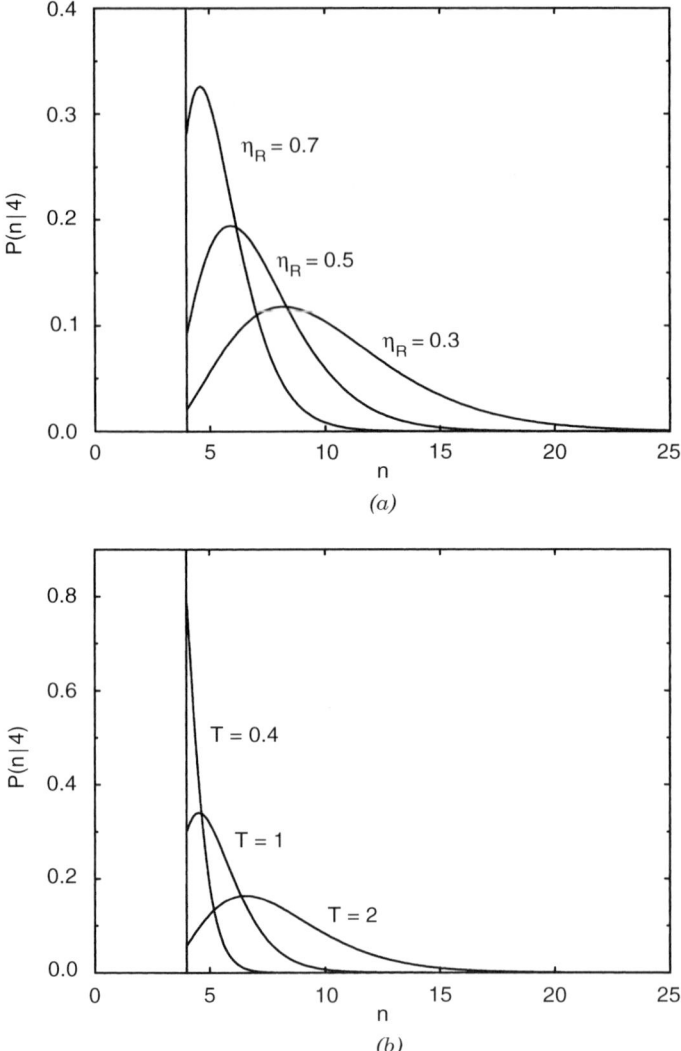

Figure 3. Conditional probability of an impinging photon number n when the detector registers $m = 4$ for (a) different efficiencies η_R at a fixed OPA gain $T = 3$ and (b) different gains T at a fixed efficiency $\eta_R = 0.3$.

A much higher counting rate is obtained by using an array of diodes following the proposal by Paul et al. [14,15]. For four photons, the count rate is now on the order of 10^{-6}, thus yielding 20–80 counts per second with a laser pulsed at an 80-MHz rate.

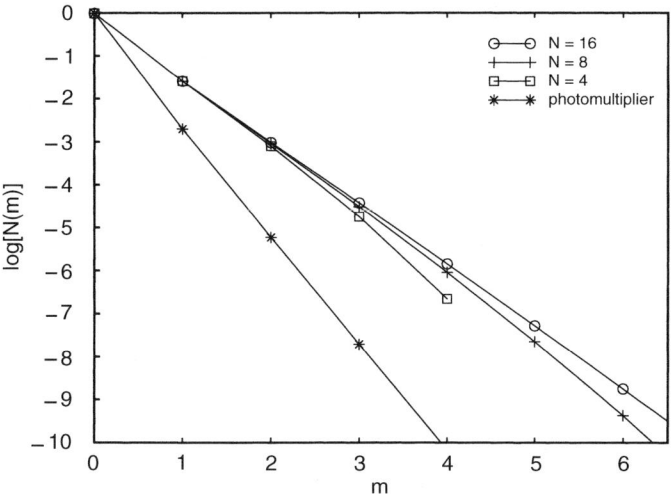

Figure 4. As Fig. 2 but with fixed $\eta_R = 0.3$, $T = 1$ (dashed line), and $T = 0.4$ (solid line).

The Y_S quadrature is measured via a homodyne detector. The mode S is superposed to a reference field at frequency ω, with appropriate phase, and we measure the intensity of the superposition.

The probability that the S detector counts N photons in the resulting field, if its efficiency is 1, is given by

$$P_o(N) = \left| \int_{-\infty}^{+\infty} \langle N|y_S\rangle \psi(y_S - A) \, dy_S \right|^2 \qquad (21)$$

Accounting for the photodetector efficiency $\eta_S < 1$, the count probability becomes

$$P_o^\eta(M) = \sum_{N \geq M} P_o(N) P(M|N) \qquad (22)$$

where

$$P(M|N) = \binom{N}{M} \eta_S^M (1 - \eta_S)^{N-M} \qquad (23)$$

In Fig. 5a we report the distributions $P_o^\eta(M)$ for some values η_S of the homodyne detector efficiency in the case of a superposition of two coherent states of opposite phase with separation $2|\alpha| = 2\sqrt{5}$. For $\eta_S = 0.8$ the fringes are barely visible.

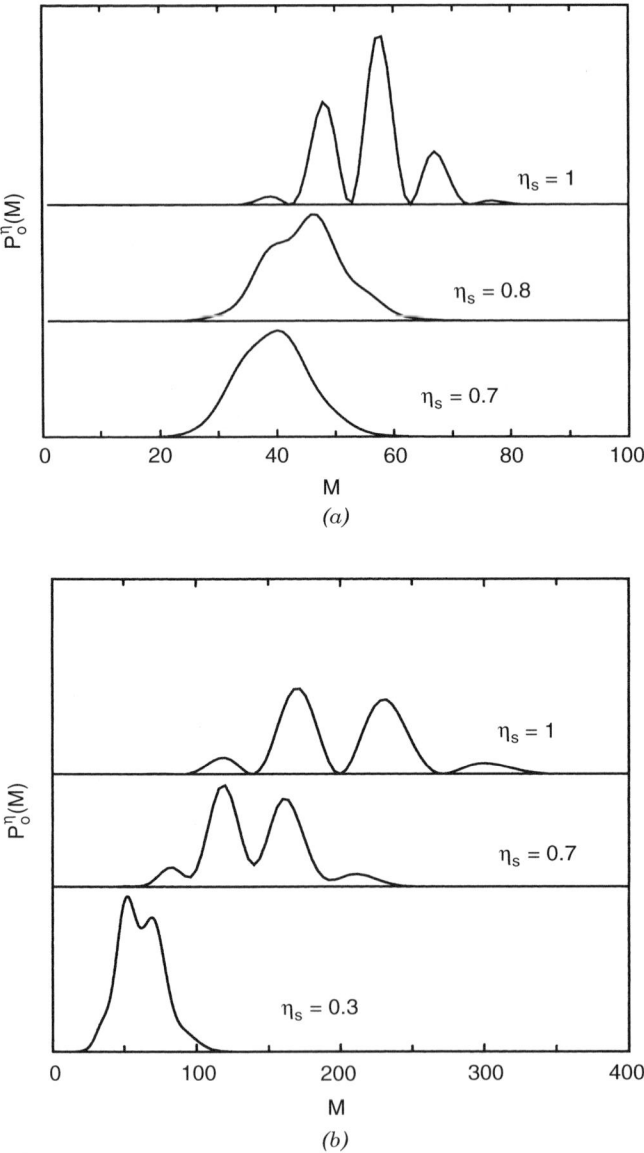

Figure 5. Photocount distribution after the homodyne detection for an input field of two coherent states $|\pm\alpha\rangle$ where $|\alpha|^2 = 5$ for different homodyne detector efficiencies η_S and with a pre-OPA set at different gains l: (a) $l = 1$ (no pre-OPA), (b) $l = 0.3$, (c) $l = 0.15$. The different horizontal scales correspond to different LO intensity for the three cases.

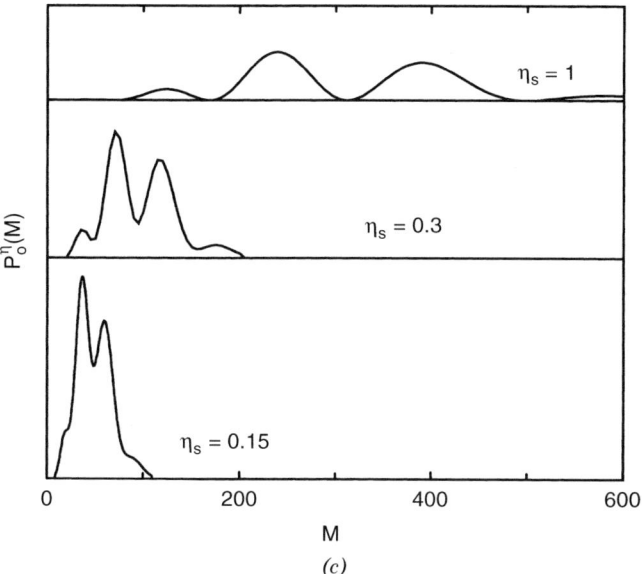

Figure 5. (*Continued*)

We expect that use of PIN diodes should provide a high detection efficiency [16]. However, we devise a way to improve the fringe visibility as if the efficiency were very close to unity.

$P(M|N)$ has the effect of rounding off the distribution $P_o(N)$. By spreading the distribution $P_o(N)$, it becomes less sensitive to this roundoff. This can be done by incorporating in the detection system a pre-OPA whose role consists in separating the fringes.

Indeed, the decoherence rate is proportional to the square root of the distance between the two states in the phase plane; thus an auxiliary OPA before the homodyne, with gain less than unity for the x_S quadrature, reduces the separation and therefore reduces the effect of the losses.

If $l(<1)$ is the shrinking factor for x_S in the auxiliary OPA, then the probability $P_o(N)$ of Eq. (21) changes to

$$P_o(N) = \left| \int_{-\infty}^{+\infty} \langle N|y_S\rangle \psi[l(y_S - A)] \, dy_S \right|^2 \qquad (24)$$

The corresponding distributions $P_o^\eta(M)$ are reported for $l = 0.3$ and 0.15 in Figs. 5b and 5c, respectively, for the case $|\alpha|^2 = 5$. With $l = 0.3$ and $\eta_S = 0.7$ the fringes are well visible, confirming the validity of the proposed strategy.

Notice that Fig. 5 is evaluated for a photon number around 100, for sake of demonstration; in fact, the experiment is carried with a much higher LO intensity.

To summarize, the opposite roles of second and third OPA consist, respectively, of putting the two states of the superposition away and then reapproaching them. This means that a measurement done in the intermediate space region would resolve two widely separate states. The setup here proposed is an optical implementation of the ideal experiment suggested for the same purpose by Wigner in the case of two spin 1/2 particles, by use of two Stern–Gerlach apparatuses [17].

III. "WHICH PATH" EXPERIMENT WITH A LARGE PHOTON NUMBER

The availability of an intermediate spatial region suggests a way of transforming the phase-space separation of the two states of the superposition into a real space separation. Precisely, we might insert a Mach–Zehnder interferometer between second and third OPA. The two inputs of the first beam splitter are fed, respectively, by the superposition state $(|\alpha\rangle + |-\alpha\rangle)/\sqrt{2}$ and by a coherent state $|\gamma\rangle$ with $|\gamma| = |\alpha|$ and adjustable phase. By a suitable choice of this phase, the two separate arms A and B of the interferometer have a field given by the superposition

$$|\beta_1\rangle_A |0\rangle_B + |0\rangle_A |\beta_2\rangle_B \tag{25}$$

where $|\beta_1| = |\beta_2| = \sqrt{2}|\alpha|$.

Thus, a photodetection performed on the two arms of the interferometer would provide a photon number $2|\alpha|^2$ on one arm and 0 on the other or viceversa; however, if no measurement is performed within the interferometer, the homodyne system at the output will detect an interference between the two alternative paths. Adjusting the two interference arm lengths, we recover the input states at interferometer output. This final measurement is a "which path" experiment, upgraded to a packet of $2|\alpha|^2$ photons. So far this experiment had been performed with only one photon, whereas in our setup it is scaled to a large photon number.

The corresponding experiment is being carried at the National Institute of Applied Optics (INOA) in Florence, Italy.

A first run, with an Nd:YAG mode locked laser at $\lambda = 1.06\,\mu m$, was hampered by the low efficiency of available avalanche Si detectors at that wavelength. An improved version, using a Ti:Sa laser at $\lambda = 800\,nm$, provides a much better matching within the peak efficiency of the Si detectors. Preliminary reconstructions of the Wigner function of the superposition state have already tested the soundness of the proposed scheme.

Acknowledgment

This work was partly supported by INFM through the Advanced Research Project CAT.

References

1. Zurek, *Phys. Rev.* **D26**, 1862 (1982); *Phys. Today* **Oct.**, 36–44 (1991) and several comments regarding this article in *Phys. Today* **April**, 13–15, 81–90 (1993).
2. A. O. Caldeira and A. J. Leggett, *Physica A* **121**, 587 (1983).
3. E. Schrödinger, *Naturwissenschaften* **23**, 807 (1935).
4. A. J. Leggett, *Prog. Theor. Phys. Supplement* **69**, 1 (1980).
5. A. J. Leggett, *Proceedings, International Symposium on Foundations of Quantum Mechanics*, Tokyo, pp. 74–82 (1983).
6. A. J. Leggett and A. K. Garg, *Phys. Rev. Lett.* **54**, 857 (1985).
7. C. Monroe, D. M. Meekhof, B. E. King, and D. J. Wineland, *Science* **272**, 1131 (1996).
8. M. Brune, E. Hagley, J. Dreyer, X. Maitre, A. Maali, C. Wunderlich, J. M. Raimond, and S. Haroche, *Phys. Rev. Lett.* **77**, 4887 (1996).
9. B. Yurke and D. Stoler, *Phys. Rev. Lett.* **57**, 13 (1986).
10. S. Song, C. M. Caves, and B. Yurke, *Phys. Rev. A* **41**, 5261 (1990).
11. A. Montina and F. T. Arecchi, *Phys Rev. A* **58**, 3472 (1998).
12. A. La Porta, R. E. Slusher, and B. Yurke, *Phys. Rev. Lett.* **62**, 28 (1989).
13. B. Yurke, W. Schleich, and D. F. Walls, *Phys. Rev. A* **42**, 1703 (1990).
14. H. Paul, P. Torma, T. Kiss, and I. Jex, *Phys. Rev. Lett.* **76**, 2464 (1996).
15. H. Paul, P. Torma, T. Kiss, and I. Jex, *Phys. Rev. A* **56**, 4076 (1997).
16. G. Breitenbach, S. Schiller, and J. Mlynek, *Nature* **387**, 471 (1997).
17. E. P. Wigner, *Am. J. Phys.* **31**, 6 (1963).

QUANTUM DECOHERENCE AND THE GLAUBER DYNAMICS FROM THE STOCHASTIC LIMIT

L. ACCARDI and S. V. KOZYREV

Centro Vito Volterra, Polymathematics, Facolta di Economica, Universita degli Studi di Roma Tor Vergata, Rome, Italy

CONTENTS

I. Introduction
II. The Model and Its Stochastic Limit
III. The Langevin Equation
IV. Dynamics for Generic Systems
V. Control of Coherence
VI. The Glauber Dynamics
VII. Evolution for Subalgebra of Local Operators
Acknowledgments
References

I. INTRODUCTION

In the present chapter we investigate a general model of quantum system interacting with a bosonic reservoir via a Hamiltonian of the form

$$H = H_0 + \lambda H_I$$

where H_0 is called the free Hamiltonian and H_I is called the interaction Hamiltonian.

The stochastic golden rules, which arise in the stochastic limit of quantum theory as natural generalizations of Fermi golden rule [1,2], provide a natural tool to associate a stochastic flow, driven by a white noise (stochastic Schrödinger) equation, with any discrete system interacting with a quantum

---

Dynamical Systems and Irreversibility: A Special Volume of Advances in Chemical Physics,
Volume 122, Edited by Ioannis Antoniou. Series Editors I. Prigogine and Stuart A. Rice.
ISBN 0-471-22291-7. © 2002 John Wiley & Sons, Inc.

field. This white noise Hamiltonian equation, when put in normal order, becomes equivalent to a quantum stochastic differential equation. The Langevin (stochastic Heisenberg) and master equations are deduced from this white noise equation by means of standard procedures that are described in Ref. 1.

We use these equations to investigate the decoherence in quantum systems.

In Ref. 3, extending previous results obtained with perturbative techniques in Ref. 9, it was shown in the example of the spin–boson Hamiltonian that the decoherence in quantum systems can be controlled by the following constants (c.f. Section 2 for the definition of the quantities involved):

$$\mathrm{Re}(g|g)_\omega^+ = \int dk\, |g(k)|^2 2\pi\delta(\omega(k) - \omega) N(k)$$

In this chapter we extend the approach of Ref. 3 from two-level systems to arbitrary quantum systems with discrete spectrum. Our results show that the stochastic limit technique gives us an effective method to control quantum decoherence.

We find that in the above-mentioned interaction, all the off-diagonal matrix elements, of the density matrix of a generic discrete quantum system, will decay exponentially if $\mathrm{Re}(g|g)$ are nonzero. In other words, we obtain the asymptotic diagonalization of the density matrix.

Moreover, we show that for the generic quantum system the off-diagonal elements of the density matrix decay exponentially as $\exp(-N\,\mathrm{Re}(g|g)t)$, with the exponent proportional to the number N of particles in the system. Therefore for generic macroscopic (large N) systems the quantum state will collapse into the classic state very quickly. This effect was built in *by hands* in several phenomenological models of the quantum measurement process. In the stochastic limit approach it is deduced from the Hamiltonian model.

This observation contributes to the clarification of one of the old problems of quantum theory: Why do macroscopic systems usually behave classically? That is, why do we observe classic states although the evolution of the system is a unitary operator described by the Schrödinger equation?

The quantum Markov semigroup we obtain leaves invariant the algebra generated by the spectral projections of the system Hamiltonian; and the associated master equation, when restricted to the diagonal part of the density matrix, takes the form of a standard classic kinetic equation. This master equation describes the convergence to equilibrium (Gibbs state) of the system, coupled with the given reservoir (quantum field).

Summing up: The convergence to equilibrium is a result of quantum decoherence.

If we can control the interaction so that some of the constants $\mathrm{Re}(g|g)$ are zero, then the corresponding matrix elements will not decay in the stochastic approximation—that is, in a time scale that is extremely long with respect to the *slow clock* of the discrete system. In this sense the stochastic limit approach provides a method for controlling quantum coherence.

The general idea of the stochastic limit (see Ref. 1) is to make the time rescaling $t \to t/\lambda^2$ in the solution of the Schrödinger (or Heisenberg) equation in the interaction picture $U_t^{(\lambda)} = e^{itH_0}e^{-itH}$, associated to the Hamiltonian H, that is,

$$\frac{\partial}{\partial t} U_t^{(\lambda)} = -i\lambda H_I(t) U_t^{(\lambda)}$$

with $H_I(t) = e^{itH_0} H_I e^{-itH_0}$. This gives the rescaled equation

$$\frac{\partial}{\partial t} U_{t/\lambda^2}^{(\lambda)} = -\frac{i}{\lambda} H_I(t/\lambda^2) U_{t/\lambda^2}^{(\lambda)} \quad (1)$$

and one wants to study the limits, in a topology to be specified,

$$\lim_{\lambda \to 0} U_{t/\lambda^2}^{(\lambda)} = U_t, \qquad \lim_{\lambda \to 0} \frac{1}{\lambda} H_I\left(\frac{t}{\lambda^2}\right) = H_t \quad (2)$$

The limit $\lambda \to 0$ after the rescaling $t \to t/\lambda^2$ is equivalent to the simultaneous limit $\lambda \to 0$, $t \to \infty$ under the condition that $\lambda^2 t$ tends to a constant (interpreted as a new slow time scale). This limit captures the dominating contributions to the dynamics, in a regime of long times and small coupling, arising from the cumulative effects, on a large time scale, of small interactions ($\lambda \to 0$). The physical idea is that, when observed from the slow time scale of the atom, the field looks like a very chaotic object: a *quantum white noise*, that is, a δ-correlated (in time) quantum field $b^*(t,k)$, $b(t,k)$ also called a *master field*.

The structure of the present chapter is as follows.

In Section II we introduce the model and consider its stochastic limit.

In Section III we derive the Langevin equation.

In Section IV we derive the master equation for the density matrix and show that for nonzero decoherence the master equation describes the collapse of the density matrix to the classic Gibbs distribution and discusses the connection of this fact with the procedure of quantum measurement.

In Section V, using the characterization of quantum decoherence obtained in Section IV and generalizing arguments of Ref. 3, we find that our general model exhibits macroscopic quantum effects (in particular, conservation of quantum coherence). These effects are controllable by the state of the reservoir (that can be controlled by filtering).

In Section VI we apply our general scheme to the model of a quantum system of spins interacting with a bosonic field and derive a quantum extension of the Glauber dynamics.

Thus our stochastic limit approach provides a microscopic interpretation, in terms of fundamental Hamiltonian models, of the dynamics of quantum spin systems. Moreover, we deduce the full stochastic equation and not only the master equation. This is new even in the case of classic spin systems.

II. THE MODEL AND ITS STOCHASTIC LIMIT

In the present paper we consider a general model, describing the interaction of a system S with a reservoir, represented by a bosonic quantum field. Particular cases of this general model were investigated in [3–5]. The total Hamiltonian is

$$H = H_0 + \lambda H_I = H_S + H_R + \lambda H_I$$

where H_R is the free Hamiltonian of a bosonic reservoir R:

$$H_R = \int \omega(k) a^*(k) a(k) \, dk$$

acting in the representation space $\mathscr{F}$ corresponding to the state $\langle \cdot \rangle$ of bosonic reservoir generated by the density matrix $\mathbf{N}$ that we take in the algebra of spectral projections of the reservoir Hamiltonian. The reference state $\langle \cdot \rangle$ of the field is a mean zero gauge invariant Gaussian state, characterized by the second-order correlation function equal to

$$\langle a(k) a^*(k') \rangle = (N(k) + 1) \delta(k - k')$$
$$\langle a^*(k) a(k') \rangle = N(k) \delta(k - k')$$

where the function $N(k)$ describes the density of bosons with frequency k. One of the examples is the (Gaussian) bosonic equilibrium state at temperature β^{-1}.

The system Hamiltonian has the following spectral decomposition:

$$H_S = \sum_r \varepsilon_r P_{\varepsilon_r}$$

where the index r labels the spectral projections of H_S. For example, for a nondegenerate eigenvalue ε_r of H_S the corresponding spectral projection is

$$P_{\varepsilon_r} = |\varepsilon_r\rangle \langle \varepsilon_r|$$

where $|\varepsilon_r\rangle$ is the corresponding eigenvector.

The interaction Hamiltonian H_I (acting in $\mathscr{H}_S \otimes \mathscr{F}$) has the form

$$H_I = \sum_j (D_j^* \otimes A(g_j) + D_j \otimes A^*(g_j)), \qquad A(g) = \int dk \bar{g}(k) a(k)$$

where $A(g)$ is a smeared quantum field with cutoff function (form factor) $g(k)$. To perform the construction of the stochastic limit, one needs to calculate the free evolution of the interaction Hamiltonian: $H_I(t) = e^{itH_0} H_I e^{-itH_0}$.

Using the identity

$$1 = \sum_r P_{\varepsilon_r}$$

we write the interaction Hamiltonian in the form

$$H_I = \sum_j \sum_{rr'} P_{\varepsilon_r} D_j^* P_{\varepsilon_{r'}} \int dk \bar{g}_j(k) a(k) + \text{h.c.} \qquad (3)$$

Let us introduce the set of energy differences (Bohr frequencies)

$$F = \{\omega = \varepsilon_r - \varepsilon_{r'} : \varepsilon_r, \varepsilon_{r'} \in \text{Spec } H_S\}$$

and the set of all energies of the form

$$F_\omega = \{\varepsilon_r : \exists \varepsilon_{r'} (\varepsilon_r, \varepsilon_{r'} \in \text{Spec } H_S) \text{ such that } \varepsilon_r - \varepsilon_{r'} = \omega\}$$

With these notations we rewrite the interaction Hamiltonian (3) in the form

$$H_I = \sum_j \sum_{\omega \in F} \sum_{\varepsilon_r \in F_\omega} P_{\varepsilon_r} D_j^* P_{\varepsilon_r - \omega} \int dk \bar{g}_j(k) a(k) + \text{h.c.}$$

$$= \sum_j \sum_{\omega \in F} E_\omega^*(D_j) \int dk \bar{g}_j(k) a(k) + \text{h.c.} \qquad (4)$$

where

$$E_\omega(X) := \sum_{\varepsilon_r \in F_\omega} P_{\varepsilon_r - \omega} X P_{\varepsilon_r} \qquad (5)$$

It is easy to see that the free volution of $E_\omega(X)$ is

$$e^{itH_S} E_\omega(X) e^{-itH_S} = e^{-it\omega} E_\omega(X)$$

Using the formula for the free evolution of bosonic fields

$$e^{itH_R}a(k)e^{-itH_R} = e^{-it\omega(k)}a(k)$$

we obtain the following for the free evolution of the interaction Hamiltonian:

$$H_I(t) = \sum_j \sum_{\omega \in F} E^*_\omega(D_j) \int dk \bar{g}_j(k) e^{-it(\omega(k)-\omega)} a(k) + \text{h.c.} \quad (6)$$

In the stochastic limit the field $H_I(t)$ gives rise to a family of quantum white noises, or master fields. To investigate these noises, let us suppose the following:

1. $\omega(k) \geq 0, \forall k$.
2. The $(d-1)$-dimensional Lebesgue measure of the surface $\{k : \omega(k) = 0\}$ is equal to zero (so that $\delta(\omega(k)) = 0$) (e.g., $\omega(k) = k^2 + m$ with $m \geq 0$).

Now let us investigate the limit of $H_I(t/\lambda^2)$ using one of the basic formulae of the stochastic limit:

$$\lim_{\lambda \to 0} \frac{1}{\lambda^2} \exp\left(\frac{it}{\lambda^2} f(k)\right) = 2\pi \delta(t) \delta(f(k)) \quad (7)$$

which shows that the term $\delta(f(k))$ in (7) is not identically equal to zero only if $f(k) = 0$ for some k in a set of nonzero $(d-1)$-dimensional Lebesgue measure. This explains condition (2) above.

The rescaled interaction Hamiltonian is expressed in terms of the rescaled creation and annihilation operators

$$a_{\lambda,\omega}(t,k) = \frac{1}{\lambda} e^{-i\frac{t}{\lambda^2}(\omega(k)-\omega)} a(k), \qquad \omega \in F$$

After the stochastic limit, every rescaled annihilation operator corresponding to any transition from $\varepsilon_{r'}$ to ε_r with the frequency $\omega = \varepsilon_r - \varepsilon_{r'}$ generates nontrivial quantum white noise

$$b_\omega(t,k) = \lim_{\lambda \to 0} a_{\lambda,\omega}(t,k) = \lim_{\lambda \to 0} \frac{1}{\lambda} e^{-i\frac{t}{\lambda^2}(\omega(k)-\omega)} a(k)$$

with the relations

$$[b_\omega(t,k), b^*_\omega(t',k')] = \lim_{\lambda \to 0} [a_{\lambda,\omega}(t,k), a^*_{\lambda,\omega}(t',k')]$$

$$= \lim_{\lambda \to 0} \frac{1}{\lambda^2} e^{-i\frac{t-t'}{\lambda^2}(\omega(k)-\omega)} \delta(k-k') \quad (8)$$

$$= 2\pi \delta(t-t') \delta(\omega(k)-\omega) \delta(k-k')$$

$$[b_\omega(t,k), b^*_{\omega'}(t',k')] = 0$$

[c.f. (7)]. This shows, in particular that quantum white noises, corresponding to different Bohr frequencies, are mutually independent.

The stochastic limit of the interaction Hamiltonian is therefore equal to

$$h(t) = \sum_j \sum_{\omega \in F} E_\omega^*(D_j) \int dk \bar{g}_j(k) b_\omega(t, k) + \text{h.c.} \tag{9}$$

The state on the master field (white noise) $b_\omega(t, k)$, corresponding to our choice of the initial state of the field, is the mean zero gauge invariant Gaussian state with correlations:

$$\langle b_\omega^*(t, k) b_\omega(t', k') \rangle = 2\pi \delta(t - t') \delta(\omega(k) - \omega) \delta(k - k') N(k)$$

$$\langle b_\omega(t, k) b_\omega^*(t', k') \rangle = 2\pi \delta(t - t') \delta(\omega(k) - \omega) \delta(k - k') (N(k) + 1)$$

and vanishes for noises corresponding to different Bohr frequences.

Now let us investigate the evolution equation in the interaction picture for our model. According to the general scheme of the stochastic limit, we obtain the (singular) white noise equation

$$\frac{d}{dt} U_t = -ih(t) U_t \tag{10}$$

whose normally ordered form is the quantum stochastic differential equation [6]

$$dU_t = (-idH(t) - G dt) U_t \tag{11}$$

where $h(t)$ is the white noise Hamiltonian (9) given by the stochastic limit of the interaction Hamiltonian and

$$dH(t) = \sum_j \sum_{\omega \in F} (E_\omega^*(D_j) dB_{j\omega}(t) + E_\omega(D_j) dB_{j\omega}^*(t)) \tag{12}$$

$$dB_{j\omega}(t) = \int dk \bar{g}_j(k) \int_t^{t+dt} b_\omega(\tau, k) \, d\tau \tag{13}$$

According to the stochastic golden rule (11) the limit dynamical equation is obtained as follows: The first term in (11) is just the limit of the iterated series solution for (1)

$$\lim_{\lambda \to 0} \frac{1}{\lambda} \int_t^{t+dt} H_I\left(\frac{\tau}{\lambda^2}\right) d\tau$$

The second term, Gdt, called the drift, is equal to the limit of the expectation value in the reservoir state of the second term in the iterated series solution for (1)

$$\lim_{\lambda \to 0} \frac{1}{\lambda^2} \int_t^{t+dt} dt_1 \int_t^{t_1} dt_2 \left\langle H_I\left(\frac{t_1}{\lambda^2}\right) H_I\left(\frac{t_2}{\lambda^2}\right) \right\rangle$$

Making in this formula the change of variables $\tau = t_2 - t_1$ we get

$$\lim_{\lambda \to 0} \frac{1}{\lambda^2} \int_t^{t+dt} dt_1 \int_{t-t_1}^{0} d\tau \left\langle H_I\left(\frac{t_1}{\lambda^2}\right) H_I\left(\frac{t_1}{\lambda^2} + \frac{\tau}{\lambda^2}\right) \right\rangle \qquad (14)$$

Computing the expectation value and using the fact that the limits of oscillating factors of the form $\lim_{\lambda \to 0} e^{ict_1/\lambda^2}$ vanish unless the constant c is equal to zero, we see that we can have a nonzero limit only when all oscillating factors of a kind e^{ict_1/λ^2} (with t_1) in (14) cancel. In conclusion we obtain

$$G = \sum_{ij} \sum_{\omega \in F} \int_{-\infty}^{0} d\tau \left(\int dk \, \overline{g_i(k)} g_j(k) e^{i\tau(\omega(k)-\omega)} (N(k)+1) E_\omega^*(D_i) E_\omega(D_j) \right.$$

$$\left. + \int dk \, g_i(k) \overline{g_j(k)} e^{-i\tau(\omega(k)-\omega)} N(k) E_\omega(D_i) E_\omega^*(D_j) \right)$$

and therefore from the formula

$$\int_{-\infty}^{0} e^{it\omega} dt = \frac{-i}{\omega - i0} = \pi \delta(\omega) - i\, \text{P.P.} \frac{1}{\omega} \qquad (15)$$

we obtain the following expression for the drift G:

$$\sum_{ij} \sum_{\omega \in F} \left(\int dk \, \overline{g_i(k)} g_j(k) \frac{-i(N(k)+1)}{\omega(k) - \omega - i0} E_\omega^*(D_i) E_\omega(D_j) \right.$$

$$\left. + \int dk \, g_i(k) \overline{g_j(k)} \frac{iN(k)}{\omega(k) - \omega + i0} E_\omega(D_i) E_\omega^*(D_j) \right)$$

$$= \sum_{ij} \sum_{\omega \in F} ((g_i|g_j)_\omega^- E_\omega^*(D_i) E_\omega(D_j) + \overline{(g_i|g_j)_\omega^+} E_\omega(D_i) E_\omega^*(D_j)) \qquad (16)$$

Let us note that for (16) we have the following Cheshire Cat effect found in Ref. 3: Even if the frequency ω is negative and therefore does not generate a quantum white noise, the corresponding values $(g|g)_\omega^\pm$ in (16) will be nonzero. In other words, negative Bohr frequencies contribute to an energy shift in the system, but not to its damping.

Remark. If F is any subset of Spec H_S and X_r are arbitrary bounded operators on $\mathcal{H}_S$, then for any $t \in R$ we have

$$e^{itH_S} \sum_{\varepsilon_r \in F} P_{\varepsilon_r} X_r P_{\varepsilon_r} = \sum_{\varepsilon_r \in F} e^{it\varepsilon_r} P_{\varepsilon_r} X_r P_{\varepsilon_r} = \sum_{\varepsilon_r \in F} P_{\varepsilon_r} X_r P_{\varepsilon_r} e^{itH_S}$$

In other words, $\sum_{\varepsilon_r \in F} P_{\varepsilon_r} X_r P_{\varepsilon_r}$ belongs to the commutant $L^\infty(H_S)'$ of the abelian algebra $L^\infty(H_S)$, generated by the spectral projections of H_S.

A corollary of this remark is that, for each $\omega \in F$, for any bounded operator $X \in L^\infty(H_S)'$ and for each pair of indices (i, j) the operators

$$E_\omega(D_i) X E_\omega^*(D_j), \qquad E_\omega^*(D_i) X E_\omega(D_j) \tag{17}$$

belong to the commutant $L^\infty(H_S)'$ of $L^\infty(H_S)$. In particular, if H_S has nondegenerate spectrum so that $L^\infty(H_S)$ is a maximal abelian subalgebra of $B(\mathcal{H}_S)$, the operators (17) also belong to $L^\infty(H_S)$.

III. THE LANGEVIN EQUATION

Now we will find the Langevin equation, which is the limit of the Heisenberg evolution, in interaction representation. Let X be an observable. The Langevin equation is the equation satisfied by the stochastic flow j_t, defined by

$$j_t(X) = U_t^* X U_t$$

where U_t satisfies Eq. (11) in the previous section, that is,

$$dU_t = (-idH(t) - Gdt)U_t \tag{18}$$

To derive the Langevin equation we consider

$$dj_t(X) = j_{t+dt}(X) - j_t(X) = dU_t^* X U_t + U_t^* X dU_t + dU_t^* X dU_t \tag{19}$$

The only nonvanishing products in the quantum stochastic differentials are

$$dB_{i\omega}(t) dB_{j\omega}^*(t) = 2\text{Re}(g_i|g_j)_\omega^- dt, \qquad dB_{i\omega}^*(t) dB_{j\omega}(t) = 2\text{Re}(g_i|g_j)_\omega^+ dt \tag{20}$$

Combining the terms in (19) and using (18), (12), (16), and (20), we get the Langevin equation

$$dj_t(X) = \sum_\alpha j_t \circ \theta_\alpha(X) \, dM^\alpha(t) = \sum_{n=-1,1; j\omega} j_t \circ \theta_{nj\omega}(X) dM^{nj\omega}(t) + j_t \circ \theta_0(X) \, dt$$

$$\tag{21}$$

where

$$dM^{-1,j\omega}(t) = dB_{j\omega}(t), \qquad \theta_{-1,j\omega}(X) = -i[X, E_\omega^*(D_j)] \qquad (22)$$

$$dM^{1,j\omega}(t) = dB_{j\omega}^*(t), \qquad \theta_{1,j\omega}(X) = -i[X, E_\omega(D_j)] \qquad (23)$$

and

$$\theta_0(X) = \sum_{ij}\sum_{\omega\in F}\Bigl(-i\,\mathrm{Im}(g_i|g_j)_\omega^-[X, E_\omega^*(D_i)E_\omega(D_j)]$$
$$+ i\,\mathrm{Im}(g_i|g_j)_\omega^+[X, E_\omega(D_i)E_\omega^*(D_j)]$$
$$+ 2\,\mathrm{Re}(g_i|g_j)_\omega^-\Bigl(E_\omega^*(D_i)XE_\omega(D_j) - \frac{1}{2}\{X, E_\omega^*(D_i)E_\omega(D_j)\}\Bigr)$$
$$+ 2\,\mathrm{Re}(g_i|g_j)_\omega^+\Bigl(E_\omega(D_i)XE_\omega^*(D_j) - \frac{1}{2}\{X, E_\omega(D_i)E_\omega^*(D_j)\}\Bigr)\Bigr) \qquad (24)$$

is a quantum Markovian generator. The structure map $\theta_0(X)$ has the standard form of the generator of a master equation [7]:

$$\theta_0(X) = \Psi(X) - \frac{1}{2}\{\Psi(1), X\} + i[H, X]$$

where Ψ is a completely positive map and H is self-adjoint. In our case $\Psi(X)$ is a linear combination of terms of the type

$$E_\omega^*(D_i)XE_\omega(D_j)$$

Remark. *A corollary of the remark at the end of Section II is that the Markovian generator θ_0 maps $L^\infty(H_S)'$ into itself. Moreover, if X in (24) belongs to the $L^\infty(H_S)$, then the Hamiltonian part of $\theta_0(X)$ vanishes and only the dissipative part remains. In particular, if H_S has nondegenerate spectrum, then $\theta_0(X)$ maps $L^\infty(H_S)$ and has the form*

$$\theta_0(X) = \sum_{ij}\sum_{\omega\in F}(2Re(g_i|g_j)_\omega^-(E_\omega^*(D_i)XE_\omega(D_j) - XE_\omega^*(D_i)E_\omega(D_j))$$
$$+ 2Re(g_i|g_j)_\omega^+(E_\omega(D_i)XE_\omega^*(D_j) - XE_\omega(D_i)E_\omega^*(D_j)))$$

for any $X \in L^\infty(H_S)$.

In Ref. 8 the following theorem was proved.

Theorem. *For any pair of operators in the system algebra X, Y, the structure maps in the Langevin equation (21) satisfy the equation*

$$\theta_\alpha(XY) = \theta_\alpha(X)Y + X\theta_\alpha(Y) + \sum_{\beta,\gamma} c_\alpha^{\beta\gamma} \theta_\beta(X)\theta_\gamma(Y)$$

where the structure constants $c_\alpha^{\beta\gamma}$ is given by the Ito table:

$$dM^\beta(t)dM^\gamma(t) = \sum_\alpha c_\alpha^{\beta\gamma} dM^\alpha(t)$$

The conjugation rules of $dM^\alpha(t)$ and θ_α are connected in such a way that formula (21) defines a $*$-flow ($* \circ j_t = j_t \circ *$).

Let us now investigate the master equation for the density matrix ρ.

We will show that if the reservoir is in the equilibrium state at temperature β^{-1}, then for the generic system with decoherence the solution of the master equation $\rho(t)$ with $t \to \infty$ tends to the classic Gibbs state with the same temperature β^{-1}. This phenomenon realizes the quantum measurement procedure: The quantum state (density matrix) collapses into the classic state.

To show this we use the control of quantum decoherence that was found in the stochastic approximation of quantum theory (see Ref. 3 and discussion below).

Let us consider the evolution of the state (positive normed linear functional on system observables) given by the density matrix ρ, $\rho(X) = \mathrm{tr}\,\hat\rho X$. The evolution of the state is defined as follows:

$$\rho_t = j_t^*(\rho) = \rho \circ j_t$$

Therefore from (21) we obtain the evolution equation:

$$d\rho_t(X) = \rho \circ dj_t(X) = \rho \circ \sum_\alpha j_t \circ \theta_\alpha(X)\,dM^\alpha(t) = \sum_\alpha \rho_t(\theta_\alpha(X)\,dM^\alpha(t))$$

Only the stochastic differential dt in this formula will survive and we obtain the master equation

$$\frac{d}{dt}\rho_t(X) = \rho_t \circ \theta_0(X) \equiv \theta_0^*(\rho_t)(X) \tag{25}$$

Let us consider the density matrix $\hat\rho = \hat\rho_S \otimes \hat\rho_R$,

$$\hat\rho_{S,t} = \sum_{\mu,\nu} \rho(\mu,\nu,t)|\mu\rangle\langle\nu|$$

where $|\mu\rangle$, $|\nu\rangle$ are eigenvectors of the system Hamiltonian H_S.

Using the form (24) of θ_0, the master equation (25) will take the form

$$\sum_{\mu,\nu}\frac{d}{dt}\rho(\mu,\nu,t)|\mu\rangle\langle\nu| = \sum_{\mu,\nu}\rho(\mu,\nu,t)$$

$$\sum_{ij}\sum_{\omega\in F}\Bigg(-i\,\mathrm{Im}(g_i|g_j)^-_\omega(|\mu\rangle\langle\nu|\chi_\omega(\varepsilon_\nu)D_i^*P_{\varepsilon_\nu-\omega}D_jP_{\varepsilon_\nu}-\chi_\omega(\varepsilon_\mu)P_{\varepsilon_\mu}D_i^*P_{\varepsilon_\mu-\omega}D_j|\mu\rangle\langle\nu|)$$

$$+2\mathrm{Re}(g_i|g_j)^-_\omega\bigg(\chi_\omega(\varepsilon_\mu+\omega)\chi_\omega(\varepsilon_\nu+\omega)P_{\varepsilon_\mu+\omega}D_i^*|\mu\rangle\langle\nu|D_jP_{\varepsilon_\nu+\omega}$$

$$-\frac{1}{2}(|\mu\rangle\langle\nu|\chi_\omega(\varepsilon_\nu)D_i^*P_{\varepsilon_\nu-\omega}D_jP_{\varepsilon_\nu}+\chi_\omega(\varepsilon_\mu)P_{\varepsilon_\mu}D_i^*P_{\varepsilon_\mu-\omega}D_j|\mu\rangle\langle\nu|)\bigg)$$

$$+i\,\mathrm{Im}(g_i|g_j)^+_\omega(|\mu\rangle\langle\nu|\chi_\omega(\varepsilon_\nu+\omega)D_iP_{\varepsilon_\nu+\omega}D_j^*P_{\varepsilon_\nu}-\chi_\omega(\varepsilon_\mu+\omega)$$

$$\times P_{\varepsilon_\mu}D_iP_{\varepsilon_\mu+\omega}D_j^*|\mu\rangle\langle\nu|)+2\mathrm{Re}(g_i|g_j)^+_\omega\bigg(\chi_\omega(\varepsilon_\mu)\chi_\omega(\varepsilon_\nu)P_{\varepsilon_\mu-\omega}D_i|\mu\rangle\langle\nu|D_j^*P_{\varepsilon_\nu-\omega}$$

$$-\frac{1}{2}(|\mu\rangle\langle\nu|\chi_\omega(\varepsilon_\nu+\omega)D_iP_{\varepsilon_\nu+\omega}D_j^*P_{\varepsilon_\nu}+\chi_\omega(\varepsilon_\mu+\omega)P_{\varepsilon_\mu}D_iP_{\varepsilon_\mu+\omega}D_j^*|\mu\rangle\langle\nu|)\bigg)\Bigg) \tag{26}$$

where $\chi_\omega(\varepsilon_\mu)=1$ if $\varepsilon_\mu\in F_\omega$ and equals 0 otherwise.

IV. DYNAMICS FOR GENERIC SYSTEMS

Let us investigate the behavior of a system with dynamics defined by (26). This dynamics will depend on the Hamiltonian of the system.

We will call the Hamiltonian H_S generic, if:

1. The spectrum Spec H_S of the Hamiltonian is nondegenerate.
2. For any Bohr frequency ω there exists a unique pair of energy levels ε, $\varepsilon'\in\mathrm{Spec}\,H_S$, such that

$$\omega=\varepsilon-\varepsilon'$$

We investigate (26) for generic Hamiltonian. We also consider the case of one test function $g_i(k)=g(k)$, although this is not important. In this case the Markovian generator θ_0 takes the form

$$\theta_0(X)=\sum_{\sigma,\sigma'}|\langle\sigma'|D|\sigma\rangle|^2\bigg(-i\,\mathrm{Im}(g|g)^-_{\sigma\sigma'}[X,|\sigma\rangle\langle\sigma|]+2\mathrm{Re}(g|g)^-_{\sigma\sigma'}$$

$$\times\bigg(|\sigma\rangle\langle\sigma|\langle\sigma'|X|\sigma'\rangle-\frac{1}{2}\{X,|\sigma\rangle\langle\sigma|\}\bigg)+i\,\mathrm{Im}(g|g)^+_{\sigma\sigma'}[X,|\sigma'\rangle\langle\sigma'|]$$

$$+2\mathrm{Re}(g|g)^+_{\sigma\sigma'}\bigg(|\sigma'\rangle\langle\sigma'|\langle\sigma|X|\sigma\rangle-\frac{1}{2}\{X,|\sigma'\rangle\langle\sigma'|\}\bigg)\bigg) \tag{27}$$

Here we use the notion

$$(g|g)_{\mu\sigma} = (g|g)_{\varepsilon_\mu - \varepsilon_\sigma}$$

Notice that the factors $\operatorname{Re}(g|g)_{\sigma\sigma'}^{\pm}$ are > 0 only for $\varepsilon_\sigma > \varepsilon_{\sigma'}$ and vanish for the opposite case.

It is easy to see that the terms in (27) of the form

$$|\sigma\rangle\langle\sigma|\langle\sigma'|X|\sigma'\rangle$$

for off-diagonal elements of the density matrix $X = |\mu\rangle\langle v|$ are equal to zero. We will show that in such case Eq. (26) will predict fast damping of the states of the kind $|\mu\rangle\langle v|$.

In the nongeneric case one can expect the fast damping of the state $|\mu\rangle\langle v|$ with different energies ε_μ and ε_v.

With the given assumptions the action of θ_0 on the off-diagonal matrix unit $|\mu\rangle\langle v|$, $\varepsilon_\mu \neq \varepsilon_v$ is equal to $A_{\mu v}|\mu\rangle\langle v|$, where the number $A_{\mu v}$ is given by the following:

$$A_{\mu v} = \sum_\sigma (-i\operatorname{Im}(g|g)_{\mu\sigma}^{-}|\langle\sigma|D|\mu\rangle|^2 + i\operatorname{Im}(g|g)_{v\sigma}^{-}|\langle\sigma|D|v\rangle|^2$$
$$+ i\operatorname{Im}(g|g)_{\sigma\mu}^{+}|\langle\mu|D|\sigma\rangle|^2 - i\operatorname{Im}(g|g)_{\sigma v}^{+}|\langle v|D|\sigma\rangle|^2 - \operatorname{Re}(g|g)_{\mu\sigma}^{-}|\langle\sigma|D|\mu\rangle|^2$$
$$- \operatorname{Re}(g|g)_{v\sigma}^{-}|\langle\sigma|D|v\rangle|^2 - \operatorname{Re}(g|g)_{\sigma\mu}^{+}|\langle\mu|D|\sigma\rangle|^2 - \operatorname{Re}(g|g)_{\sigma v}^{+}|\langle v|D|\sigma\rangle|^2)$$
(28)

The map θ_0 multiplies off-diagonal matrix elements of the density matrix $\hat{\rho}_S$ by a number $A_{\mu v}$. Let us note that

$$\operatorname{Re} A_{\mu v} \leq 0$$

We will prove in the present section that for generic Hamiltonian the map θ_0^* mixes diagonal elements of the density matrix but does not mix diagonal and off-diagonal elements (the action of θ_0^* on-diagonal element is equal to the linear combination of diagonal elements). Therefore

$$j_t^*(|\mu\rangle\langle v|) = \exp(A_{\mu v} t)|\mu\rangle\langle v|$$

We see that if any of the $\operatorname{Re}(g|g)|\langle\beta|D|\alpha\rangle|^2$ in (28) is nonzero, then the corresponding off-diagonal matrix element of the density matrix decays. We obtain an effect of the diagonalization of the density matrix. This gives an effective criterium for quantum decoherence in the stochastic approximation: The system will exhibit decoherence if the constants $\operatorname{Re}(g|g)^{\pm}$ are nonzero.

Now we estimate the velocity of decay of the density matrix $|\mu\rangle\langle\nu|$ for a quantum system with N particles. The eigenstate $|\mu\rangle$ of the Hamiltonian of such a system can be considered as a tensor product over degrees of freedom of the system of some substates. Let us estimate from below the number of degrees of freedom of the system by the number of particles that belong to the system (for each particle we have few degrees of freedom). To obtain the estimate from below for the velocity of decay we assume that $|\langle\sigma|D|\mu\rangle|^2$ in (28) is nonzero only if the state σ differs from the state μ only for one degree of freedom.

Then the summation over ω (or equivalently over σ) in (28) can be estimated by the summation over the degrees of freedom, or over particles belonging to the system. If we have total decoherence, [i.e., all $\mathrm{Re}(g|g)$ are nonzero] then, taking all corresponding $|\langle\sigma|D|\mu\rangle|^2 = 1$, we can estimate (28) as $-N\mathrm{Re}(g|g)$, where N is the number of particles in the system, or

$$j_t^*(|\mu\rangle\langle\nu|) = \exp(-N\mathrm{Re}(g|g)t)|\mu\rangle\langle\nu| \tag{29}$$

The off-diagonal element of the density matrix decays exponentially, with the exponent proportional to the number of particles in the system. Therefore for macroscopic (large N) systems with decoherence the quantum state will collapse into the classic state very quickly.

This observation clarifies why macroscopic quantum systems usually behave classically. Equation (29) describes such a type of behavior, predicting that the quantum state damps at least as quickly as $\exp(-N\mathrm{Re}(g|g)t)$. Therefore a macroscopic system (large N) will become classic in a time of order $(N\mathrm{Re}(g|g))^{-1}$.

In the final section of this chapter we will illustrate the collapse phenomenon (29) using the quantum extension of the Glauber dynamics for a system of spins.

We see that the stochastic limit predicts the collapse of a quantum state into a classic state and, moreover, allows us to estimate the velocity of the collapse (29). One can consider (29) as a more detailed formulation of the Fermi golden rule: The Fermi golden rule predicts exponential decay of quantum states; formula (29) also relates the speed of the decay to the dimensions (number of particles) of the system.

Consider now the system density matrix $\hat{\rho}_S \in \mathscr{C}$, where $\mathscr{C}$ is the algebra generated by the spectral projections of the system Hamiltonian H_S, and consider the master equation (27) (we consider the generic case). We will find that the evolution defined by this master equation will conserve the algebra $\mathscr{C}$ and therefore will be a classic evolution. We will show that this classic evolution in fact describes quantum phenomena.

For $\hat{\rho}_{S,t} \in \mathscr{C}$ we define the evolved density matrix of the system

$$\hat{\rho}_{S,t} = \sum_\sigma \rho(\sigma,t)|\sigma\rangle\langle\sigma|$$

For this density matrix the master equation (25) takes the form

$$\frac{d}{dt}\rho(\sigma,t) = \sum_{\sigma'}(\rho(\sigma',t)(2\mathrm{Re}(g|g)^-_{\sigma'\sigma}|\langle\sigma|D|\sigma'\rangle|^2 + 2\mathrm{Re}(g|g)^+_{\sigma\sigma'}|\langle\sigma'|D|\sigma\rangle|^2)$$
$$- \rho(\sigma,t)(2\mathrm{Re}(g|g)^+_{\sigma'\sigma}|\langle\sigma|D|\sigma'\rangle|^2 + 2\mathrm{Re}(g|g)^-_{\sigma\sigma'}|\langle\sigma'|D|\sigma\rangle|^2)) \tag{30}$$

Let us note that if $\rho(\sigma,t)$ satisfies the detailed balance condition

$$\rho(\sigma,t)2\mathrm{Re}(g|g)^-_{\sigma\sigma'} = \rho(\sigma',t)2\mathrm{Re}(g|g)^+_{\sigma\sigma'} \tag{31}$$

then $\rho(\sigma,t)$ is the stationary solution for (30).

Let us investigate (30) and (31) for the equilibrium state of the field. In this case

$$2\mathrm{Re}(g|g)^-_{\sigma\sigma'} = 2\pi\int dk\,|g(k)|^2\delta(\omega(k) + \varepsilon_{\sigma'} - \varepsilon_\sigma)\frac{1}{1 - e^{-\beta\omega(k)}}$$
$$= 2\pi\int dk\,|g(k)|^2\delta(\omega(k) + \varepsilon_{\sigma'} - \varepsilon_\sigma)\frac{1}{1 - e^{-\beta(\varepsilon_\sigma - \varepsilon_{\sigma'})}}$$
$$= \frac{C_{\sigma\sigma'}}{1 - e^{-\beta(\varepsilon_\sigma - \varepsilon_{\sigma'})}} \quad 2\mathrm{Re}(g|g)^+_{\sigma\sigma'} = \frac{C_{\sigma\sigma'}}{e^{\beta(\varepsilon_\sigma - \varepsilon_{\sigma'})} - 1}$$

Equation (30) takes the form

$$\frac{d}{dt}\rho(\sigma,t)e^{\beta\varepsilon_\sigma} = \sum_{\sigma'}\frac{C_{\sigma\sigma'}|\langle\sigma'|D|\sigma\rangle|^2 - C_{\sigma'\sigma}|\langle\sigma|D|\sigma'\rangle|^2}{1 - e^{-\beta(\varepsilon_\sigma - \varepsilon_{\sigma'})}}$$
$$\times (\rho(\sigma',t)e^{\beta\varepsilon_{\sigma'}} - \rho(\sigma,t)e^{\beta\varepsilon_\sigma}) \tag{32}$$

Let us note that $C_{\sigma\sigma'}$ are nonzero (and therefore positive) only if denominators in (32) are positive, and $C_{\sigma'\sigma}$ are nonzero only if the corresponding denominators are negative.

If the system possesses decoherence, then $C_{\sigma\sigma'}$ and $C_{\sigma'\sigma}$ are nonzero and the solution of Eq. (32) for $t \to \infty$ tends to the stationary solution given by the detailed balance condition (31)

$$\frac{\rho(\sigma,t)}{1 - e^{-\beta(\varepsilon_\sigma - \varepsilon_{\sigma'})}} = \frac{\rho(\sigma',t)}{e^{\beta(\varepsilon_\sigma - \varepsilon_{\sigma'})} - 1}$$

or

$$\rho(\sigma,t)e^{\beta\varepsilon_\sigma} = \rho(\sigma',t)e^{\beta\varepsilon_{\sigma'}}$$

This means that the stationary solution (31) of (30) describes the equilibrium state of the system

$$\rho(\sigma, t) = \frac{e^{-\beta\varepsilon_\sigma}}{\sum_{\sigma'} e^{-\beta\varepsilon_{\sigma'}}}$$

For a system with decoherence the density matrix will tend, as $t \to \infty$, to the stationary solution (31) of (30). In particular, as $t \to \infty$, the density matrix collapses to the classic Gibbs distribution.

The phenomenon of a collapse of a quantum state into a classic state is connected with the quantum measurement procedure. The quantum uncertainty will be concentrated at the degrees of freedom of the quantum field and vanishes after the averaging procedure. One can speculate that the collapse of the wavefunction is a property of open quantum systems: We can observe the collapse of the wavefunction of the system averaging over the degrees of freedom of the reservoir interacting with the system. Usually the collapse of a wavefunction is interpreted as a projection onto a classic state (the von Neumann interpretation). The picture emerging from our considerations is more general: The collapse is a result of the unitary quantum evolution and conditional expectation (averaging over the degrees of freedom of quantum field). This is a generalization of the projection: It is easy to see that every projection P generates a (nonidentity preserving) conditional expectation $E_P(X) = PXP$; more generally, a set of projections P_i generates the conditional expectation

$$\sum_i \alpha_i E_{P_i}, \qquad \alpha_i \geq 0$$

but not every conditional expectation could be given in this way.

We have found the effect of the collapse of density matrix for $\rho(t) = \langle U_t \rho U_t^* \rangle$, where $U_t = \lim_{\lambda \to 0} e^{itH_0} e^{-itH}$ is the stochastic limit of interacting evolution. The same effect of collapse will be valid for the limit of the full evolution e^{-itH}, because the full evolution is the composition of interacting and free evolution. The free evolution leaves invariant the elements of diagonal subalgebra and multiplies the above-considered nondiagonal element $|\sigma'\rangle\langle\sigma|$ by the oscillating factor $e^{it(\varepsilon_{\sigma'} - \varepsilon_\sigma)}$. Therefore for the full evolution we obtain the additional oscillating factor, and the collapse phenomenon will survive.

V. CONTROL OF COHERENCE

In this section we generalize the approach of Ref. 3 and investigate different regimes of qualitative behavior for the considered model.

The master equation (30) at first sight looks completely classic. In this chapter we derived this equation using quantum arguments. Now we will show that (30) in fact describes a quantum behavior. To show this we consider the following example.

Let us rewrite (30) using the particular form (16) of $(g|g)^\pm$. Using (15) and (16), we obtain

$$\frac{d}{dt}\rho(\sigma,t) = \sum_{\sigma'} 2\pi \int dk\, |g(k)|^2 ((N(k)+1)(\rho(\sigma',t)\delta(\omega(k)+\varepsilon_\sigma-\varepsilon_{\sigma'})|\langle\sigma|D|\sigma'\rangle|^2$$
$$- \rho(\sigma,t)\delta(\omega(k)+\varepsilon_{\sigma'}-\varepsilon_\sigma)|\langle\sigma'|D|\sigma\rangle|^2)$$
$$+ N(k)(\rho(\sigma',t)\delta(\omega(k)+\varepsilon_{\sigma'}-\varepsilon_\sigma)\langle\sigma'|D|\sigma\rangle|^2$$
$$- \rho(\sigma,t)\delta(\omega(k)+\varepsilon_\sigma-\varepsilon_{\sigma'})|\langle\sigma|D|\sigma'\rangle|^2) \qquad (33)$$

The first term (integrated with $N(k)+1$) on the right-hand side of this equation describes the emission of bosons, and the second term (integrated with $N(k)$) describes the absorption of bosons. For the emission term the part with $N(k)$ describes the induced emission, and the part with 1 describes the spontaneous emission of bosons.

Let us note that the Einstein relation for probabilities of emission and absorption of bosons with quantum number k

$$\frac{\text{probability of emission}}{\text{probability of absorption}} = \frac{N(k)+1}{N(k)}$$

is satisfied in the stochastic approximation.

Formula (33) describes a macroscopic quantum effect. To show this let us take the spectrum of a system Hamiltonian (the set of system states $\Sigma = \{\sigma\}$) as follows: Let Σ contain two groups Σ_1 and Σ_2 of states with the energy gap between these groups (or, for simplicity, two states σ_1 and σ_2 with $\varepsilon_{\sigma_2} > \varepsilon_{\sigma_1}$). This type of Hamiltonian was considered in different models of quantum optics; for a review see Ref. 10 (for the case of two states we get the spin–boson Hamiltonian investigated in [3] using the stochastic limit). Let the state $\langle\cdot\rangle$ of the bosonic field be taken in such a way that the density $N(k)$, of quanta of the bosonic field, has support in a set of momentum variables k such that

$$0 < \omega(k) < \omega_0 < |\varepsilon_{\sigma_1} - \varepsilon_{\sigma_2}|, \qquad k \in \operatorname{supp} N(k) \qquad (34)$$

This means that high-energetic bosons are absent. It is natural to consider the state $\langle\cdot\rangle$ as a sum of an equilibrium state at temperature β^{-1} and an nonequilibrium part. Therefore the density $N(k)$ will be nonzero for small k because the equilibrium state satisfies this property.

Under the considered assumption (34) the integral of δ-function $\delta(\omega(k) + \varepsilon_{\sigma_1} - \varepsilon_{\sigma_2})$ with $N(k)$ in (33) is identically equal to zero. Therefore the right-hand side of (33) will be equal to

$$\sum_{\sigma'} 2\pi \int dk \, |g(k)|^2 (\rho(\sigma',t)\delta(\omega(k) + \varepsilon_\sigma - \varepsilon_{\sigma'})|\langle\sigma|D|\sigma'\rangle|^2$$
$$- \rho(\sigma,t)\delta(\omega(k) + \varepsilon_{\sigma'} - \varepsilon_\sigma)|\langle\sigma'|D|\sigma\rangle|^2)$$

It is natural to consider this value (corresponding to the spontaneous emission of bosons by the system) as small with respect to the induced emission (for $N(k) \gg 1$). In this case the density matrix $\rho(\sigma,t)$ will be almost constant in time. This is an effect of conservation of quantum coherence: In the absence of bosons with the energy $\omega(k)$ equal to $\varepsilon_{\sigma_1} - \varepsilon_{\sigma_2}$ the system cannot jump between the states σ_1 and σ_2 (or, at least, this transition is very slow), because in the stochastic limit such a jump corresponds to quantum white noise that must be on a mass shell.

At the same time, the transitions between states inside the groups Σ_1 and Σ_2 are not forbidden by (34), because these transitions are connected with the soft bosons (with small k) that are present in the equilibrium part of $\langle \cdot \rangle$. In the above, the assumptions equation (33) describes the transition of the system to intermediate equilibrium, where the transitions between groups of states Σ_1 and Σ_2 are forbidden.

If the state $\langle \cdot \rangle$ does not satisfy the property (34), then the system undergoes fast transitions between states σ_1 and σ_2. We can switch on such a transition by switching on the bosons with the frequency $\omega(k) = \varepsilon_{\sigma_2} - \varepsilon_{\sigma_1}$.

In conclusion, Eq. (33) describes a macroscopic quantum effect controlled by the distribution of bosons $N(k)$ which can be physically controlled, for example, by filtering.

VI. THE GLAUBER DYNAMICS

In the present section we apply the master equation (30) to the derivation of the quantum extension of the classic Glauber dynamics. The Glauber dynamics is a dynamics for a spin lattice with nearest-neighbor interaction (see Refs. 11 and 12). We will prove that the Glauber dynamics can be considered as a dynamics generated by the master equation of the type (30) derived from a stochastic limit for a quantum spin system interacting with a bosonic quantum field.

We take the bosonic reservoir space $\mathscr{F}$ corresponding to the bosonic equilibrium state at temperature β^{-1}. Thus the reservoir state is Gaussian with mean zero and correlations given by

$$\langle a^*(k)a(k')\rangle = \frac{1}{e^{\beta\omega(k)} - 1}\delta(k - k')$$

For simplicity we only consider the case of a one-dimensional spin lattice, but our considerations extend without any change to multidimensional spin lattices.

The spin variables are labeled by integer numbers Z; and for each finite subset $\Lambda \subseteq Z$ with cardinality $|\Lambda|$, the system Hilbert space is

$$\mathcal{H}_S = \mathcal{H}_\Lambda = \otimes_{r \subset \Lambda} C^2$$

and the system Hamiltonian has the form

$$H_S = H_\Lambda = -\frac{1}{2} \sum_{r,s \in \Lambda} J_{rs} \sigma_r^z \sigma_s^z$$

where σ_r^x, σ_r^y, σ_r^z are Pauli matrices ($r \in \Lambda$) at the rth site in the tensor product

$$\sigma_r^i = 1 \otimes \cdots \otimes 1 \otimes \sigma^i \otimes 1 \otimes \cdots \otimes 1$$

For any $r, s \in \Lambda$

$$J_{rs} = J_{sr} \in R, \qquad J_{rr} = 0$$

We consider for simplicity the system Hamiltonian that describes the interaction of spin with the nearest neighbors (Ising model):

$$J_{rs} = J_{r,r+1}$$

The interaction Hamiltonian H_I (acting in $\mathcal{H}_S \otimes \mathcal{F}$) has the form

$$H_I = \sum_{r \in \Lambda} \sigma_r^x \otimes \psi(g_r), \quad \psi(g) = A(g) + A^*(g), \quad A(g) = \int dk\, \bar{g}(k) a(k)$$

where ψ is a field operator, and $A(g)$ is a smeared quantum field with cutoff function (form factor) $g(k)$.

The eigenvectors $|\sigma\rangle$ of the system Hamiltonian H_S can be labeled by spin configurations σ (sequences of ± 1), which label the natural basis in $\mathcal{H}_S$ consisting of tensor products of eigenvectors of σ_r^z (spin-up and spin-down vectors $|\varepsilon_r\rangle$, corresponding to eigenvalues $\varepsilon_r = \pm 1$)

$$|\sigma\rangle = \otimes_{r \in \Lambda} |\varepsilon_r\rangle$$

In the present section we denote ε_r the energy of the spin at site r and denote $E(\sigma)$ the energy of the spin configuration σ:

$$E(\sigma) = -\frac{1}{2} \sum_{r,s \in \Lambda} J_{rs} \varepsilon_r \varepsilon_s$$

The action of the operator σ_r^x on the spin configuration σ is defined using the action of σ_r^x on the corresponding eigenvector $|\sigma\rangle$, so the operator σ_r^x flips the spin at the rth site in the sequence σ (i.e., it maps the vector $|\varepsilon_r\rangle$ in the tensor product into the vector $|-\varepsilon_r\rangle$). From the form of H_S and H_I it follows that in (30) the matrix element $\langle\sigma|D|\sigma'\rangle$ of any two eigenvectors, corresponding to the spin configurations σ, σ', will be nonzero only if the configurations σ, σ' differ exactly at one site. If the configurations σ, σ' differ exactly at one site, then $\langle\sigma|D|\sigma'\rangle = 1$.

The (Classic) Glauber dynamics will be given by the master equation for the density matrix after using the algebra of spectral projections of the system Hamiltonian (30):

$$\frac{d}{dt}\rho(\sigma,t) = \sum_{r\in\Lambda}(\rho(\sigma_r^x\sigma,t)(2\mathrm{Re}(g|g)^-_{\sigma_r^x\sigma,\sigma} + 2\mathrm{Re}(g|g)^+_{\sigma,\sigma_r^x\sigma})$$
$$- \rho(\sigma,t)(2\mathrm{Re}(g|g)^+_{\sigma_r^x\sigma,\sigma} + 2\mathrm{Re}(g|g)^-_{\sigma,\sigma_r^x\sigma})) \qquad (35)$$

which gives the Glauber dynamics of a system of spins (see Refs. 11 and 12). Here

$$2\mathrm{Re}(g|g)^-_{\sigma,\sigma_r^x\sigma} = 2\pi\int dk\,|g(k)|^2\delta(\omega(k) - J_{r-1,r}\varepsilon_{r-1} - J_{r,r+1}\varepsilon_{r+1})\frac{1}{1-e^{-\beta\omega(k)}} \qquad (36)$$

and analogously all the other $(g|g)^\pm$.

Up to now we have investigated the dynamics for the diagonal part of the density matrix. The master equation for the off-diagonal part of the density matrix (25) will give the quantum extension of the Glauber dynamics. We consider now this off-diagonal part:

$$\sum_{\mu\neq\nu}\rho(\mu,\nu,t)|\mu\rangle\langle\nu|$$

From (25) and (28) we obtain the equation for the off-diagonal elements of the density matrix:

$$\frac{d}{dt}\rho(\mu,\nu,t) = A_{\mu\nu}\rho(\mu,\nu,t) \qquad (37)$$

$$A_{\mu\nu} = \sum_{r\in\Lambda}(-i\,\mathrm{Im}(g|g)^-_{\mu,\sigma_r^x\mu} + i\,\mathrm{Im}(g|g)^-_{\nu,\sigma_r^x\nu} + i\,\mathrm{Im}(g|g)^+_{\sigma_r^x\mu,\mu}$$
$$- i\,\mathrm{Im}(g|g)^+_{\sigma_r^x\nu,\nu} - \mathrm{Re}(g|g)^-_{\mu,\sigma_r^x\mu} - \mathrm{Re}(g|g)^-_{\nu,\sigma_r^x\nu}$$
$$- \mathrm{Re}(g|g)^+_{\sigma_r^x\mu,\mu} - \mathrm{Re}(g|g)^+_{\sigma_r^x\nu,\nu}) \qquad (38)$$

Equations (35), (37), and (38) describe the quantum extension of the classic Glauber dynamics (35). As was already noted in Section IV, the coefficient $A_{\mu\nu}$ in (38) is proportional to $|\Lambda|$ (the number of particles in the system). Due to the summation on $r \in \Lambda$ the coefficient $A_{\mu\nu}$ will diverge for large $|\Lambda|$ (the real part of $A_{\mu\nu}$ will tend to $-\infty$). Therefore the density matrix will collapse to the diagonal subalgebra (the classic distribution function) very quickly.

Let us consider now the particular case of a one-dimensional system with a translationally invariant Hamiltonian:

$$J_{rs} = J_{r,r+1} = J > 0$$

The translationally invariant Hamiltonian does not satisfy the generic nondegeneracy conditions on the system spectrum that we have used in the derivation of Eqs. (35) and (37), and therefore we cannot apply these equations to describe the dynamics for this Hamiltonian.

However, in the translation invariant one-dimensional case we can investigate these equations by direct methods.

In this case the $(g|g)^{\pm}$, given by (36), are nonzero only if $\varepsilon_{r-1} = \varepsilon_{r+1} = 1$, and we get for (36)

$$2\,\text{Re}(g|g)^{-}_{\sigma,\sigma_r^x\sigma} = 2\pi \int dk\, |g(k)|^2 \delta(\omega(k) - 2J) \frac{1}{1 - e^{-2\beta J}} = \frac{C}{1 - e^{-2\beta J}} \quad (39)$$

Therefore for one-dimensional translation invariant Hamiltonians we obtain for (35) (compare with Refs. 11 and 12)

$$\frac{d}{dt}\rho(\sigma,t) = \frac{C}{1 - e^{-2\beta J}} \left(\sum_{r \in \Lambda; E(\sigma) > E(\sigma_r^x\sigma)} (e^{-2\beta J}\rho(\sigma_r^x\sigma,t) - \rho(\sigma,t)) \right.$$
$$\left. + \sum_{r \in \Lambda; E(\sigma) < E(\sigma_r^x\sigma)} (\rho(\sigma_r^x\sigma,t) - e^{-2\beta J}\rho(\sigma,t)) \right) \quad (40)$$

The detailed balance stationary solution of (40) satisfies the following: For two spin configurations σ, $\sigma_r^x\sigma$ that differ by the flip of spin at site r, the energy of corresponding configurations differ by $2J$. The expectation $\rho(\mu)$, $\mu = \sigma, \sigma_r^x\sigma$ of configuration with the higher energy will be $e^{-2\beta J}$ times less.

For the off-diagonal part of the density matrix for the case of one-dimensional translation invariant Hamiltonian, the terms in the imaginary part of (38) cancel and using (39) we get for (38)

$$A_{\mu\nu} = -\sum_{r \in \Lambda} \left(\frac{2C}{1 - e^{-2\beta J}} + \frac{2C}{e^{2\beta J} - 1} \right) = -2C \sum_{r \in \Lambda} \frac{1 + e^{-2\beta J}}{1 - e^{-2\beta J}}$$

This sum, over r, of equal terms diverges with $|\Lambda| \to \infty$. Therefore the off-diagonal elements of the density matrix that satisfy (37) will decay very quickly; and for sufficiently large t, $|\Lambda|$ the dynamics of the system will be given by classic Glauber dynamics.

For the master equation considered above we used the master equation for generic (nondegenerate) Hamiltonian. This gives us the Glauber dynamics. But the translation invariant Hamiltonian is degenerate. Therefore in the translation invariant case we will obtain some generalization, of the Glauber dynamics. To derive this generalization, let us consider the general form (26) of the master equation. For the considered spin system this gives

$$\sum_{\mu,\nu} \frac{d}{dt}\rho(\mu,\nu,t)|\mu\rangle\langle\nu| = \sum_{\mu,\nu}\rho(\mu,\nu,t)\sum_{a,b\in\Lambda}\Bigg(-i(|\mu\rangle\langle P_{E(\nu)}\sigma_a^x P_{\nu-}\sigma_b^x\nu|$$
$$-|P_{E(\mu)}\sigma_a^x P_{\mu-}\sigma_b^x\mu\rangle\langle\nu|) + \frac{C}{1-e^{-2\beta J}}\Big(|P_{E(\mu)+J}\sigma_a^x\mu\rangle\langle P_{E(\nu)+J}\sigma_b^x\nu|$$
$$-\frac{1}{2}(|\mu\rangle\langle P_{E(\nu)}\sigma_a^x P_{E(\nu)-J}\sigma_b^x\nu| + |P_{E(\mu)}\sigma_a^x P_{E(\mu)-J}\sigma_b^x\mu\rangle\langle\nu|)\Big)$$
$$+i(|\mu\rangle\langle P_{E(\nu)}\sigma_a^x P_{\nu+}\sigma_b^x\nu| - |P_{E(\mu)}\sigma_a^x P_{\mu+}\sigma_b^x\mu\rangle\langle\nu|) + \frac{C}{e^{2\beta J}-1}$$
$$\times\Big(|P_{E(\mu)-J}\sigma_a^x\mu\rangle\langle P_{E(\nu)-J}\sigma_b^x\nu| - \frac{1}{2}(|\mu\rangle\langle P_{E(\nu)}\sigma_a^x P_{E(\nu)+J}\sigma_b^x\nu|$$
$$+|P_{E(\mu)}\sigma_a^x P_{E(\mu)+J}\sigma_b^x\mu\rangle\langle\nu|)\Big)\Bigg) \qquad (41)$$

Here C is given by (39), operator $P_{E(\mu)}$ is a projector onto the states with the energy $E(\mu)$, and operator $P_{\nu-}$ is given by

$$P_{\nu-} = \text{Im}(g|g)_{-1}^- P_{E(\nu)-J} + \text{Im}(g|g)_0^- P_{E(\nu)} + \text{Im}(g|g)_1^- P_{E(\nu)+J}$$

$$(g|g)_a^- = \text{P.P.}\int dk\,|g(k)|^2 \frac{1}{\omega(k)+aJ}\frac{1}{1-e^{-\beta\omega(k)}}, \qquad a=-1,0,1$$

For the operator $P_{\nu+}$ we obtain the analogous expression

$$P_{\nu+} = \text{Im}(g|g)_{-1}^+ P_{E(\nu)-J} + \text{Im}(g|g)_0^+ P_{E(\nu)} + \text{Im}(g|g)_1^+ P_{E(\nu)+J}$$

with the coefficients $(g|g)_a^+$:

$$(g|g)_a^+ = \text{P.P.}\int dk\,|g(k)|^2 \frac{1}{\omega(k)-aJ}\frac{1}{e^{\beta\omega(k)}-1}, \qquad a=-1,0,1$$

Equation (41) gives the quantum generalization of the Glauber dynamics. The matrix elements $\rho(\mu, \nu, t)$ of the density matrix corresponding to the states μ, ν with different energies will decay quickly. But for the translation invariant Hamiltonian there exist different μ, ν with equal energies. Corresponding matrix elements will decay with the same speed as the diagonal elements of the density matrix. Moreover, one can expect nonergodic behavior for this model. Therefore the generalization (41) of the Glauber dynamics is nontrivial.

VII. EVOLUTION FOR SUBALGEBRA OF LOCAL OPERATORS

In this section, to compare with the results of [4] we consider the dynamics of spin systems, described in the previous section, for Hamiltonian with not necessarily a finite set of spins Λ but for local observable X.

The observable X is local if it belongs to the local algebra—that is, UHF algebra (uniformly hyperfinite algebra):

$$\mathscr{A} = \bigcup_{\Lambda \text{ is finite}} \mathscr{A}_\Lambda$$

where $\mathscr{A}_\Lambda$ is the $*$-algebra generated by the elements

$$\otimes_i X_i, \qquad X_i = 1 \quad \text{for} \quad i \notin \Lambda$$

Consider now the action of θ_0 on local X:

$$\theta_0(X) = \sum_{ij} \sum_{\omega \in F} \Bigg(-i \operatorname{Im}(g_i|g_j)_\omega^- [X, E_\omega^*(D_i) E_\omega(D_j)]$$
$$+ i \operatorname{Im}(g_i|g_j)_\omega^+ [X, E_\omega(D_i) E_\omega^*(D_j)]$$
$$+ 2 \operatorname{Re}(g_i|g_j)_\omega^- \left(E_\omega^*(D_i) X E_\omega(D_j) - \frac{1}{2} \{X, E_\omega^*(D_i) E_\omega(D_j)\} \right)$$
$$+ 2 \operatorname{Re}(g_i|g_j)_\omega^+ \left(E_\omega(D_i) X E_\omega^*(D_j) - \frac{1}{2} \{X, E_\omega(D_i) E_\omega^*(D_j)\} \right) \Bigg) \quad (42)$$

For $E_\omega(D_i)$ we obtain

$$E_\omega(D_i) = \sum_{E(r) \in F_\omega} P_{E(r)-\omega} D_i P_{E(r)}$$
$$= 1 \otimes |\varepsilon_{i-1}\rangle\langle\varepsilon_{i-1}| \otimes |-\varepsilon_i\rangle\langle\varepsilon_i| \otimes |\varepsilon_{i+1}\rangle\langle\varepsilon_{i+1}| \otimes 1 + 1 \otimes |-\varepsilon_{i-1}\rangle$$
$$\times \langle-\varepsilon_{i-1}| \otimes |\varepsilon_i\rangle\langle-\varepsilon_i| \otimes |-\varepsilon_{i+1}\rangle\langle-\varepsilon_{i+1}| \otimes 1 \quad (43)$$

with the frequency ω of the following form:

$$\omega = J_{i-1,i}\varepsilon_{i-1} + J_{i,i+1}\varepsilon_{i+1}$$

Therefore the operator E_ω given by (43) is local; moreover, the corresponding map θ_0 given by (42) maps $\mathscr{A}$ into itself.

Formula (43) explains the physical meaning of the operator $E_\omega(D_i)$. For positive ω it flips the spin at site i along the direction of the mean field of its neighbors (for negative ω it flips the same spin into the opposite direction).

Acknowledgments

Sergei Kozyrev is grateful to Luigi Accardi and Centro Vito Volterra (where this work was done) for their kind hospitality. The authors are grateful to I. V. Volovich for discussions. This work was partially supported by INTAS 9900545 grant. Sergei Kozyrev was partially supported by RFFI 990100866 grant.

References

1. L. Accardi, Y. G. Lu, and I. V. Volovich, *Quantum Theory and Its Stochastic Limit*, Springer-Verlag, New York, 2000.
2. L. Accardi, Y. G. Lu, and I. V. Volovich, *Interacting Fock Spaces and Hilbert Module Extensions of the Heisenberg Commutation Relations*, Publications of IIAS, Kyoto, 1997.
3. L. Accardi, S. V. Kozyrev, and I. V. Volovich, "Dynamics of dissipative two-level system in the stochastic approximation," *Phys. Rev. A* **57**, N3 (1997).
4. L. Accardi and S. V. Kozyrev, Glauber dynamics from stochastic limit, in *White Noise Analysis and Related Topics*, volume in honor of T. Hida, CESNAM, Kyoto, 1999.
5. L. Accardi and S. V. Kozyrev, Stochastic dynamics of lattice systems in stochastic limit, *Chaos, Solitons and Fractals*, to appear.
6. L. Accardi, Y. G. Lu, and I. V. Volovich, *A White Noise Approach to Classical and Quantum Stochastic Calculus*, preprint of Centro Vito Volterra, no. 375, Rome, July 1999.
7. G. Lindblad, *Commun. Math. Phys.* **48**, 119–130 (1976).
8. L. Accardi and S. V. Kozyrev, The stochastic limit of quantum spin system, invited talk at the 3rd Tohwa International Meeting on Statistical Physics, Tohwa University, Fukuoka, Japan, November 8–11, 1999. to appear in the proceedings
9. A. Caldeira and A. J. Leggett, *Physica A* **121**, 587 (1983).
10. D. F. Walls and G. J. Milburn, *Quantum Optics*, Springer-Verlag, New York, 1994.
11. R. J. Glauber, *J. Math. Phys.* **4**, 263 (1963).
12. K. Kawasaki, in *Phase Transitions and Critical Phenomena*, Vol. 2, C. Domb and M. S. Green, eds., Academic Press, New York, 1972.

CP VIOLATION AS ANTIEIGENVECTOR-BREAKING

KARL GUSTAFSON

*Department of Mathematics, University of Colorado, Boulder, CO, U.S.A.;
and International Solvay Institutes for Physics and Chemistry,
University of Brussels, Brussels, Belgium*

CONTENTS

I. Introduction
II. Operator Trigonometry
    A. Brief History of the Operator Trigonometry
    B. Essentials of Operator Trigonometry
    C. Extended Operator Trigonometry
III. Quantum Probability
    A. Bell's Inequality
    B. A General Triangle Inequality
    C. Classic and Quantum Probabilities
IV. CP Symmetry Violation
    A. Elementary Particle Physics
    B. Strangeness Total Antieigenvectors
    C. CP Violation as Antieigenvector-Breaking
V. Conclusion
Acknowledgments
References

I. INTRODUCTION

At my presentation at Solvay XXI, I showed that my operator trigonometry provided a natural geometry for the analysis of quantum probabilities in certain spin systems—for example, as related to Bell's inequality. These results have since been published in recent papers [1,2]. Therefore here I will only summarize those results, with further details available from those papers. Then I want to move forward and present my more recent finding that the operator trigonometry may also be intrinsically connected to certain Kaon systems. I only briefly mentioned this fact at the end of Refs. 1 and 2. In particular here I will show that CP violation may be seen as a certain slightly unequal weighting of the CP antieigenvectors. The results presented here are preliminary, and it is hoped that their refinements may be pursued elsewhere.

II. OPERATOR TRIGONOMETRY

My operator trigonometry came into being 35 years ago in a question about multiplicative perturbation of contraction semigroup infinitesimal generators. The setting was the functional analysis of partial differential equation initial value problems. Although that setting was one of abstract mathematics, in hindsight it may be seen that one of the questions which I addressed at that time, that of multiplicative perturbation BA of a given semigroup generator A, naturally led to both of the key entities of the operator trigonometry: the $\cos \phi(A)$ and the $\sin \phi(A)$ of an arbitrary linear operator A. Here $\phi(A)$ denotes what I called the angle of A: the maximum (or supremum) angle through which the operator A may turn vectors in its domain $\mathscr{D}(A)$.

A. Brief History of the Operator Trigonometry

For more details see Refs. 3 and 4, where the developments of the operator trigonometry are brought up to 1995, and see Ref. 5 for an update to 2000. Also I will refer to Ref. 6 for some other recent developments of the operator trigonometry.

The origins of the operator trigonometry can be traced back to 1966 and three theorems characterizing contraction semigroup generators.

Theorem 1 (Hille–Yosida–Phillips–Lumer). *In a Banach space X, a densely defined linear operator A is the infinitesimal generator of a contraction semigroup e^{tA} iff it is dissipative and $\mathscr{R}(I - A) = X$.*

Theorem 2 (Rellich–Kato–Sz.Nagy–Gustafson). *If A is a contraction semigroup generator and B is A-small, then $A + B$ is a generator iff $A + B$ is dissipative.*

Theorem 3 (Gustafson). *If A is a generator and B is bounded and strongly accretive, then BA is a generator iff BA is dissipative.*

Of course there are a number of generalizations of these results to bounded semigroups, left and right perturbations, and so on (see the literature). Recall that A dissipative means that $\operatorname{Re}[Ax, x] \leq 0$ for all x in the domain of A, for some given semi-inner product, A accretive means $-A$ is dissipative, B m-accretive (also called B strongly accretive) means that $\text{Re}\,[Bx, x] \geq m\|x\|^2$ for some $m > 0$, and B is A-small means $\|Bx\| \leq a\|x\| + b\|Ax\|$ for some $b < 1$.

Theorem 1 may be usefully viewed in the control theory context. Generators of stable evolutions must have their spectra in the left half-complex plane. Theorem 2 is, as is well known, essential to the proof of the spectral completeness of the Hamiltonians of quantum physics. For example, you may take $A = -\Delta$ and $B = -1/r$—that is, A the self-adjoint Laplacian diffusion operator and B the Coulomb potential. Then $i(A + B)$ is the hydrogen operator, here arranged with its spectrum placed on the imaginary axis, which still counts as being dissipative, although in the nonstrict sense. Theorem 3 also has physical applications (e.g., to stochastic time changes), but it is the least important of the above three theorems in the abstract semigroup generator theory. Nonetheless, it was Theorem 3 which led to the operator trigonometry.

Theorem 2 may be proved from Theorem 1, and Theorem 3 may be proved from Theorem 2 (see Ref. 3 and 4. What is missing from Theorem 3 is a criteria for BA to be dissipative when A is dissipative and B is accretive. This question is the origin of the operator trigonometry. Consider in particular the question of when bounded operators A and B are both m-accretive and you want BA to be accretive. From that question in 1966–1968, I found the following four results.

Theorem 4. *For bounded A and B which are both m-accretive on a Banach space X, a sufficient condition for BA to be accretive is*

$$\sin \phi(B) \leq \cos \phi(A) \tag{1}$$

In (1), $\phi(A)$ denotes the angle of an operator A, defined through the first antieigenvalue μ of A

$$\mu = \cos \phi(A) = \inf_{x \neq 0} \frac{\text{Re}[Ax, x]}{\|Ax\|\|x\|} \tag{2}$$

and $\sin \phi(B)$ denotes the entity

$$\nu = \sin \phi(B) = \inf_{\epsilon > 0} \|\epsilon B - I\| \tag{3}$$

That ν deserves to be called $\sin \phi(B)$ follows from the following important result.

Theorem 5. *For any bounded strongly accretive operator B on a Hilbert space X,*

$$\nu^2(B) + \mu^2(B) = 1 \tag{4}$$

I mention that (4) is no longer true for X a general Banach space. I refer to Theorem 5 as the min–max theorem of the operator trigonometry because I proved it that way.

Theorem 6. *For any bounded strictly positive selfadjoint or Hermitian operator A on a Hilbert space X,*

$$\cos\phi(A) = \frac{2\sqrt{mM}}{m+M}, \qquad \sin\phi(A) = \frac{M-m}{M+m} \qquad (5)$$

In (5) $M = \|A\|$ = the upper bound of A and $m = \|A^{-1}\|^{-1}$ = the lower bound of A. One more important result from the early days is the following Euler equation of the variational functional (2).

Theorem 7. *For A a strongly accretive bounded operator on a Hilbert space, the functional in (2) has Euler equation*

$$2\|Ax\|^2\|x\|^2(\operatorname{Re} A)x - \|x\|^2\operatorname{Re}\langle Ax, x\rangle A^*Ax - \|Ax\|^2\operatorname{Re}\langle Ax, x\rangle x = 0 \qquad (6)$$

The Euler equation (6) is satisfied by the first antieigenvectors (see below). When A is self-adjoint, Hermitian, unitary, or normal, all of A's eigenvectors also satisfy (6). Thus (6) generalizes the Rayleigh–Ritz variational characterization of eigenvectors to also include antieigenvectors. The antieigenvectors minimize the functional (2), whereas the eigenvectors maximize it.

All of the above became known in the period 1966–1968. However, we were able to show that for unbounded accretive operators A in a Hilbert space, $\cos\phi(A) = 0$. This in itself is an interesting geometrical characterization of the topological distinction between bounded and unbounded strongly accretive operators A: The unbounded operators can turn vectors as close to 90° as you like, whereas the bounded ones cannot. However, from the point of view of differential equations in Hilbert spaces (e.g., see Ref. 7), the most interesting contraction semigroup generators (e.g., heat equation, Schrödinger equation, etc.) are unbounded. For unbounded operators, criterion (1) is not very interesting. So I lost interest in the operator trigonometry for awhile.

Later (the 1980s) my interests turned to computational fluid dynamics (e.g., see Refs. 3 and 7). One learns quickly that in those applications about 80% of your computer time is used doing linear solvers $Ax = b$. Here A is a very large sparse matrix, often symmetric positive definite (called SPD in the following). As an important widely computed example, A will be a discretized Laplacian or related potential operator. One set of methods for solving very large linear systems $Ax = b$ are the so-called iterative methods. Classic versions of these are the Jacobi, Gauss–Seidel, and Successive Overrelation schemes, but there are interesting faster, more recent algorithms too. See my discussion in Ref. 5. Often the convergence rates of these algorithms depend significantly, although not totally on A's largest and smallest eigenvalues λ_n and λ_1. In a series of papers (see Ref. 5) in the 1990s I was able to provide an alternate, new theory of convergence of many of these methods, in terms of my operator trigonometric entities $\sin\phi(A)$ and $\cos\phi(A)$. Such a geometrical understanding of linear solver iterative convergence, notwithstanding the intensive development for industry of these computational methods over the last 30 years, had not been evident before the operator trigonometry came along.

B. Essentials of Operator Trigonometry

It is possible that the abstract operator trigonometry theory discussed above will find applications elsewhere in operator-theoretic abstract models. However, in this section I restrict A to be a real SPD $n \times n$ matrix with simple eigenvalues. In this way we can fix ideas and summarize the essentials of the operator trigonometry. Besides, A (an SPD matrix) is probably the most important situation for applications. Remember that the operator trigonometry is concerned with an operator's critical turning angles, analogous to the standard spectral theory which is concerned with an operator's critical stretchings—that is, a matrix A's eigenvalues and corresponding eigenvectors. Accordingly, I gave the names antieigenvalues and their corresponding antieigenvectors to A's critical angles (cosines thereof) and the vectors which are turned those amounts, respectively. For a symmetric $n \times n$ positive definite matrix A with eigenvalues $0 < \lambda_1 < \lambda_2 \cdots < \lambda_n$ and corresponding normalized eigenvectors $x_1, \ldots, x_n$, we then have [3,4] the following fundamental matrix-trigonometric entities:

$$\mu_1 = \cos\phi(A) = \min_{x\neq 0} \frac{\langle Ax, x\rangle}{\|Ax\|\|x\|} = \frac{2\sqrt{\lambda_1 \lambda_n}}{\lambda_1 + \lambda_n}$$

$$\nu_1 = \sin\phi(A) = \min_{\epsilon > 0} \|\epsilon A - I\| = \frac{\lambda_n - \lambda_1}{\lambda_n + \lambda_1} \quad (7)$$

$$\phi(A) = \text{the (maximum turning) angle of } A$$

$$x_\pm^1 = \pm\left(\frac{\lambda_n}{\lambda_1 + \lambda_n}\right)^{1/2} x_1 + \left(\frac{\lambda_1}{\lambda_1 + \lambda_n}\right)^{1/2} x_n$$

The first antieigenvectors x_+^1 and x_-^1 are turned by A the maximal turning angle $\phi(A)$, and μ_1 is called the corresponding first antieigenvalue. Higher antieigenvalues and antieigenvectors were originally defined variationally, but later it was seen that it is more natural to define them as

$$x_\pm^i = \pm\left(\frac{\lambda_i}{\lambda_i + \lambda_{n-i+1}}\right)^{1/2} x_{n-i+1} + \left(\frac{\lambda_{n-i+1}}{\lambda_i + \lambda_{n-i+1}}\right)^{1/2} x_i \quad (8)$$

These are the vectors most turned by A on the reducing subspaces $\text{sp}\{x_i, \ldots, x_{n-i+1}\}$ as one excludes the preceding antieigenvectors already accounted for. The cases when the λ_i are not simple eigenvalues pose no problem, and then the eigenvector components of the corresponding antieigenvectors just inherit those higher eigenvector multiplicities (see Ref. 6). The important early result (1968) was that in Hilbert space one has the relationship (4), written more graphically as $\cos^2 \phi(A) + \sin^2 \phi(A) = 1$. The same relationship holds on the reducing subspaces for the reduced turning angles $\phi_i(A)$. Without (4), in my opinion, there is no operator trigonometry.

I have normalized the antieigenvectors in (7) and (8) to norm one, but clearly any scalar multiple of an antieigenvector is also an antieigenvector. Obviously any (real, nonzero) scalar multiple of A or A^{-1} has the same operator trigonometry as A. However, it is important to note that the two antieigenvectors $x_\pm^i$ do not determine a corresponding antieigenspace. Notice that $\text{sp}\{x_+^i, x_-^i\}$ is just the span $\text{sp}\{x_{n-i+1}, x_i\}$ of the two constituent eigenvectors.

Early on in the operator trigonometry I defined a "total" operator turning angle for arbitrary operators A by

$$\cos \phi_{\text{total}}(A) = \inf_{Ax \neq 0} \frac{|\langle Ax, x \rangle|}{\|Ax\| \|x\|} \tag{9}$$

For A a diagonalized normal operator, the total antieigenvectors are the same as in (7) and (8) except that the λ_i must be replaced by $|\lambda_i|$ throughout those expressions. One property of $\phi_{\text{total}}(A)$ that I want to point out for use later in this chapter is the following. Letting A be a unitary operator U, it follows directly from (9) that $\cos \phi_{\text{total}}(U) = \min_{\lambda \in W(U)} |\lambda|$, where $W(U)$ denotes the numerical range of U. (See Ref. 4.) For U, a unitary operator, its numerical range $W(U)$ is just the closure of the convex hull of its spectrum. Thus for a strongly accretive unitary matrix U, $\cos\phi_{\text{total}}(U)$ is the distance from the origin of the complex plane to the nearest point of $W(U)$—or, stated another way, the distance to the nearest chord between U's eigenvalues on the unit circle.

C. Extended Operator Trigonometry

With a view toward extending the rather complete operator trigonometry known for A (an SPD matrix) to arbitrary invertible $n \times n$ matrices A for the computational linear solver problem $Ax = b$, recently in Ref. 6, I extended the operator trigonometry to arbitrary (i.e., beyond A positive self-adjoint) invertible operators A via A's polar decomposition $A = U|A|$ via the fact that $|A|$ is positive self-adjoint. Then I set $\phi_e(A) = \phi(|A|)$. This is not entirely satisfactory because I ignore the unitary part U of A's polar factorization. Of course I was originally interested in the critical *relative* turnings, just as in the eigenvalue theory one is interested in the critical *relative* stretchings. So the uniform turnings that one would find in a unitary operator U were not what was sought. I will return to this point later in this chapter.

In my extended operator trigonometry, one changes definition (3) to

$$\sin \phi_e(A) = \inf_\epsilon \|\epsilon A - U\| \tag{10}$$

Then, considering, for example, A to be arbitrary $n \times n$ nonsingular matrix with singular values $\sigma_1 \geqq \sigma_2 \geqq \cdots \geqq \sigma_n > 0$, we obtain from (7) that

$$\sin \phi_e(A) = \min_{\epsilon > 0} \|\epsilon |A| - I\| = \frac{\sigma_1(A) - \sigma_n(A)}{\sigma_1(A) + \sigma_n(A)} \tag{11}$$

One may check that the key min–max identity (4) is then satisfied if one modifies (2) to

$$\cos \phi_e(A) = \min_{x \neq 0} \frac{\langle |A|x, x \rangle}{\||A|x\| \|x\|} \tag{12}$$

Cos $\phi_e(A)$ is then given as in (7) with λ_1 and λ_n replaced by σ_n and σ_1, respectively.

Because more details on the extended operator trigonometry are available in the recent paper [6], I will just summarize the main points in the following theorem.

Theorem 8. *Let A be an arbitrary invertible operator on a complex Hilbert space X. From $A = U|A|$ polar form define the angle $\phi(A)$ according to (10) and $\cos \phi(A)$ according to (12). Then the min–max identity*

$$\sin^2 \phi_e(A) + \cos^2 \phi_e(A) = 1 \tag{13}$$

holds, and a full operator trigonometry of relative turning angles obtains for A from that of $|A|$. In this extended operator trigonometry, the Euler equation (6) is replaced by the simpler expression, with $\|x\| = 1$ for simplicity:

$$\frac{|A|^2 x}{\langle |A|^2 x, x \rangle} - \frac{2|A|x}{\langle |A|x, x \rangle} + x = 0 \tag{14}$$

Now a few remarks. As concerns my extended operator trigonometry, surely one could have instead used the singular value decomposition of A, left polar form, or other factorizations of A. However, after I tried several alternatives, I preferred the right polar form. Remember that this exists for arbitrary closed operators in a Hilbert space. However, that choice comes at the price of ignoring any trigonometric effects of the unitary factor of A. Until recently I had little need for some finer operator trigonometry of unitary operators, since intuitively one can view them as block uniform rotations or block mirror symmetries. However, when considering quantum computing, it became clear [5] that I now need a better operator trigonometry for unitary operators. This also became clear as concerns the treatment of quantum spin probabilities [1,2]. It also becomes clear from this chapter. I am not going to try to remedy this lacuna in my theory here, but I hope to do so elsewhere.

The operator trigonometry is at once both generally defined and quite incomplete. Even in the A SPD case where I have a "complete" theory, one finds new things. For example, I showed in Ref. 6 that the angle $\phi_{\pm}$ between the two first antieigenvectors x_+^1 and x_-^1 is always $\phi(A) + \pi/2$. This finding led to other new facts and questions [5,6]. As another example, recently I wondered: Can I make the antieigenvectors $x_{\pm}^1$ orthogonal? That would make them more resemble the setting for eigenvectors. Let me observe here that the answer is yes, and quite immediate. Define a new inner product $\langle x, y \rangle_A = \langle Ax, y \rangle$. Then you may quickly verify that

$$\langle x_+^1, x_-^1 \rangle_A = \langle Ax_+^1, x_-^1 \rangle = \left(\frac{\lambda_n}{\lambda_1 + \lambda_n}\right)\lambda_1 - \left(\frac{\lambda_1}{\lambda_1 + \lambda_n}\right)\lambda_n = 0 \quad (15)$$

This orthogonality extends to various combinations of the higher antieigenvectors as well. For example, one may confirm that $\langle x_+^i, x_-^j \rangle_A = 0$ for any $1 \leq i \leq n/2$, $1 \leq j \leq n/2$, by a similar computation. However, for $i \neq j$ that fact was already a consequence of x_+^i and x_-^j being in orthogonal reducing subspaces of A. The same pairwise orthogonality holds for the extended operator antieigenvectors using the $|A|$ inner product. However, it is not a property of the total antieigenvectors.

III. QUANTUM PROBABILITY

About 5 years ago (1996) I noticed a connection between my operator trigonometry and certain issues occurring in or related to quantum spin system probabilities. Essentially, the result became that certain inequalities obtained by Bell and Wigner for consideration of hidden variables in quantum mechanics, and certain inequalities obtained by Accardi and Fedullo for the existence or nonexistence of quantum or classic probability logics, can all be embedded naturally into my operator trigonometry, which then also extends those earlier results. See Refs. 1 and 2 for full details.

In this section I will refer to the papers [1,2] for all bibliographical citations that I mention here. Also I will be brief. For the purposes of this chapter, this section may be regarded as the bridge between the operator trigonometry of Section II and its new application to CP violation in Section IV. It was my finding [1,2] that my operator trigonometry provides a natural and more general geometry for these quantum probability spin systems that prompted me to look further into quantum elementary particle theory and, in particular, into the CP-violation models.

A. Bell's Inequality

The following is well known, so I will summarize. In 1935 Einstein, Podolsky, and Rosen created a gedanken experiment which they claimed demonstrated that conventional quantum mechanics cannot provide a complete description of reality. In 1951 David Bohm addressed this issue and formulated his version of a hidden variable theory which could conceivably provide a more complete description within quantum mechanics. In 1964 Bell presented his famous inequality and exhibited certain quantum spin measurement configurations whose expectation values could not satisfy his inequality. Bell's analysis assumes that physical systems (e.g., two measuring apparatuses) can be regarded as physically totally separated, in the sense of being free of any effects one from the other. Thus his inequality could provide a "test" that could be failed by measurements performed on correlated quantum systems. In particular, he argued that local realistic hidden variable theories could not hold. As is well known, the 1982 physical experiments of Aspect et al. demonstrated that beyond any reasonable doubt the Bell inequalities are violated by certain quantum systems, and papers continue to appear with further demonstrated violations. However, it should be stressed that these experiments show the violation of certain locality assumptions and do not really deal with the hidden varable issues.

In a 1970 paper, Wigner simplified and clarified in several ways the argument of Bell. Wigner assumed that all possible measurements are predetermined, even if they involve incompatible observables; moreover, any measurement on one of two apparatuses does not change the present outcomes of measurements on the other apparatus. Thus the meanings of locality and realism are made more clear, and both assumptions are present in the model setup. It is helpful to imagine, for example, that the "hidden variable" is just the directional orientation of each of the two apparatuses, each of which can be thought of as just a three-dimensional possibly skew coordinate system. Then two spin $1/2$ particles are sent to the apparatuses, each to one, both coming simultaneously from a common atomic source, with perfect anticorrelation and singlet properties. Nine measurements are then needed to simultaneously measure the direction vectors $\omega_1, \omega_2, \omega_3$ of the two spins. Each spin has two possible values $1/2 \equiv +$, $-1/2 \equiv -$, so each measurement can permit four relative results: $++, --, +-, -+$. Therefore there are 4^9 possible outcomes. Wigner then assumes that the spins are not affected by the orientation of the particular measuring apparatus. This reduces the outcomes to 2^6 possibilities. For example, if the hidden variables are in the possibility domain $(+,-,-;-+-)$, then the measurement of the spin component of the first particle in the ω_1 direction will yield value spin $= +$, no matter what direction the spin of the second particle is measured.

I will come back to Bell's original inequality below. However, it is less interesting than Wigner's version, because Bell's considerations evolved from thinking in terms of Kolmogorovian classic probabilities, whereas Wigner placed the question squarely in quantum mechanical Hilbert space. To continue, Wigner reformulates Bell's setup and reduces the outcomes to 2^6 possibilities—for example, the instance in which the hidden variables are in the domain $(+,-,-;-,+,-)$ that I mentioned above. Then he shows that these 64 possibilities can be grouped by sixteens, with most terms canceling. In the first of the four resulting spin measurement possibilities—that is, that of $++$ for the first particle in direction ω_1 and for the second particle in direction ω_3—Wigner then arrives at the conclusion that the hidden parameters can reproduce the quantum mechanical probabilities only if the three directions $\omega_1, \omega_2, \omega_3$ in which the spins are measured are so situated that

$$\frac{1}{2}\sin^2\frac{1}{2}\theta_{23} + \frac{1}{2}\sin^2\frac{1}{2}\theta_{12} \geq \frac{1}{2}\sin^2\frac{1}{2}\theta_{31} \tag{16}$$

Then to make the point very clear, he specializes to the case in which the three directions $\omega_1, \omega_2, \omega_3$ in 3 space are coplanar and with ω_2 bisecting the angle between ω_1 and ω_3. Then $\theta_{12} = \theta_{23} = \theta_{31}/2$ and inequality (16) becomes

$$\sin^2\left(\frac{1}{2}\theta_{12}\right) \geq \frac{1}{2}\sin^2(\theta_{12}) = 2\sin^2\left(\frac{1}{2}\theta_{12}\right)\cos^2\left(\frac{1}{2}\theta_{12}\right) \tag{17}$$

from which $\cos^2(\frac{1}{2}\theta_{12}) \leq 1/2$ and hence $\theta_{31} = 2\theta_{12} \geq \pi$. Thus condition (16), which is necessary for appropriate quantum mechanical spin probabilities for the hidden variable theories, is violated for all $\theta_{31} < \pi$. Wigner then asserts (without giving the details) that the same conclusion may be drawn for all coplanar directions. He also similarly considers the other three spin measurement configurations and in each case reduces them to inequalities analogous to (16), with similar conclusions. I treat all of these in [1,2], so we will look only at this first of the four spin configuration possibilities here.

B. A General Triangle Inequality

In the early operator trigonometry we needed the following general triangle inequality (e.g., see Refs. 3 and 4). It does not seem to be well known. Its "truth" is geometrically "self-evident," but we never found a proof so we had to provide one ourselves.

Let x, y, z be any three vectors in a real or complex Hilbert space of any dimension. For convenience take them to be of norm 1, although that is not necessary. Let $\langle x, y \rangle = a_1 + ib_1$, $\langle y, z \rangle = a_2 + ib_2$, $\langle x, z \rangle = a_3 + ib_3$. Define the angles $\phi_{xy}, \phi_{yz}, \phi_{xz}$ in $[0, \pi]$ by $\cos \phi_{xy} = a_1$, $\cos \phi_{yz} = a_2$, and $\cos \phi_{xz} = a_3$.

Theorem 9. *There holds the general triangle inequality*

$$\phi_{xz} \leq \phi_{xy} + \phi_{yz} \tag{18}$$

The easiest proof of (18) seems to arrive via the Gram matrix

$$G = \begin{bmatrix} \langle x,x \rangle & \langle x,y \rangle & \langle x,z \rangle \\ \langle y,x \rangle & \langle y,y \rangle & \langle y,z \rangle \\ \langle z,x \rangle & \langle z,y \rangle & \langle z,z \rangle \end{bmatrix} \tag{19}$$

A Gram matrix is positive semidefinite in any number of dimensions, and it is definite iff the given vectors are linearly independent. Let me prove (18) here so that you may get the feel of Theorem 9 and, more importantly, so that it will become clear that Wigner's results are a special case of (18).

It suffices to show

$$\cos \phi_{xz} \geq \cos(\theta_{xy} + \phi_{yz}) \tag{20}$$

which by the sum formula for cosines is equivalent to

$$\sqrt{1 - a_1^2} \sqrt{1 - a_2^2} \geq a_1 a_2 - a_3 \tag{21}$$

The desired result (18) follows trivially when the right-hand side of (21) is negative. In the other case we need

$$(1 - a_1^2)(1 - a_2^2) \geq (a_1 a_2 - a_3)^2 \tag{22}$$

which is equivalent to

$$1 - a_1^2 - a_2^2 - a_3^2 + 2 a_1 a_2 a_3 \geq 0 \tag{23}$$

But for unit vectors the determinant of the Gram matrix (19) becomes (using complex cancellations)

$$|G| = \begin{vmatrix} 1 & a_1 & a_3 \\ a_1 & 1 & a_2 \\ a_3 & a_2 & 1 \end{vmatrix} = 1 + 2 a_1 a_2 a_3 - (a_1^2 + a_2^2 + a_3^2) \geq 0 \tag{24}$$

which is equivalent to (23).

Let us now look at Wigner's inequalities (16) and (17) above from the operator trigonometric perspective. The Gram determinant G of (24) vanishes if and only if the three directions are coplanar, no matter what their frame of reference. Then we may write the equality (24) as follows:

$$(1 - a_1^2) + (1 - a_2^2) - (1 - a_3^2) = 2a_3(a_3 - a_1 a_2) \tag{25}$$

or in the terminology of Wigner

$$\sin^2\left(\frac{1}{2}\theta_{12}\right) + \sin^2\left(\frac{1}{2}\theta_{23}\right) - \sin^2\left(\frac{1}{2}\theta_{13}\right)$$
$$= 2\cos\left(\frac{1}{2}\theta_{13}\right)\left[\cos\left(\frac{1}{2}\theta_{13}\right) - \cos\left(\frac{1}{2}\theta_{12}\right)\cos\left(\frac{1}{2}\theta_{23}\right)\right] \tag{26}$$

Wigner's situation for avoidance of probability violation in the coplanar case is equivalent to the right-hand side of (26) being nonnegative. Because all half-angles in (26) do not exceed $\pi/2$, except for the trivial case when $\frac{1}{2}\theta_{13} = \pi/2$, the nonnegativity of (26) means that of its second factors. By choosing the direction ω_2 to be the "one in between" among the half-angles, we can without loss of generality assume that $\frac{1}{2}\theta_{12} + \frac{1}{2}\theta_{23} = \frac{1}{2}\theta_{13}$. The required nonnegativity of (26) then reduces by the elementary cosine sum formula to

$$\cos\left(\frac{\theta_{12} + \theta_{23}}{2}\right) \geq \frac{1}{2}\left[\cos\left(\frac{\theta_{12} + \theta_{23}}{2}\right) + \cos\left(\frac{\theta_{12} - \theta_{23}}{2}\right)\right] \tag{27}$$

that is, $\cos((\theta_{12} + \theta_{23})/2) \geq \cos((\theta_{12} - \theta_{23})/2)$, which is false for positive θ_{23}. This completes Wigner's argument and is the meaning of coplanar quantum probability violation.

Theorem 10. *The Wigner hidden variable geometric considerations are special cases of the operator trigonometry.*

C. Classic and Quantum Probabilities

Motivated partially by Bell's and Wigner's inequalities and also by related classic versus quantum probability issues, Accardi and Fedullo (and others) developed certain key inequalities upon which their results as to whether certain spin systems could support Kolmogorov classic probabilities or Hilbert-space quantum probabilities depended. I don't want to reproduce all the details of how I showed that these inequalities may also be seen as special cases of the general triangle inequality (18), so see Refs. 1 and 2 for those details. However, to make the point clear here, note that the key inequality of Accardi and Fedullo (1982), namely

$$\cos^2\alpha + \cos^2\beta + \cos^2\gamma - 1 \leq 2\cos\alpha\cos\beta\cos\gamma \tag{28}$$

a necessary and sufficient condition for the angles α, β, γ of a quantum spin model in a two-dimensional complex Hilbert space, is precisely the same as the operator trigonometry relation (23) for the real cosines a_1, a_2, a_3 of the angles between arbitrary unit vectors in any complex Hilbert space. The angles of inequality (28) are related to transition probability matrices $P(A|B), P(B|C), P(C|A)$ for three observables A, B, C which may take two values. See Refs. 1 and 2 for more specifics.

Theorem 11. *The Accardi–Fedullo probability geometric considerations are special cases of the operator trigonometry.*

There is one comment I would like to add here. It is that the embedding of the Accardi–Fedullo et al. results within my operator trigonometry is to this date only established for the case of three observables that may take on two values. Via Pauli spin matrices, in that case the issues concerning quantum probabilities then reduce to questions about 2×2 matrices. Moreover, in that situation the necessary inequalities for the existence of classic Kolmogorov probabilities may be seen to be included within the same inequalities needed for the quantum case. If one goes to, say, two observables with four possible values, one has an open question. Obviously there are many other related open questions. I hope to address some of those matters in a later publication.

Finally, let's look at Bell's Inequality, as I promised above. There is of course a huge literature on Bell-type inequalities, and I have cited some sources for that in Refs. 1 and 2. However, keep in mind that our real interest in this chapter is with quantum probabilities, whereas Bell's model evolved from Kolmogorovian probability concepts. The original Bell inequality is

$$|P(a,b) - P(a,c)| \leq 1 + P(b,c) \tag{29}$$

Here $P(a,b)$ is the expectation value of the product of the spin components along unit vector directions a and b. Bell showed that if one considers expectation values, for which $P(a,b) = -\cos\theta_{ab}$, then some direction combinations will violate (29). Let us cast that situation into the operator trigonometric frame of this chapter. Let $a_3 = \cos\theta_{ac}$, $a_1 = \cos\theta_{ab}$, $a_2 = \cos\theta_{bc}$. Square Bell's inequality (29), which then becomes the two cases

$$\begin{aligned} a_1^2 + a_2^2 + a_3^2 + 2[a_3 a_2 - a_3 a_1 - a_2 a_1] \leq 1 \\ a_1^2 + a_2^2 + a_3^2 + 2[a_2 a_1 - a_3 a_1 - a_3 a_2] \leq 1 \end{aligned} \tag{30}$$

Consider the first case, and the second may be similarly examined. Insert the Grammian expression (24) into (30) to obtain

$$1 - (a_1^2 + a_2^2 + a_3^2) + 2a_1a_2a_3 - 2a_1a_2a_3 - 2a_3a_2 + 2a_3a_1 + 2a_2a_1 \geqq 0 \quad (31)$$

that is,

$$|G| \geqq 2[a_1a_2a_3 + a_3a_2 - a_3a_1 - a_2a_1] \quad (32)$$

When the directions a, b, c are coplanar, $|G| = 0$ and the right-hand side of (32) will tell you exactly the violating directions—that is, those for which the right-hand side is positive. Similar violating and nonviolating directional delineations may in principle be obtained for the general noncoplanar case $|G| > 0$ in the same way. Such Bell inequality constraints appear to give rise to some interesting new trigonometric algebraic–geometry inequalities, even in the low-dimensional case of three space dimensions.

IV. CP SYMMETRY VIOLATION

The principal goal of this chapter is to report the interesting natural occurrence of the operator trigonometry within the CP elementary particle theory. More details as well as investigation of possible extensions of this finding to more general elementary particle considerations will be pursued elsewhere. Also for the reasons that will become clear below, I hope the elementary particle physics may help indicate appropriate extensions of the operator trigonometry in the future.

I first noticed this intrinsic connection of the operator trigonometry to the CP-violation developments when reading the recent review [8] where Bertram Schwarzschild discusses the most recent [9] experimental results that apparently finally remove all doubts about physical demonstration that CP violation exists. See the original experimental finding of CP violation in Ref. 10, the early theoretical models [11,12], and the extensive experimental work and results [13,14] about 10 years ago as cited in Ref. 8. Here I want only to attempt a "thumbnail" sketch for the nonspecialist of some of the issues involved. For that I will lean heavily on Refs. 8–14, and I have also consulted the treatments in Refs. 15–20.

A. Elementary Particle Physics

The standard model (Glashow–Weinberg–Salam–Yang–Mills et al.) of elementary particle physics is based upon the gauge group $S(U(2) \times U(3))$, which may be represented as a subset of the 5×5 unitary matrices. The goal is to understand all of the interactions of matter which are seen in the large particle accelerators. The famous CPT theorem [15] states the necessary invariance of any quantum field theory under the combined action of charge conjugation C, parity reversal P, and time reversal T. That is, if you reverse the sign of charge, the sign of space, and the sign of time, in any order, the local field theory remains valid. From a scattering theory point of view, through S-matrix analysis continuation from positive to negative energies, one is led to formulate the existence of antiparticles, that correspond to an exact interchange of the outgoing states with the incoming states.

CP denotes the combined operation of charge conjugation (the replacement of particles by their antiparticles) and parity inversion (mirror symmetry in space). The only two undisputed physical manifestations of CP symmetry violation in nature to date are (a) the predominance of matter over antimatter in the visible universe and (b) decay of neutral K mesons. The latter was first demonstrated in Ref. 10. There the K_L° meson in a helium gas was shown to decay into two charged pions, symbolically $K_L^\circ \to \pi^+\pi^-$, in about 2×10^{-3} of the experiments. Here K_L° denotes the longer-lived of the two neutral kaon eigenstates of lifetime and mass, and K_S° denotes the short-lived eigenstate. Let K° and $\bar{K}^\circ$ be the two (degenerate) mass eigenstates of strangeness. Strangeness is an added quantum number that is conserved by the strong and electromagnetic interactions, although not by the weak interaction. For parity conservation, one needs pair pure states so K° and $\bar{K}^\circ$ have assigned eigenvalues of $\lambda_1 = +1$ and $\lambda_2 = -1$, respectively. The landmark paper [10] then concluded on the basis of the experimental evidence that K_L° could not be a pure CP eigenstate. Let

$$K_1 = \frac{K^\circ + \bar{K}^\circ}{\sqrt{2}}, \qquad K_2 = \frac{K^\circ - \bar{K}^\circ}{\sqrt{2}} \tag{33}$$

be the pure CP eigenstates, with eigenvalues $+1$ and -1, respectively. Then before CP violation was demonstrated, it was assumed that $K_L^\circ = K_2$ and, by symmetry, also $K_S^\circ = K_1$. The result of Ref. 10 is that K_L° could not be a pure CP eigenstate due to the presence of a two-pion decay mode. From their measurements they then proposed that K_L° is a mixture of K_2 (predominant) and a small amount of K_1, specifically [10]

$$\begin{aligned} K_L^\circ &= K_2 + \epsilon K_1 \\ &= 2^{-1/2}[(K^\circ - \bar{K}^\circ) + \epsilon(K^\circ + \bar{K}^\circ)] \end{aligned} \tag{34}$$

where the small CP-violation mixing parameter is about $\epsilon = 0.0023$.

Shortly after Ref. 10, in Ref. 11 it was proposed that CP violation might be due to a very, very weak interaction—say, 10^{-7} of the standard weak interaction. See also the discussion of other models in Ref. 12, where the need to introduce new interacting weak fields to implement the CP-violating decays is put forth. However, according to the recent summary [8], the experiments [13,14] and especially the most recent experiments [9] support the assertion that, at least in part, the kaon CP violation may be explained within the standard model without the need to postulate the existence of some new superweak force.

B. Strangeness Total Antieigenvectors

Let me consider here the more common occurrence of CP symmetry—that is, no CP violation. Then in the neutral kaon situation we have

$$K_L^\circ = K_2 = \frac{K^\circ - \bar{K}^\circ}{\sqrt{2}} = \text{1st strangeness total antieigenvector}$$
$$K_S^\circ = K_1 = \frac{K^\circ + \bar{K}^\circ}{\sqrt{2}} = \text{2nd strangeness total antieigenvector} \quad (35)$$

where here for convenience I have just moved the $\pm$ in the earlier antieigenvector expression (7) to $\mp$ in the second term; that is, a normal operator's total first antieigenvectors are given by [3,4]

$$x_\mp^1 = \left(\frac{|\lambda_n|}{|\lambda_1| + |\lambda_n|}\right)^{1/2} x_1 \mp \left(\frac{|\lambda_1|}{|\lambda_1| + |\lambda_n|}\right)^{1/2} x_n \quad (36)$$

Because $n = 2$ and $|\lambda_1| = |\lambda_2| = 1$ for the mass eigenstates K° and $\bar{K}^\circ$, (35) says that the longer-lived kaon mass eigenstate K_L°, which is the focus of the experimental studies reported in Refs. 8–10, 13, and 14, would be the first strangeness total antieigenvector. This is my first observation.

Theorem 12. *The two CP eigenstates K_2 and K_1 are exactly the two strangeness total antieigenvectors.*

Next I want to recall the notion of quark mixing matrix. In Ref. 12 by appropriate phase convention arguments it was shown that the 4×4 representation matrix of the $SU_{\text{weak}}(2)$ interaction was of the form $\Lambda = K \begin{bmatrix} 0 & U \\ 0 & 0 \end{bmatrix} K^{-1}$; and by ignoring the gauge field, one could focus on the 2×2 unitary matrix

$$U = \begin{bmatrix} \cos\theta & \sin\theta \\ -\sin\theta & \cos\theta \end{bmatrix} \quad (37)$$

Such matrices and their generalizations are commonly referred to as CKM (Cabibbo–Kobayashi–Maskawa) matrices or, more commonly, just as the *quark mixing matrix*. Their purpose is to relate the quark mass eigenstates to the (e.g., weak) interaction eigenstates. Thus I may write (35) in the form $U\begin{bmatrix}\bar{K}^\circ\\K^\circ\end{bmatrix}=\begin{bmatrix}K_1\\K_2\end{bmatrix}$ — that is, specifically,

$$\frac{1}{\sqrt{2}}\begin{bmatrix}1 & 1\\-1 & 1\end{bmatrix}\begin{bmatrix}\bar{K}^\circ\\K^\circ\end{bmatrix}=\begin{bmatrix}K_1\\K_2\end{bmatrix}=\begin{bmatrix}K_S^\circ\\K_L^\circ\end{bmatrix} \quad (38)$$

in terms of the quark mixing matrix with a phase parameter $\theta = \pi/4$.

A simple computation reveals that the quark mixing matrix U in (37) has eigenvalues $\lambda_1 = e^{i\theta}$ and $\lambda_2 = e^{-i\theta}$, corresponding eigenvectors $x_1 = \frac{1}{\sqrt{2}}\begin{bmatrix}1\\i\end{bmatrix}$ and $x_2 = \frac{1}{\sqrt{2}}\begin{bmatrix}1\\-i\end{bmatrix}$, and total antieigenvectors $x_- = \begin{bmatrix}0\\i\end{bmatrix}$ and $x_+ = \begin{bmatrix}1\\0\end{bmatrix}$, respectively, according to (36). Of course, U is a uniform rotation matrix and all real vectors are rotated an angle $-\theta$, but I want to single out the two antieigenvectors $x_\mp$ here. Simple computations verify that

$$\frac{|\langle Ux_\pm^1, x_\pm^1\rangle|}{\|Ux_\pm^1\|\|x_\pm^1\|} = \frac{\text{Re}\langle Ux_\pm^1, x_\pm^1\rangle}{\|Ux_\pm^1\|\|x_\pm^1\|} = \cos\theta \quad (39)$$

Thus from (2), (9), (36), and (39) we have verified the following theorem.

Theorem 13. *The quark mixing matrix (37) has total antieigenvectors (36). These are also its usual (real) antieigenvectors. U's antieigenvalues are $\mu_1 = \cos\theta \equiv \mu_1^{total}$ and U's operator angles are $\phi(U) \equiv \phi_{\text{total}}(U) = \theta$.*

Reference 12 also presented an interesting 3×3 unitary quark mixing matrix (which I won't reproduce here) that results from assuming that there is another field interaction in the model. The later formulation and finding of top and bottom quarks and the extension of quartet models to octet models justified this extension of the CP-violation theory to larger unitary quark mixing matrices. I don't want to go into "all that" here, but there seem to be many still unresolved related issues [16–20]. In principle, we could compute the antieigenvectors and operator trigonometry for such octet models should that appear to be interesting.

C. CP Violation as Antieigenvector-Breaking

Turning now to the CP-violation case and accepting the Christenson et al. [10] ansatz that $K_L^\circ = K_2 + \epsilon K_1$ is a mixed CP state [see (34) above], immediately we see that the effect of the mixing parameter $\epsilon \neq 0$ is that the kaon longer-lived mass eigenstate K_L° is no longer a strangeness antieigenvector in the sense of (36). I don't know what this really means physically, but from the operator trigonometry point of view it means that CP violation not only breaks a fundamental physical symmetry but it also somehow reduces a maximum turning angle of a strangeness operator. I will call this phenomena *antieigenvector-breaking*. This is my second main observation.

Theorem 14. *CP violation with nonvanishing mixing parameter ϵ implies that the mass eigenstates K_L° and K_S° are no longer the proper strangeness antieigenvectors.*

Let me provide some feeling for Theorem 14 by a small computation. If I match the $K_L^\circ = K_2 + \epsilon K_1$ ansatz [10] with a corresponding assumption that $K_S^\circ = K_1 - \epsilon K_2$, we may see the perturbative effect of the CP-violating mixing parameter. I hasten to say that to my knowledge the K_S° mixing parameter need not exist, need not be ϵ, and so on, but the computation can proceed anyway. Then we arrive at the quark mixing relation

$$\frac{1}{\sqrt{2}}\left\{\begin{bmatrix} 1 & 1 \\ -1 & 1 \end{bmatrix} + \epsilon \begin{bmatrix} 1 & -1 \\ 1 & 1 \end{bmatrix}\right\}\begin{bmatrix} \bar{K}^\circ \\ K^\circ \end{bmatrix} = \begin{bmatrix} K_S^\circ \\ K_L^\circ \end{bmatrix} \qquad (40)$$

The CP-symmetry quark mixing matrix U has become $U + \epsilon U^T$. The CP-violating quark mixing matrix $U + \epsilon U^T$ has eigenvalues $\lambda_{1,2} = (1+\epsilon)/\sqrt{2} \pm i(1-\epsilon)/\sqrt{2}$. Although these eigenvalues lie on a $(1+\epsilon^2)^{1/2}$ radius and the matrix is no longer unitary or orthogonal, this broken-symmetry quark mixing matrix M_ϵ of (40) is still a normal operator. Thus its antieigenvectors could be computed according to the known operator trigonometry, and thus one could express the CP-violating eigenstates as perturbed antieigenvectors of strangeness. This perturbed quark mixing matrix has a slightly smaller operator turning angle than that of the CP-symmetry quark mixing matrix. This can be ascertained by plotting the two eigenvalues, drawing the chord between them, and then using the property I pointed out just after (9), that the $\cos \phi_{\text{total}}(M_\epsilon)$ is the distance from the origin to this chord. This antieigenvalue property holds for normal operators as well. From this one finds that instead of $\phi_{\text{total}}(U) = \theta = \pi/4$, one now has turning angle $\phi_{\text{total}}(M_\epsilon) = \tan^{-1}((1-\epsilon)/(1+\epsilon))$. For the experimental $\epsilon = 0.0023$, this turning angle is about 44.868°. Finally I comment that for both U and M_ϵ one obtains the antieigenvalue at real phase; that is, the midpoint of the chord lies exactly on the real axis.

V. CONCLUSION

A complete treatment of this new connection of the operator trigonometry to elementary particle physics would necessitate further extension of my operator trigonometry and will hopefully be pursued elsewhere. Also, there is a lot to learn about the theoretical physical elementary particle models, both standard and nonstandard. Moreover, the big business that these experimental efforts really constitute must be recognized. Note that there are 49 authors of Ref. 13 and 29 authors of Ref. 14 and 33 additional acknowledged staff workers in Ref. 13, a total of 111 scientists. The large group [9] that produced the latest CP-violation measurements is so far unnamed but represents a continuation of a 15 year competition between Fermilab and CERN on the CP-violation question.

Acknowledgments

I would like to thank the Solvay Institute for the invitation and financial support enabling my participation in this Congress. In particular, I would like to thank Professors Ilya Prigogine and Ioannis Antoniou for organizing this Congress and for many pleasant and fruitful collegial discussions over the years. I would also like to thank the Japanese sponsors of the Congress. In particular, I believe a hearty statement of appreciation should go to Professor Kazuo Kitahara and his team, who carried the burdens of local arrangements and who also collected in an extremely conscientious and impressive manner all of the taped or written comments associated with the lectures at the Congress. I would like to acknowledge several useful conversations with Luigi Accardi concerning Kolmogorov and Quantum probability models. Finally, I would like to thank Professor Takehisa Abe of the Shibaura Institute of Technology, Professor Mitsuharu Otani of Waseda University, and Dr. Mei Kobayashi of IBM—Japan for hosting me in Tokyo for two lectures as I unjetlagged there a couple of days just prior to the Congress.

References

1. K. Gustafson, Quantum trigonometry. *Infinite Dimensional Analysis, Quantum Probab. Relat. Topics* **3**, 33–52 (2000).
2. K. Gustafson, Probability, geometry, and irreversibility in quantum mechanics. *Chaos, Solitons & Fractals*, **12**, 2849–2858 (2001).
3. K. Gustafson, *Lectures on Computational Fluid Dynamics, Mathematical Physics, and Linear Algebra*, World Scientific, Singapore, 1997.
4. K. Gustafson and D. Rao, *Numerical Range*, Springer, Berlin, 1997.
5. K. Gustafson, An unconventional computational linear algebra: Operator trigonometry, in *Unconventional Models of Computation, UMC'2K*, I. Antoniou, C. Calude, and M. Dinneen, eds., Springer, London, 2001, pp. 48–67.
6. K. Gustafson, An extended operator trigonometry. *Lin. Algebra & Appl.* **319**, 117–135 (2000).
7. K. Gustafson, *Partial Differential Equations and Hilbert Space Methods*, Dover, New York, 1999.
8. B. Schwarzschild, Comments on CP violation in kaon decay. *Physics Today*, **May**, 17–19 (1999).
9. P. Shawhan, KTeV Report, Fermilab seminar, 24 February, 1999.
10. J. Christenson, J. Cronin, V. Fitch, and R. Turlay, Evidence for the 2π decay of the K_2^0 meson. *Phys. Rev. Lett.* **13**, 138–140 (1964).
11. L. Wolfenstein, Violation of CP invariance and the possibility of very weak interactions. *Phys. Rev. Lett.* **13**, 562–564 (1964).
12. M. Kobayashi and T. Maskawa, CP-violation in the renormalizable theory of weak interaction. *Prog. Theor. Phys.* **49**, 652–657 (1973).
13. G. Barr et al., A new measurement of direct CP violation in the neutral kaon system. *Phys. Lett.* **B317**, 233–242 (1993).
14. L. Gibbons et al., Measurement of the CP-violation parameter $Re(\epsilon'/\epsilon)$. *Phys. Rev. Lett.* **70**, 1203–1206 (1993).

15. G. Luders, Proof of the TCP Theorem. *Ann. Phys.* **2**, 1–15 (1957).
16. C. Jarlskog, *CP Violation*, World Scientific, Singapore, 1989.
17. L. Mathelitsch and W. Plessas, *Broken Symmetries*. Springer, Berlin, 1998.
18. I. Bigi and A. Sanda, *CP Violation*, Cambridge Press, Cambridge, 2000.
19. C. Froggatt and H. Nielson, *Origin of Symmetries*, World Scientific, Singapore, 1991.
20. A. Afriat and F. Selleri, *The Einstein, Podolsky, and Rosen Paradox*, Plenum Press, New York, 1999.

PART FOUR

EXTENSION OF QUANTUM THEORY AND FIELD THEORY

DYNAMICS OF CORRELATIONS. A FORMALISM FOR BOTH INTEGRABLE AND NONINTEGRABLE DYNAMICAL SYSTEMS*

I. PRIGOGINE

Center for Studies in Statistical Mechanics and Complex Systems, The University of Texas, Austin, Texas, U.S.A.; and International Solvay Institutes for Physics and Chemistry, Free University of Brussels, Brussels, Belgium

CONTENTS

I. Introduction
II. Poincaré's Theorem
III. Extension of Unitary Transformations
IV. Dynamics of Dissipative Systems
V. Concluding Remarks
Acknowledgments
References

I. INTRODUCTION

It is well known that classic dynamics and quantum mechanics leads to time reversible and deterministic laws. Still in many fields we discover situations where this picture of nature is not applicable. There are two obvious examples: kinetic theory (or nonequilibrium statistical mechanics) and thermodynamics. Kinetic theory deals with probabilities. As in thermodynamics, it includes a broken time symmetry. You find in many books that kinetic theory and thermodynamics are based on approximations. But that is difficult to accept. Indeed to quote only one example, kinetic theory as developed by Boltzmann and

---

*This chapter is a revised version of the paper presented at the Solvay conference in 1998.

Dynamical Systems and Irreversibility: A Special Volume of Advances in Chemical Physics, Volume 122, Edited by Ioannis Antoniou. Series Editors I. Prigogine and Stuart A. Rice.
ISBN 0-471-22291-7. © 2002 John Wiley & Sons, Inc.

others leads to predictions of transport coefficients for dilute gases, which are in complete agreement with observation. Also nonequilibrium thermodynamics leads to the prediction of coherent structures, which are in quantitative agreement with experiment. For this reason, our group has always been interested to formulate dynamics in such a way that probabilities and time symmetry breaking are included in the microscopic description.

Classic or quantum mechanics are generally formulated in the Hilbert space formalism. For such systems, dynamics can be reduced to a set of noninteracting modes by a canonical or unitary transformation. On the other hand, we know from Poincaré that most systems are nonintegrable. As we shall see, these systems break time symmetry and cannot be described by unitary evolution in the Hilbert space. Integrable systems and nonintegrable systems seem to obey quite different laws. But we have now obtained a unified formulation of dynamics, which applies to both (a) the integrable systems for which there exists a unitary operator U leading to a diagonalization of the Hamiltonian and (b) nonintegrable systems outside the Hilbert space. As we shall see, this involves the construction of a nonunitary transformation operator Λ that corresponds to an analytic extension of U. Therefore we see that kinetic theory and thermodynamics, far from being based on any "falsification" of dynamics, are well-defined extensions of classic or quantum dynamics. For reasons we shall explain later, our form of dynamics can be called a dynamics of "correlations."

Many years ago, in my monograph of 1962 [1], I introduced the idea of "dynamics of correlations." However, it is only in recent years that, thanks mainly to the work of Professor Tomio Petrosky and Dr. Gonzalo Ordonez, the generality of this approach was made explicit. In a sense, the main point is that we replace interactions (potential energy) by correlations. For integrable systems, we shall show that these concepts are equivalent. But for nonintegrable systems we can now formulate dynamics in terms of correlations, replacing the potential energy. In this chapter, we shall try to give a simple introduction to the physical ideas. Calculations can be followed in the original papers [2–6].

II. POINCARÉ'S THEOREM

We consider systems with Hamiltonian

$$H = H_0 + \lambda V \qquad (1)$$

where H_0 is the unperturbed Hamiltonian describing noninteracting particles, and V is the interaction. The coupling constant λ is assumed to be dimensionless. For integrable systems, Poincaré has shown that the invariants associated with H_0 can be extended to H. The interaction V can be eliminated. But Poincaré has also shown that for most classic systems there appear divergences in the construction of invariants of motion. We call these systems "nonintegrable in the sense of Poincaré." More precisely, the divergences occur in the perturbation expansion (i.e., expansion in λ^n with $n \geq 0$) of invariants of motion other than functions of the Hamiltonian. The divergences are due to vanishing denominators, which occur when the frequencies of the system obey relations called Poincaré resonances. Then the interactions cannot be eliminated by unitary (or canonical) analytical transformations.

An essential condition for the appearance of Poincaré resonances is that the frequencies are continuous functions of the momenta. This implies that quantum systems in a finite volume are integrable because they have discrete spectra. However, the situation changes when we consider systems in the limit of infinite volume.[1] Then we have a continuous spectrum and resonances.

We can still deal with the vanishing denominators if we interpret them as distributions; for example,

$$\frac{1}{w} \Rightarrow \frac{1}{w \pm i\epsilon} = \mathscr{P}\frac{1}{w} \mp \pi i \delta(w) \qquad (2)$$

where $\epsilon > 0$ is an infinitesimal. We shall come back later to the choice of $\mp i\epsilon$, which is crucial to obtain well-defined perturbation expansions. For such situations, the nonunitary transformation Λ leads to new units of systems that cannot be reduced to trajectory or wavefunction descriptions.

For nonintegrable systems we have no more "certainty." We come to a different description. Indeed, once the regularization of Poincaré's divergences is achieved, we find two unexpected new elements: the breaking of time symmetry and the appearance of intrinsic probabilities. We come to new units or modes, which are no longer invariants. They obey irreversible kinetic processes describing their mutual interactions.

The basic problem of classic mechanics or quantum mechanics is the diagonalization of H. This corresponds to the introduction of a unitary operator U, such that H is diagonal. For integrable systems, we can always diagonalize H. For nonintegrable systems, this is generally not possible. Our interest is in nonintegrable systems.

There is a second aspect, which is clearly exhibited on the level of probability or distribution functions ρ in the Liouville–von Neumann space. As is well known, we have

$$i\frac{\partial}{\partial t}\rho = L_H \rho, \qquad L_H \equiv [H, \ ] \qquad (3)$$

---

[1] This avoids the introduction of boundary conditions. We never have isolated dynamical systems as correlations cross the boundaries.

where we use a unit $\hbar = 1$. L_H is a "superoperator" (in quantum mechanics, it acts on the density operator ρ). Because L_H is a Hermitian operator, Eq. (3) has only real eigenvalues in the Hilbert space for ρ. To describe dissipative processes, we need an extension of L_H outside the Hilbert space [7]. Therefore the formulation of dynamics for nonintegrable systems involves an extension of the statistical description outside the Hilbert space.

The idea that interactions may kick the system out from the Hilbert space was already introduced by Dirac in the frame of field theory [8].

III. EXTENSION OF UNITARY TRANSFORMATIONS

In order to have a common formulation for both integrable and non-integrable systems, we study the dynamics in the Liouville space. We shall consider quantum mechanics. The dynamics is given by the Liouville–von Neumann equation (3). As for the Hamiltonian (1) we have $L_H = L_0 + \lambda L_V$.

Let us consider first the case of noninteracting particles, with $\lambda = 0$. We decompose the density operator ρ into independent components

$$\rho = \sum_\nu P^{(\nu)} \rho \qquad (4)$$

where $P^{(\nu)}$ are projectors to orthogonal eigenspaces of L_0 [see Eq. (20)]. We have

$$L_0 P^{(\nu)} = P^{(\nu)} L_0 = w^{(\nu)} P^{(\nu)} \qquad (5)$$

where $w^{(\nu)}$ are real eigenvalues, which are in quantum mechanics the differences of energies between the ket and bras of a dyatic operator [e.g., see (24)].

The unperturbed Liouville equation is then decomposed into a set of independent equations,

$$i \frac{\partial}{\partial t} P^{(\nu)} \rho = w^{(\nu)} P^{(\nu)} \rho \qquad (6)$$

We associate the diagonal component of ρ with $\nu = 0$. We have $w^{(0)} = 0$; that is the diagonal density matrices are invariants of motion in the unperturbed case. The off-diagonal components with $\nu \neq 0$ simply oscillate with frequencies $w^{(\nu)}$.

Next we consider the interacting case, with $\lambda \neq 0$. We first assume that the system is integrable in the sense of Poincaré. This means that we can construct, by perturbation expansion or otherwise, a superoperator U that puts the dynamics into the same form as in the unperturbed case. We have

$$i \frac{\partial}{\partial t} U\rho = (U L_H U^{-1}) U\rho \qquad (7)$$

This corresponds to a change of representation,

$$\rho \Rightarrow \bar{\rho} \equiv U\rho, \qquad L_H \Rightarrow \bar{\Theta} \equiv UL_HU^{-1} \qquad (8)$$

(hereafter we use overbars to denote operators defined for integrable systems). The defining property of this transformation is that the transformed Liouville operator is diagonal in the unperturbed basis; that is, we have

$$\bar{\Theta} P^{(\nu)} = P^{(\nu)} \bar{\Theta} \equiv \bar{\theta}^{(\nu)} = \bar{w}^{(\nu)} P^{(\nu)} \qquad (9)$$

where $\bar{w}^{(\nu)}$ are the real eigenvalues of $\bar{\Theta}$, corresponding to $w^{(\nu)}$ shifted by the interaction. As a consequence, the dynamics is reduced to the set of equations

$$i\frac{\partial}{\partial t} P^{(\nu)} \bar{\rho} = \bar{w}^{(\nu)} P^{(\nu)} \bar{\rho} \qquad (10)$$

Let us note that for the integrable case the problem of diagonalization of L_H is reducible to the problem of diagonalization of the Hamiltonian H. As is well known in quantum mechanics, U is factorizable as

$$U = u \times u^{-1} \qquad (11)$$

where u is the transformation that diagonalizes H (we use the notation $(A \times B)\rho = A\rho B$ to denote factorizable superoperators). Due to the factorizability of U, for $\nu = 0$ we still have $\bar{w}^{(0)} = 0$. This means that there is a one-to-one correspondence (through the transformation U) between the unperturbed and the perturbed invariants of motion.

In the nonintegrable case we can no longer construct U, due to Poincaré's resonances. However, as mentioned before, we can extend the construction of U if we interpret the denominators as distributions with suitable analytic continuations. We shall present an example below.

Our new construction leads to a nonunitary operator Λ, which is an extension of U to nonintegrable systems. We have now the transformations

$$\rho \Rightarrow \tilde{\rho} = \Lambda\rho, \qquad L_H \Rightarrow \tilde{\Theta} \equiv \Lambda L_H \Lambda^{-1} \qquad (12)$$

The transformed Liouville equation is then

$$i\frac{\partial}{\partial t} \tilde{\rho} = \tilde{\Theta} \tilde{\rho} \qquad (13)$$

The transformed Liouvillian $\tilde{\Theta}$ (now called the "collision operator" in kinetic theory) is no longer diagonal in the unperturbed basis of projectors (this would bring us back to the integrable case). However, we may still require the commutation with the unperturbed projectors, because

$$\tilde{\Theta} P^{(v)} = P^{(v)} \tilde{\Theta} \equiv \tilde{\theta}^{(v)} \tag{14}$$

This means that $\tilde{\Theta}$ is block-diagonal in the unperturbed basis of $P^{(v)}$. It leads to transitions *inside* each $P^{(v)}$ subspace.

We can introduce the complete set of $P^{(v)}$ projectors and express Λ and $\tilde{\Theta}$ in this set:

$$\Lambda = \sum_v P^{(v)} \Lambda, \qquad \tilde{\Theta} = \sum_v \tilde{\theta}^{(v)} \tag{15}$$

Let us also note that Eq. (14) can be written as

$$L_H \Pi^{(v)} = \Pi^{(v)} L_H \tag{16}$$

where

$$\Pi^{(v)} = \Lambda^{-1} P^{(v)} \Lambda \tag{17}$$

The projectors $\Pi^{(v)}$ have been used extensively in our approach [9]. The dynamics is decomposed into a set of independent "subdynamics."

Instead of obtaining a set of equations with invariant or oscillating solutions, we now obtain a set of Markovian kinetic equations

$$i \frac{\partial}{\partial t} P^{(v)} \tilde{\rho} = \tilde{\Theta}^{(v)} P^{(v)} \tilde{\rho} \tag{18}$$

There is an important point: In the construction of Λ (instead of U), time symmetry is broken as a consequence of the analytic continuation of the denominators. As a consequence, the operator $\tilde{\Theta}$ is not Hermitian and leads to complex eigenvalues. It becomes a dissipative operator that describes dissipative processes such as decay or diffusion. To lowest order in the coupling constant, it reduces to Pauli's collision operator of quantum mechanics. In classic mechanics it leads to the Fokker–Planck equation. From the similitude relation of the operator $\tilde{\Theta}$ with the Liouville operator, we can show that L_H has the same eigenvalues as $\tilde{\Theta}$. This is only possible if ρ is not in the Hilbert space. To introduce time irreversibility, we need to go outside the Hilbert space: ρ is then a "distribution" (generalized function). Furthermore, the appearance of kinetic equations means that the wavefunction description is not preserved by the transformation Λ. Indeed, in contrast to U, Λ is a nonfactorizable superoperator. In this representation for nonintegrable systems, dynamics is described in the Liouville–von Neumann space and not in terms of wavefunctions.

In both integrable and nonintegrable cases, we transform the Liouville equation into a set of independent equations. But as already mentioned in the integrable case, the meaning is quite different from the nonintegrable case. Indeed, Eq. (7) corresponds to oscillations, while Eq. (18) corresponds to Markovian kinetic equations. Note that all non-Markovian memory effects are eliminated in the representation $\tilde{\rho} = \Lambda \rho$, which describes interacting dressed particles or modes [10].

Before we go further, let us define more precisely the meaning of the projection operators $P^{(v)}$. As an example, consider a model of a particle interacting with a field, the Friedrichs model with Hamiltonian

$$H = \omega_1 |1\rangle\langle 1| + \sum_k \omega_k |k\rangle\langle k| + \lambda \sum_k V_k(|k\rangle\langle 1| + |1\rangle\langle k|) \qquad (19)$$

The state $|1\rangle$ represents the bare particle (or atom) in its excited level and no field present, while the state $|k\rangle$ represents a bare field mode of momentum k together with the particle in its ground state. The interaction describes transitions between these states, corresponding to absorption and emission processes.

For density matrices the diagonal elements provide the probability to find the particle in the state $|1\rangle$ or the field in a mode $|k\rangle$, while the off-diagonal elements give information on the quantum correlations between particle and field, or among field modes. The interaction changes the state of the correlations. Hence, in the density matrix formulation, there appears naturally a "dynamics of correlations" [1]. To formulate this more precisely, let us first introduce the concept of the "vacuum-of-correlations subspace" that is the set of diagonal dyads $|\alpha\rangle\langle\alpha|$ with $\alpha = 1, k$. We then introduce an integer d that specifies the degree of correlation. This is defined as the minimum number d of successive interactions λL_V by which a given dyadic state can reach the vacuum of correlation. For example, the dyadic states $|1\rangle\langle k|$ and $|k\rangle\langle 1|$ corresponding to particle–field correlations have $d = 1$, while the dyads $|k\rangle\langle k'|$ corresponding to field–field correlations have $d = 2$. For the Friedrichs model, $d = 2$ is the maximum value of the degree of correlation.

The projection operators are written explicitly $P^{(v)}$,

$$\begin{aligned} P^{(0)} &\equiv \sum_{\alpha=1,k} |\alpha;\alpha\rangle\rangle\langle\langle\alpha;\alpha| \\ P^{(\alpha\beta)} &\equiv |\alpha;\beta\rangle\rangle\langle\langle\alpha;\beta| \qquad (\alpha \neq \beta) \end{aligned} \qquad (20)$$

which are orthogonal and complete:

$$P^{(\mu)} P^{(\nu)} = P^{(\mu)} \delta_{\mu\nu}, \qquad \sum_\nu P^{(\nu)} = 1 \qquad (21)$$

with $(\nu) = (0)$ or $(\alpha\beta)$. Here we have introduced a double-ket notation $|\rho\rangle\rangle$ as a double-bra $\langle\langle A|$ for the operators ρ and A with a definition of an inner product:

$$\langle\langle A|\rho\rangle\rangle = \text{Tr}(A^+ \rho) \qquad (22)$$

and we have represented a dyadic operator as $|\alpha; \beta\rangle\rangle \equiv |\alpha\rangle\langle\beta|$. The projector $P^{(0)}$ corresponds to the vacuum of correlations subspace, while the projectors $P^{(k1)}$ and $P^{(1k)}$ correspond to the $d = 1$ subspace and $P^{(kk')}$ to the $d = 2$ subspace. The complement projectors $Q^{(\nu)}$ are defined by

$$P^{(\nu)} + Q^{(\nu)} = 1 \qquad (23)$$

They are orthogonal to $P^{(\nu)}$ (i.e., $Q^{(\nu)} P^{(\nu)} = P^{(\nu)} Q^{(\nu)} = 0$) and satisfy $[Q^{(\nu)}]^2 = Q^{(\nu)}$. As mentioned before, the bare projectors $P^{(\nu)}$ commute with L_0 and they are eigenprojectors of L_0,

$$[P^{(\nu)}, L_0] = 0, \qquad L_0 P^{(\nu)} = w^{(\nu)} P^{(\nu)} \qquad (24)$$

where $w^{(\nu)}$ are the eigenvalues,

$$w^{(0)} = 0, \qquad w^{(\alpha\beta)} = \omega_\alpha - \omega_\beta \qquad (25)$$

IV. DYNAMICS OF DISSIPATIVE SYSTEMS

We come now to what we may call the backbone of our approach. For integrable systems, we have seen that the central problem is the construction of the unitary operator U in the Liouville–von Neuman space. For nonintegrable systems, we introduced in our previous papers [5–7] new operators $C^{(\nu)}$, $D^{(\nu)}$, $\chi^{(\nu)}$ corresponding to the dynamics of correlations. The superoperator $C^{(\nu)}$ is an "off-diagonal" superoperator, because it describes off-diagonal transitions $C^{(\nu)} = Q^{(\nu)} C^{(\nu)} P^{(\nu)}$ from the $P^{(\nu)}$ correlation subspace to the $Q^{(\nu)}$ subspace. By operating $C^{(\nu)}$ on the ν correlation subspace $P^{(\nu)}$, this operator creates correlations other than the ν correlation. In particular, $C^{(0)}$ creates higher correlations from the vacuum of correlations. For this reason the $C^{(\nu)}$ are generally called "creation-of-correlations" superoperators, or creation operators in short. Conversely, the $D^{(\nu)} = P^{(\nu)} D^{(\nu)} Q^{(\nu)}$ are called destruction operators.

The superoperator $\chi^{(\nu)} = P^{(\nu)} \chi^{(\nu)} P^{(\nu)}$ is "diagonal," because it describes a diagonal transition between states belonging to the same subspace $P^{(\nu)}$.

In terms of these operators, we may indeed consider dynamics as a dynamics of correlations. U is expressed in terms of the kinetic operators C, D, and χ. The kinetic operators for integrable systems are given by definition by the relations

$$\bar{\chi}^{(\nu)} \equiv P^{(\nu)} U^{-1} P^{(\nu)}$$
$$\bar{C}^{(\nu)} \bar{\chi}^{(\nu)} \equiv Q^{(\nu)} U^{-1} P^{(\nu)} \tag{26}$$

We also have the Hermitian conjugate components

$$[\bar{\chi}^{(\nu)}]^\dagger \equiv P^{(\nu)} U P^{(\nu)}$$
$$[\bar{\chi}^{(\nu)}]^\dagger \bar{D}^{(\nu)} \equiv P^{(\nu)} U Q^{(\nu)} \tag{27}$$

where $\bar{D}^{(\nu)} \equiv [\bar{C}^{(\nu)}]^\dagger$. The diagonalization of the Hamiltonian starting with the projectors $P^{(\nu)}$ is equivalent to the dynamics of correlations. We have

$$U^{-1} P^{(\nu)} = (P^{(\nu)} + \bar{C}^{(\nu)}) \bar{\chi}^{(\nu)}$$
$$P^{(\nu)} U = [\bar{\chi}^{(\nu)}]^\dagger (P^{(\nu)} + \bar{D}^{(\nu)}) \tag{28}$$

As we shall see later, the "kinetic operators" for integrable systems form the basis used to perform the analytic continuation to nonintegrable systems.

For the integrable case the transformation U relates bare eigenfunctions of L_0 to dressed eigenfunctions of L_H. The eigenfunctions of L_H are given by the dyads of eigenstates of H as $|\bar{\phi}_\alpha; \bar{\phi}_\beta\rangle\rangle$, where

$$|\bar{\phi}_\alpha; \bar{\phi}_\beta\rangle\rangle = |\bar{\phi}_\alpha\rangle\langle\bar{\phi}_\beta| = [\langle\langle\bar{\phi}_\alpha; \bar{\phi}_\beta|]^\dagger \tag{29}$$

and $H|\bar{\phi}_\alpha\rangle = \bar{\omega}_\alpha |\bar{\phi}_\alpha\rangle$. The states $|\alpha\rangle$ are eigenstates of H_0, while $|\bar{\phi}_\alpha\rangle$ are dressed eigenstates of H. In terms of the unitary operators U in the wavefunction space and U^{-1} in the Liouville–von Neumann space we have, respectively [c.f. Eq. (11)],

$$|\bar{\phi}_\alpha\rangle = u^{-1}|\alpha\rangle, \qquad |\bar{\phi}_\alpha; \bar{\phi}_\beta\rangle\rangle = U^{-1}|\alpha; \beta\rangle\rangle \tag{30}$$

One can also write the eigenfunctions of L_H in terms of the kinetic operators. Here we use the notation [5,7]

$$|\bar{F}_\alpha^0\rangle\rangle \equiv |\bar{\phi}_\alpha; \bar{\phi}_\alpha\rangle\rangle, \qquad |\bar{F}^{\alpha\beta}\rangle\rangle \equiv |\bar{\phi}_\alpha; \bar{\phi}_\beta\rangle\rangle \quad (\alpha \neq \beta) \tag{31}$$

Then we have

$$L_H|\bar{F}_j^v\rangle\rangle = \bar{w}^{(v)}|\bar{F}_j^v\rangle\rangle \tag{32}$$

where $\bar{w}^{(\alpha\beta)} \equiv \bar{\omega}_\alpha - \bar{\omega}_\beta$ and $\bar{w}^{(0)} \equiv 0$. From Eq. (28) we obtain

$$|\bar{F}_j^v\rangle\rangle = (P^{(v)} + \bar{C}^{(v)})|f_j^v\rangle\rangle, \quad \langle\langle \bar{F}_j^v| = \langle\langle f_j^v|(P^{(v)} + \bar{D}^{(v)}) \tag{33}$$

where $|f_j^v\rangle\rangle \equiv \bar{\chi}^{(v)}|v_j\rangle\rangle$ where j is a degenerary index [see an example $j = \alpha$ in (31)].

Note that $P^{(v)}|\bar{F}_j^v\rangle\rangle = |f_j^v\rangle\rangle$ and $Q^{(v)}|\bar{F}_j^v\rangle\rangle = \bar{C}^{(v)}|f_j^v\rangle\rangle$. Hence the $Q^{(v)}$ component of $|\bar{F}_j^v\rangle\rangle$ is a functional of the $P^{(v)}$ component,

$$Q^{(v)}|\bar{F}_j^v\rangle\rangle = \bar{C}^{(v)}P^{(v)}|\bar{F}_j^v\rangle\rangle \tag{34}$$

Similarly, for the left eigenstates of L_H we have

$$\langle\langle \bar{F}_j^v|Q^{(v)} = \langle\langle \bar{F}_j^v|P^{(v)}\bar{D}^{(v)} \tag{35}$$

The functional relations expressed by Eqs. (34) and (35) are highly nontrivial. It is at the basis of the construction of dynamics of nonintegrable systems.

The above construction allows us to give an explicit expression to the collision operator in (9) in terms of the kinetic operators [7]

$$\bar{\theta}^{(v)} = P^{(v)}w^{(v)} + [\bar{\chi}^{(v)}]^{-1}\lambda L_V \bar{C}^{(v)}\bar{\chi}^{(v)} \tag{36}$$

The main result is that the dyadic formulation of quantum mechanics can be expressed in terms of the kinetic operators. This is the starting point for our transition from integrable to nonintegrable systems.

The time evolution of the density matrix in the unperturbed representation depends on the correlations. These correlations replace here the interaction V [see Eq. (8)]. The situation changes for integrable systems when we go to the unitary representation. Then both interactions and correlations are eliminated [as the kinetic equation reduces to Eq. (6)]. The elimination of the interaction or the correlations are equivalent problems for integrable systems (see Ref. 2).

We go now to nonintegrable systems. We have then to eliminate Poincaré's divergences. This is done by analytic continuation of the resonances which appear in the kinetic operators C, D, χ. The key point is to choose the sign of $i\epsilon$ depending on whether we have a transition to higher, equal, or lower correlations, in each term of the perturbation expansion [4,7,9]. We use the same formal expression (28) for Λ

$$\Lambda^{-1}P^{(\nu)} = (P^{(\nu)} + C^{(\nu)})\chi^{(\nu)}$$
$$P^{(\nu)}\Lambda = [\chi^{(\nu)}]^*(P^{(\nu)} + D^{(\nu)}) \quad (37)$$

However, the analytic continuation breaks the unitarity of the transformation. Λ has a new property called star-unitarity [9], which is an extension of unitarity. We have $\Lambda^{-1} = \Lambda^*$, where $*$ denotes star conjugation. Star conjugation means Hermitian conjugation plus a change in the role of higher and lower correlations.

Instead of $U\rho$, we now consider $\Lambda\rho$, which satisfies the same equation as $U\rho$ but now $\tilde{\Theta}$ is the "collision operator" of kinetic theory [compare Eq. (10) and Eq. (18)].

Let us consider an example. We consider again the Friedrichs model (19). We assume $\omega_k \geq 0$. The state $|1\rangle$ is either unstable or stable, depending on whether its energy ω_1 is above or below a certain positive threshold energy, respectively [5]. This threshold depends on the coupling constant and the potential. We first restrict ourselves to situations where $\omega_1 < 0$. This condition ensures that the state $|1\rangle$ is stable and also that all terms in the perturbation expansion are well-defined (i.e., we have integrability in the sense of Poincaré [11]).

In addition to the bare states, we can construct dressed states $|\bar{\phi}_\alpha\rangle$ that are eigenstates of H. For example, for the particle state we have

$$H|\bar{\phi}_1\rangle = \bar{\omega}_1|\bar{\phi}_1\rangle \quad (38)$$

where $\bar{\omega}_1$ is the (real) shifted energy of the discrete state.

The exact state $|\bar{\phi}_1\rangle$ is known and is expandable in perturbation series (see, e.g., Ref. 11). To first order in λ we have

$$|\bar{\phi}_1\rangle = |1\rangle - \sum_k \frac{\lambda V_k}{\omega_k - \omega_1}|k\rangle + O(\lambda^2) \quad (39)$$

which is a superposition of states $|1\rangle$ and $|k\rangle$.

The dressing of the bare particle may be written as the result of a unitary transformation in the Hilbert space,

$$u^{-1}|1\rangle = |\bar{\phi}_1\rangle \quad (40)$$

To the dressed particle state we associate the density operator

$$u^{-1}|1\rangle\langle 1|u = |\bar{\phi}_1\rangle\langle\bar{\phi}_1| \quad (41)$$

which is an invariant of motion. We can write the transformation (4.16) as

$$U^{-1}|1;1\rangle\rangle = |\bar{\phi}_1;\bar{\phi}_1\rangle\rangle \tag{42}$$

The remarkable point is that, as we have already noticed, U can be expressed in terms of the kinetic operators ($\bar{C}, \bar{D}, \bar{\chi}$) we have introduced. This leads to a closed link between quantum mechanics (or classic mechanics) and kinetic theory.

Now we turn to the case where the energy of the bare particle ω_1 is above its threshold of stability. In this case the state $|1\rangle$ becomes unstable and decays emitting photons. For this case, one can show that the state $|\bar{\phi}_1\rangle$ disappears due to "Poincaré resonances" at $\omega_k = \omega_1$ in Eq. (39) [3,11]. In other words, there is no eigenstate of H that can be obtained by a unitary transformation acting on the bare state $|1\rangle$. The disappearance of $|\bar{\phi}_1\rangle$ may be interpreted as the disappearance of one of the invariants of motion (i.e., $|\bar{\phi}_1\rangle\langle\bar{\phi}_1|$). The system is non-integrable in the sense of Poincaré. We come here to an unsolved problem of quantum mechanics [12,13]: how to define a *dressed unstable* state. We have of course the state $|1;1\rangle\rangle$ as well as the dressed states $|\bar{\phi}_1;\bar{\phi}_1\rangle\rangle$ for integrable systems. In spite of the considerable literature, this problem is not solved. However, on the level of the Liouville–von Neumann dynamics we can introduce a *dressed* particle state through the nonunitary transformation Λ obtained by the analytic continuation of U:

$$\Lambda^{-1}|1;1\rangle\rangle = \rho^p_{11} \tag{43}$$

ρ^p_{11} corresponds to the *dressed* unstable particle defined in the Liouville space, and the superscript "p" stands for the perturbed state. This state is outside the Hilbert space. This is in agreement with our remark that dissipation is only meaningful outside the Hilbert space. The properties of ρ^p_{11} have been studied in a recent paper [5], where the analytic continuation of Λ is given. ρ^p_{11} has a strict exponential decay, while wavefunctions present deviations from exponential behavior. This deviation is difficult to accept because this would destroy indiscernibility. Note that our method separates effects due to the preparation of the unstable state from the decay. According to the preparation we have different short time behavior. This corresponds to the "Zeno time" [14] as well as other effects. In contrast, the behavior of ρ^p_{11} is universal. Note also that ρ^p_{11} no longer leads to the well-known Lorentz shape, but to a distribution of photons with finite dispersion.

In Ref. 5 we have presented the exact form of U and Λ in all orders of λ. Here we present, as an example, specific components of U and Λ, up to second order in λ

$$\langle\langle k;k|U^{-1}|1;1\rangle\rangle = \frac{\lambda^2 V_k^2}{(\omega_1 - \omega_k)^2} + O(\lambda^4) \tag{44}$$

for the integrable case and present

$$\langle\langle k;k|\Lambda^{-1}|1;1\rangle\rangle = \frac{1}{2}\left[\frac{\lambda^2 V_k^2}{(\omega_1 - \omega_k + i\epsilon)^2} + \text{c.c.}\right] + O(\lambda^4) \qquad (45)$$

for the nonintegrable case. Equation (44) corresponds to the well-known Rayleigh–Schrödinger expansion (in Liouville space), while Eq. (45) corresponds to an extension of the Rayleigh–Schrödinger expansion to nonintegrable dynamical systems. Note that if we insist on keeping a unitary transformation for the nonintegrable case, we would obtain a diverging distribution as

$$\langle\langle k;k|U^{-1}|1;1\rangle\rangle_{\text{nonint}} = \frac{\lambda^2 V_k^2}{|\omega_1 - \omega_k + i\epsilon|^2} + O(\lambda^4)$$

$$\propto \frac{1}{\epsilon}\delta(\omega_1 - \omega_k) \propto \frac{1}{\epsilon}\tilde{\theta}^{(0)} \qquad (46)$$

This is an example of Poincaré's divergences. It occurs at the resonance $\omega_k = \omega_1$. It is related to the collision operator $\tilde{\theta}^{(0)}$ [3]. However, for the nonunitary transformation Eq. (45) we avoid the divergence by a suitable choice of analytic continuation.

The fourth-order corrections are given by

$$\langle\langle k;k|U^{-1}|1;1\rangle\rangle_4 = -\frac{\lambda^2 V_k^2}{(\omega_1 - \omega_k)^2}\sum_l \frac{\lambda^2 V_l^2}{(\omega_1 - \omega_l)^2}$$

$$- 2\frac{\lambda^2 V_k^2}{(\omega_1 - \omega_k)^3}\sum_l \frac{\lambda^2 V_l^2}{\omega_1 - \omega_l} \qquad (47)$$

for the integrable case and by

$$\langle\langle k;k|\Lambda^{-1}|1;1\rangle\rangle_4 = -\frac{1}{2}\left[\frac{\lambda^2 V_k^2}{(\omega_1 - \omega_k + i\epsilon)^2}\sum_l \frac{\lambda^2 V_l^2}{(\omega_1 - \omega_l + i\epsilon)^2} + \text{c.c.}\right]$$

$$- \left[\frac{\lambda^2 V_k^2}{(\omega_1 - \omega_k + i\epsilon)^3}\sum_l \frac{\lambda^2 V_l^2}{\omega_1 - \omega_l + i\epsilon} + \text{c.c.}\right]$$

$$+ \frac{1}{8}\left[\frac{\lambda^2 V_k^2}{(\omega_1 - \omega_k + i\epsilon)^2} - \text{c.c.}\right]\left[\sum_l \frac{\lambda^2 V_l^2}{(\omega_1 - \omega_l + i\epsilon)^2} - \text{c.c.}\right] \qquad (48)$$

for the nonintegrable case. The expansion in terms of λ can be pursued to all orders in λ, but the radius of convergence of the series is generally not known.

However, we have an exact compact expression for any value of λ for the Friedrichs model considered here [5]. As already mentioned, the analytic continuation has been done by separating the transitions from higher correlations from the transitions to lower correlations [4,9]. The analytical continuation is not unique. In each term we could replace $i\varepsilon$ by $-i\varepsilon$. The possibility of two different extensions corresponds to the inversion between past and future and is the basis for dissipative processes. Our method leads to a separation of processes that lead to equilibrium in the future from processes that lead to equilibrium in the past. The main point is that we can separate these processes in terms of two different "semigroups." Which semigroup to choose is a question of coherence. In the universe as known to us, all dissipative processes have the same direction. This is *by definition* the direction from past to future. Anyway the irreversible processes appear as a result of analytic continuation, and not due to any falsification. However, the mathematics of irreversible processes is highly nontrivial. For example, we have in the integrable case

$$UH^2 = (UH)^2 \tag{49}$$

The operator U is "distributive." In contrast, Λ is nondistributive:

$$\Lambda H^2 \neq (\Lambda H)^2 \tag{50}$$

This difference indicates that there are no more discrete levels, and that we have an uncertainty relation between energy and lifetime:

$$\Delta E \Delta t \geq 1/2 \tag{51}$$

where Δt is the lifetime and ΔE is given precisely by the difference [4,5]

$$(\Delta E)^2 = \langle \Lambda H^2 \rangle - \langle (\Lambda H)^2 \rangle \tag{52}$$

The time energy radiation takes here a clear meaning. The energy shift ΔE differs slightly from the usual value [5]. Our approach could be verified measuring the line shape that differs radically from the classic Lorenz shape.

V. CONCLUDING REMARKS

The dynamical description of dissipative systems has previously led to some contradictions. On the level of Fokker–Planck or Pauli equations, one retains $\lambda^2 t$ terms. In this limit, $\Lambda \approx U$. If this would be the case, irreversibility would indeed be only the result of approximations. But Λ can never be exactly equal to U. The $\lambda^2 t$ approximation corresponds only to an asymptotic expansion. It is true that for large t, equilibrium would be achieved and there would be no more dissipation. But for finite λ, *however small*, we have $\Lambda \neq U$, and we go outside the Hilbert space.

Note also that our method also applies to thermodynamic systems with N particles in a volume V in the limit $N \to \infty$, $V \to \infty$ with the density $c = N/V$ finite. In general, these systems are nonintegrable. The same applies to field theory. Free fields are integrable systems. But, in general, interacting fields are not integrable. The interactions between fields lead again to dissipation and require an extension of dynamics outside the Hilbert space [15].

Acknowledgments

I want to thank Dr. G. Ordonez for helping me to structure this chapter. I also want to thank Professor T. Petrosky, Dr. E. Karpov, and Mr. S. Kim for their contributions to the subject of this chapter. I acknowledge the International Solvay Institutes for Physics and Chemistry, the Engineering Research Program of the Office of Basic Energy Sciences at the U.S. Department of Energy, Grant No DE-FG03-94ER14465, the Robert A. Welch Foundation Grant F-0365, The European Commission ESPRIT Project 28890 NTGONGS, the National Lottery of Belgium, and the Communauté Française de Belgique for supporting this work.

References

1. I. Prigogine, *Nonequilibrium Statistical Mechanics*, Wiley Interscience, New York, 1962.
2. I. Prigogine, C. George, and J. Rae, *Physica* **56**, 25 (1971); I. Prigogine and C. George, *Physica* **56**, 329 (1971).
3. T. Petrosky and I. Prigogine, *Physica A* **147**, 439 (1988).
4. I. Prigogine and T. Petrosky, *Physica A* **147**, 461 (1988).
5. G. Ordonez, T. Petrosky and I. Prigogine, *Phys. Rev. A* **63**, 052106 (2001).
6. T. Petrosky, G. Ordonez and I. Prigogine, *Phys. Rev. A* **64**, 062101 (2001).
7. T. Petrosky and I. Prigogine, *Adv. Chem. Phys.* **99**, 1 (1997).
8. P. A. M. Dirac, *Lectures on Quantum Field Theory*, Belfer Graduate school, Yeshiva University, New York, 1966.
9. I. Prigogine, C. George, F. Henin, and L. Rosenfeld, *Chem. Scripta* **4**, 5 (1973).
10. I. Prigogine, *From Being to Becoming*, Freeman, New York, 1980.
11. T. Petrosky, I. Prigogine and S. Tasaki, *Physica A* **173**, 175 (1991).
12. W. Heitler, *The Quantum Theory of Radiation*, 3rd edition, Clarendon Press, Oxford, 1957.
13. P. A. M. Dirac, *Principles of Quantum Mechanics*, 4th edition, Oxford University Press, New York, 1962.
14. B. Misra and E. C. G. Sudarshan, *J. Math. Phys.* **18**, 756 (1977).
15. E. Karpov, G. Ordonez, T. Petrosky, and I. Prigogine, to be published.

GENERALIZED QUANTUM FIELD THEORY*

E. C. G. SUDARSHAN and LUIS J. BOYA[†]

Center for Particle Physics, Department of Physics,
The University of Texas, Austin, Texas, U.S.A.

CONTENTS

I. Introduction
II. Second Quantization
III. Continuum Spectrum
IV. Resonances and Poles
V. Spin-Statistics and CPT
Literature

I. INTRODUCTION

Quantum Field Theory was initiated by P. Jordan and developed by P. A. M. Dirac and by W. Heisenberg and W. Pauli to apply the quantum rules to arbitrary systems (including fields) and in particular to cope with the problems of radiation and matter. There have always been questions about mathematical consistency and physical interpretation. An essential difference with ordinary nonrelativistic quantum mechanics appeared very soon, when von Neumann proved the uniqueness of the irreducible representations of the commutation rules for a system with *finite* numbers of degrees of freedom, whereas the theorem was definitely false for infinite systems—that is, for quantum fields. In the first, finite

---

*Contributed to the XXI Solvay Conference. Keihanna Interaction Plaza, (Nara), Japan, November 1–5, 1998. Presented by Luis J. Boya.
[†]*Permanent address*: Department of Theoretical Physics, Faculty of Science, University of Zaragoza, Zaragoza, Spain.

Dynamical Systems and Irreversibility: A Special Volume of Advances in Chemical Physics,
Volume 122, Edited by Ioannis Antoniou. Series Editors I. Prigogine and Stuart A. Rice.
ISBN 0-471-22291-7. © 2002 John Wiley & Sons, Inc.

case, the commutation rules

$$[q_i, p_j] = i\hbar \delta_{ij} \quad \text{or} \quad [a_i, a_j^\dagger] = \delta_{ij}$$

would imply that there exists an essentially unique vacuum state $|0\rangle$ with

$$a_i |0\rangle = 0$$

The existence of different, inequivalent vacua, or even the nonexistence of vacuum at all in the case of infinite degrees of freedom, was pointed out later and elaborated by Friedrichs, Wightman and Schweber, Segal, van Hove, and others. This has nothing to do, of course, with the separability of the underlying Hilbert space, a fact that we physicists assume, but which is not mathematically necessary; even with a countable infinite number of degrees of freedom, there are a myriads of inequivalent irreducible representations of the commutation relations.

Von Neumann's theorem as such guaranteed that, for example, Heisenberg's matrix mechanics and the Schrödinger equation would produce the same results, even when they departed from the naive Bohr–Sommerfeld old quantum theory results, for example in the $\frac{1}{2}\hbar\omega$ value for the zero point energy for the oscillator: In fact, Heisenberg's treatment stressed the *energy* representation, whereas Schrödinger used the *coordinate* representation; that the wave equation was nothing but a particular representation of Heisenberg–Born–Jordan commutation rules was something subtle, which even Niels Bohr never accepted completely.

Dirac's (and simultaneously Jordan's) work on the transformation theory (1927) was the culmination of the axiomatic treatment of Quantum Mechanics (as distinct from Q.F.T.). Dirac himself was more proud of this development than any other of his (many) contributions to quantum mechanics. Transformation theory emphasizes the geometric, frame-independent and invariant description of quantum mechanics. The underlying quantum reality, somewhat cryptic, can only be revealed through a particular representation; whichever one takes, however, breaks the representation independence. One cannot but think that exactly the same is true in general relativity, where coordinate independence is dialectically opposed to the necessity of a particular frame to state a concrete problem and to exhibit the solutions. In modern times, one assumes that the M-theory is still provisional because we do not know yet the representation-independence mode of the description of the former string theory; we only know six corners of the domain, the five allowable string theories, and the "crown jewel," namely, supergravity in 11 dimensions.

Dirac's infinite sea of filled negative energy states was a constructive answer to the negative side of von Neumann's theorem for the relativistic electron field;

the "natural" representation, with all positive and negative levels empty, did not describe reality, but was in agreement with the (anti-)commutation relations all the same. But Dirac's choice was also in agreement with the irreducible representation which has the lowest energy, because all the negative-energy levels were filled.

II. SECOND QUANTIZATION

Quantum Field Theory was connected from the beginning with (special) relativity; in fact, there is a widespread belief that there is no consistent relativistic description of a single particle because of the phenomenon of pair creation, the many-body interpretation of the negative energy states, and so on. However, this is a point of view that R. P. Feynman would have challenged. One can go a long way with a relativistic quantum mechanics of a particle, with antiparticles, for example, understood as particles going backwards in time; this is especially true if one is interested only in an S-matrix description.

However, there is a beautiful description of many-body problems, even in nonrelativistic mechanics, by means of quantum field theory. This is the second quantization of Dirac (again!) and Jordan, O. Klein, and E. P. Wigner. Starting, for example, with the Schrödinger equation (with $2m = \hbar = 1$)

$$\psi''(x) + E\psi(x) = V(x)\psi(x)$$

one gets a complete set of eigenfunctions; there is first the (possible) discrete spectrum, bounded from below:

$$u_n(x) = \langle x|n \rangle, \quad E(n) = E_n, \quad n = 1, \ldots, N$$

and improper "eigenfunctions" of the continuum part

$$u_k(x) = \langle x|k \rangle \quad 0 \leq k^2 = E < \infty$$

So the spectrum is continuous, $0 \leq E < \infty$, with possibly several bound states of negative energy, $E_n < 0$.

Now the "free" second quantized field is described by an expansion

$$\Phi(x) = \sum u_n(x) a_n + u_k(x) a_k$$

Of course, one can use two real (hermitean) fields instead of the complex one,

$$\Phi(x) = \Phi_1(x) + i\Phi_2(x) = \{\Phi_1(x), \Phi_2(x)\}$$

which is very convenient for many reasons, as emphasized by J. Schwinger and others.

The time-dependent form of the Schrödinger equation in terms of real components is

$$i\frac{\partial}{\partial t}\begin{pmatrix}\psi_1(t)\\\psi_2(t)\end{pmatrix}=\begin{pmatrix}0 & -\nabla^2\\+\nabla^2 & 0\end{pmatrix}\begin{pmatrix}\psi_1(t)\\\psi_2(t)\end{pmatrix}$$

where $\psi = \psi_1 + i\psi_2$. Here ψ_1 and ψ_2 contain both signs of frequencies, though ψ and ψ^* contains positive and negative frequencies, respectively. ψ_1 and ψ_2 are real, and only they should be considered as the basic ingredient when we quantize.

Jordan and Klein determined the conmutation relations to be

$$[a_n, a_{n'}^\dagger] = \delta_{n,n'}, \qquad [a_k, a_{k'}^\dagger] = \delta(k-k')$$

if there is no restriction on the spectrum of the number operator $N_k = a_k^\dagger a_k$, which fits well with the concept of "ordinary" classic particles. Anticommutation relations for spinor particles (fields) were soon advanced in the subsequent work of Jordan and Wigner (1928).

The approach of Jordan went counterintuitive to the eminently "particle" point of view prevailing at the time in the orthodox quarters of Gottingen and Copenhagen; let us recall that the literal "wavelike" interpretation of the Schrödinger equation was eventually rejected by the inventor himself, when it was clear that it makes no sense to think of the electron smeared out throughout the space: As H. Weyl proved very soon, the probability of simultaneous detection of "bits" of the electron at two separated points is zero, because it would correspond to the product of two commuting but different projection operators. Physically it is clear that the electron is indivisible!

III. CONTINUUM SPECTRUM

Let us mention two perennial conundrums of nonrelativistic quantum mechanics that show up also in the theory of quantized fields. We refer to the treatment of the continuum spectrum and the collapse of the wave packet. We shall not have too much to say about the second topic, because of the fact that for the virtual, potential, or propensity interpretation of the wavefunction, rooted in the Aristotelian philosohy and favored by Heisenberg and K. Popper among others, there is nothing real which collapses. In any case this belongs to the interpretative program of quantum mechanics, which is beyond the modest aim of this chapter.

The so called "improper" states, necessary to describe scattering, resonance decay, and so on, are essential. They do have an irreproachable description, again due to von Neumann. He completed the proof, due to Hilbert and Schmidt (1906), of the spectral resolution for self-adjoint operators to the unbounded case: The corresponding resolution of the identity is

$$1 = \sum_n |n\rangle\langle n| + \int_E P(E)\, d\mu(E)$$

where $P(E)$ is a family of projection-valued measures with $P(-\infty) = 0$, $P(+\infty) = 1$. Likewise, the Hamiltonian resolves as

$$H = \sum_n E_n |n\rangle\langle n| + \int_E E\, dP(E)$$

valid even in the case (which is usual) where H is an unbounded operator, as long as it is self-adjoint (hypermaximal in the now-obsolete von Neumann terminology); here $|n\rangle\langle n|$ are the *bona fide* projectors onto the bound states $|n\rangle$, whereas $P(E)$ is a projector measure, associated to the continuum spectrum for all values less than E. The theorem guarantees that the usual description with "eigenfunctions" on the continuum with well-defined energy will go through to describe the scattering situation. In fact, Dirac's physical way to circumvent the continuum spectrum problems with tricks like the delta "function," separated bra's and ket's, and so on, although horrifying the mathematicians at the time, is seen today as just an anticipation of the perfectly respectable *distribution theory* of L. Schwartz (1950). What Dirac does is just to write down

$$P(E) = |E\rangle\langle E|$$

pretending that $|E\rangle$ is a state. It is not, but the average $\sum_\Delta |E\rangle\langle E|\, d\mu(E)$ always exists as a *bona fide* operator.

Nevertheless, there are alternatives; one might as well enlarge the Hilbert space by introducing the topological dual of a physical subspace, which will extend itself beyond the limits of the original Hilbert space, in particular beyond the square-integrable functions in the Lebesgue realization of an abstract Hilbert space. This is the *rigged Hilbert space* formalism, pioneered by I. M. Gelfand for other reasons. In recent times A. Bohm, M. Gadella, and coworkers have used this consistently for the scattering and resonant states.

The treatment of resonances in quantum theory has a long history, starting with Dirac's time-dependent perturbation theory (1926), Gamow's tunnel effect description of α-decay in radioactive nuclei, the Weisskopf–Wigner treatment of radiation damping and line breadth, and the resonance formulae of Breit and Wigner, and later Kapur and Peierls (developed in the mid-1930) and subsequently Peierls himself. The Kapur–Peierls formalism has been "rediscovered" recently in connection with the rigged Hilbert space approach.

One might reasonably ask, Do resonances correspond to complex energy eigenvalues? The simple answer is: Yes, if you take the inverse lifetime of the unstable state as measured by the imaginary part of the "complex" energy. No, if you remember from your school days that Hermitian operators have only real eigenvalues. So?

IV. RESONANCES AND POLES

The temptation to study the Schrödinger equation with complex eigenvalues or alternatively the analytic continuation of the analytic states and the scattering amplitude to the complex region has been irresistible. The improper understanding of either approach has led to a lot of nonsense in the physical literature.

Let us first make a simple remark. For a free particle, the spectrum is continuous and bounded from below. But we can ask the question, How are the solutions of the Schrödinger equation for negative energies, or for that matter for complex values? The mathematical solutions are there, of course; there are two of them, linearly independent. Are they of any use? Of course! For example, to construct all the transparent potentials [first found by Kay and Moses (1955), by the inverse scattering method] one starts with these mock (= unphysical) negative-energy solutions for the free particle (e.g., $\psi(x) = \cosh(x)$), applies the double Darboux isospectral method (developed in 1885!), and constructs the solitonic potentials by iteration, which are perfectly well-defined and very important; for example, in the first step one just obtains $V(x) = -2\,\mathrm{sech}^2(x)$, the paradigm of reflectionless potentials.

To study rigorously the complex continuation of the wavefunction, let us write the most general solution of the time-dependent Schrödinger equation,

$$\psi(t) = \sum c_n \exp(-iE_n t) + \int_0^\infty g(E) \exp(-iEt)\, dE$$

On the real axis there are only the poles of any bound states plus a branch cut for the continuum; of course, in the complex k-plane there is no branch cut, so one can get around the cut in the energy plane by considering a two-sheeted Riemann surface, the physical and the unphysical. In the simplest case of a finite-range potential, in the k-plane there are only "B"-type singularities, corresponding to bound states, which lie in the positive imaginary axis, "A"-type singularities, in the negative imaginary axis, called sometimes antibound states, and "C"-type singularities, associated to resonances, which are complex and appear as pairs in $\mathrm{Im}\, k < 0$.

There is nothing to prevent us from modifying the integration contour in the complex energy plane, as long as we do not cross any singularity when we move off the real axis. Needless to say, the Hamiltonian operator, assumed to be analytic, is also continued.

Of course, if we insist on deforming the contour even crossing a new pole (that should be the analytic concept of resonance), the formula is still correct, as long as we duly include the residue contribution of that pole. Therefore:

The contributions to the integral on the real axis, which do not include the resonance contribution, are the same as the contribution on a deformed contour crossing some poles (resonances), as long as they are properly taken into account.

What is the time evolution of resonances? If the time evolution is given by

$$|\psi(t)\rangle = \exp(-iHt)|\psi(0)\rangle$$

then the survival amplitude is the overlap

$$A(t) = \langle \psi^\wedge(0)|\psi(t)\rangle$$

where $\langle \psi^\wedge|$ is the dual function; then

$$\langle \psi^\wedge(t),|\psi(t)\rangle = 1 \qquad \text{independent of } t$$

Of course, for the analytic continuation the Fourier component should be written as

$$g(E) \rightarrow g^*(E^*)$$

as it is clear from the transformation formula.

Now the survival probability is the square of the modulus of the overlap $\langle \psi^\wedge(t_1)|\psi(t_2)\rangle$. Of course, it does not make sense at all to say that "the state" decays, in the sense of becoming smaller. Because the resonance is a discrete eigenstate of the analytically continued Hamiltonian, it is *stationary*. The state has unit norm and keeps it! We can say that the probability diminishes in a phenomenological approximation if we overlook the states in which the decaying systems go into, looking only to the part of the original state that persists; it is like saying that energy is not conserved in friction because we do not keep track of the energy lost on the rubbed body!

Physically it is useful to understand the resonances to be associated with poles. The Hamiltonian evolution preserves scalar products both in the Hilbert space and in the more general dual space. Note that the "Hamiltonian" that acts in the dual complex space is *not*, strictly speaking, the original (Hermitian) one, but is its continuation with complex parameters; this reconciles the "incongruence" of the very "same" Hermitian operator getting *complex* eigenvalues!

Furthermore, the process is even *reversible* in time [so we *do not* subscribe to the philosophy that resonance decay in an essentially irreversible proces; absorption is as good as spontaneous emission; that much was clear from a very early discussion between Einstein and Ritz (1908)]. The state has constant (unit, in fact) norm, and hence the survival probability cannot exceed one. So if one looks back in time, it does not make sense to say that the resonance was "arbitrarily big" in the past. In fact, the time evolution is symmetric under time reversal, if one properly takes into account the transition to the dual state of the decaying state. Reversibility holds, of course, only if the Hamiltonian itself is time-reversal invariant, which is usually the case but by no means always.

If $\mathcal{H}$ is the original space, $\mathcal{H}'$ is the dual, and $\mathcal{K}, \mathcal{K}'$ are the continued spaces in the complex domain, only a dense set in $\mathcal{H}$ maps into $\mathcal{H}'$; a delta function maps into a distribution in $\mathcal{H}'$, not in $\mathcal{H}$. In $\mathcal{H}'$ it is the Weisskopf–Wigner resonance written as a Lorentz line shape:

$$(E - z_0)^{-1} = (E - E_0 - i\Gamma/2)^{-1}$$

V. SPIN-STATISTICS AND CPT

Prigogine has raised the question of the statistics of the unstable particles. We are able to give a precise answer to this, in the dual space formalism. That is, the continued operators $a(z)$ and $a^\dagger(z)$ satisfy the same canonical commutation relations, namely $[a(z), a^\dagger(z')] = \delta_{zz'}$. So, for example, in the two-resonance state we have $a_1^\dagger a_2^\dagger |0\rangle = a_2^\dagger a_1^\dagger |0\rangle$; scalar resonances obey Bose statistics.

The disentanglement of quantum field theory from it's relativistic heritage is shown also in that some of the crowning results of the relativistic situation, the connection between spin and statistics and the CPT theorem, can be seen to remain valid also in the nonrelativistic situation. Here we are going only to show some results.

The key issue in the spin-statistics theorem turns out to be merely the structure of the equations of motion under the three-dimensional rotation group, with the attendant spin (= double-valued) representations. The stable homotopy of the fundamental groups starts precisely at dimension three, namely,

$$\pi_1(SO(n)) = Z_2, \quad n > 2$$

This is naturally true of the rotation group itself, SO(3); it is also true of its inclusion in the Galilei group or, for that matter, in the Lorentz or the Poincaré group. This has been shown in detail only recently, but was hinted at already by Bacry and others.

The projective nature of the space of rays of quantum mechanics makes it possible for the half-integral spin representations to exist. The spin-statistics theorem then merely signals the price: They have to have the antisymmetric property, with the attendant Pauli "Auschliessprinzip," which is the key to understand structures in the Universe; were it not for this principle, the matter would appear amorphous, with all the particles in the ground state—no chemical valence, no chemical bond, no molecules, no life, no ourselves, nothing!

The spin-statistics connection holds also for arbitrary dimension (greater than $2+1$); witness the modern treatment of supersymmetric charges and transformations with Grassmann numbers.

It applies, *mutatis mutandis*, to the CPT theorem. There is no problem in defining P and T for any spacetime situation, whether relativistic or not. One writes routinely, for *any* field

$$P\Phi(\vec{x},t)P^{-1} = \eta_p \beta \, \Phi(-\vec{x},t)$$

for the parity operation, where, for example, β is needed for Fermi fields; and

$$T\Phi(\vec{x},t)T^{-1} = \exp(i\pi J_2)\,\Phi^*(\vec{x},-t)$$

recalling the antilinear nature of time reversal, as discovered by Wigner in 1932.

As for charge conjugation, for any quantized field one can define

$$C\Phi(x)C^{-1}$$

from the expansion of $\Phi(x)$ in terms of creation and annihilation operators as

$$Ca_k C^{-1} = b_k, \qquad Ca_k^\dagger C^{-1} = b_k^\dagger$$

for a general complex field. Or, more to the root the matter, for purely Hermitian fields the natural definitions are

$$Ca_k C^{-1} = \pm a_k, \qquad \text{etc.}$$

Recall that the C operation is antilinear in the first quantized formalism, but linear in second quantization, whereas time reversal is *always* antilinear. With these definitions, it is not difficult to show that any Hermitian (= real) Lagrangian for nonrelativistic fields as well as for relativistic quantized fields has to be CPT invariant. Again, the result holds with generality because P and T do not have a relativistic origin; as for C, it is characteristic of the quantum theory.

Literature

P. A. M. Dirac, *Proc. R. Soc.* **A114**, 243 (1927).

W. Heisenberg and W. Pauli, *Z. Phys.* **56**, 1 (1929); *ibid.* **59**, 168 (1930).

J. von Neumann, *Grundlagen der Quantenmechanik*, 1932; English transl., Princeton University Press, Princeton, NJ, 1955.

P. A. M. Dirac, *Proc. R. Soc.* **A112**, 661 (1926).

P. Jordan, *Zeit. Phys.* **44**, 473 (1927).

G. Gamow, *Zeit. Phys.* **48**, 500 (1928).

K. Friedrichs, *Mathematical Aspects of Quantum Theory of Fields*, Interscience, New York, 1954.

G. Luders, *Dansk. Mat. Fys. Med.* **28**, 5 (1954).

A. S. Wightman and S. Schweber, *Phys. Rev.* **101**, 860 (1955).

G. Luders and B. Zumino, *Phys. Rev.* **110**, 1450 (1958).

N. Burgoyne, *Il Nuovo Cim.* **VIII**, 607 (1958).

R. E. Streater and A. S. Wightman, *PCT, Spin-Statistics and All That*, Benjamin, New York, 1964.

E. C. G. Sudarshan, C. B. Chiu, and V. Gorini, *Phys. Rev.* **D18**, 2914 (1978).

E. C. G. Sudarshan, C. B. Chiu, and G. Bhamathi, *Physica* **A202**, 540 (1994).

I. Duck and E. C. G. Sudarshan, *Pauli and the Spin-Statistics Theorem*, World Scientific, Singapore, 1998.

I. Duck and E. C. G. Sudarshan, *Am. J. Phys.* **66**, 284 (1998).

L. J. Boya, Contribution to the XXII ICGTMP, Tasmania (Australia), July 1998 (unpublished).

AGE AND AGE FLUCTUATIONS IN AN UNSTABLE QUANTUM SYSTEM

G. ORDONEZ, T. PETROSKY, and E. KARPOV

Center for Studies in Statistical Mechanics and Complex Systems, University of Texas, Austin, Texas, U.S.A.; and International Solvay Institutes for Physics and Chemistry, Free University of Brussels, Brussels, Belgium

CONTENTS

I. Introduction
II. Construction of the Time Superoperator
III. Average Age of an Excited State
IV. Age Fluctuations
V. Entropy Superoperator
VI. Concluding Remarks
Acknowledgments
References

I. INTRODUCTION

As is well known (see, e.g., Ref. 1) in quantum mechanics, there is no Hermitian time operator $\hat{t}$ conjugate to a Hamiltionian H bounded from below, such that $[H, \hat{t}] = -i$. Indeed, the operator $\hat{t}$ would imply the existence of a displacement operator $\exp(i\Delta E \partial/\partial E)$ that could shift the energy of physical states to arbitrary negative values.

As noted by Prigogine [2], the nonexistence of $\hat{t}$ in the usual formulation of quantum theory is connected to the dual role of the Hamiltonian, both as energy and time translation operator. This "degeneracy" can be lifted by going to the Liouvillian formulation of quantum dynamics.

Dynamical Systems and Irreversibility: A Special Volume of Advances in Chemical Physics, Volume 122, Edited by Ioannis Antoniou. Series Editors I. Prigogine and Stuart A. Rice. ISBN 0-471-22291-7. © 2002 John Wiley & Sons, Inc.

In the Liouvillian formulation the fundamental objects are density operators, and the generator of motion is the Liouville superoperator $L_H = [H,]$. We use the term "superoperator" to indicate that L_H acts on ordinary quantum-mechanical operators [2].

Assuming that the spectrum of H is in the positive real line, the spectrum of the Liouville superoperator will be in all the real line, because the eigenvalues of L_H are differences of energies. As a result, one can introduce [2,3] a Hermitian superoperator T that satisfies

$$[L_H, T] = -i \qquad (1)$$

In other words, the time superoperator can be constructed if there is a continuous spectrum of energy and no discrete energy levels. This situation occurs, for example, in systems having resonances associated with unstable states.

As shown by Misra, Prigogine, and Courbage [4,5] (see also Refs. 6 and 7) in classic mechanics, the existence of a time operator may be linked to the appearance of instabilities in dynamics, leading to dissipative processes. Once the time operator is constructed, one can define a microscopic "entropy" operator M as a monotonic function of T (this is a Lyapounov function analogous to Boltzmann's H function). The entropy operator indicates the "distance" of the system to its final asymptotic state, which emerges as a result of dissipation. Both T and M can only be defined in terms of ensembles, and not in terms of trajectories.

As in classic mechanics, in quantum mechanics we can construct an entropy superoperator once we have T. Both the time and entropy superoperators do not preserve the purity of states, and hence they have to be formulated in the space of density operators. In short, for both classic and quantum mechanics the existence of instabilities, resonances, or dissipation is connected to the existence of time and entropy operators acting on ensembles [2].

The time operator has been previously constructed for classic systems such as the baker map [2,5], free relativistic fields with energy unbounded from below [6,8], and nonconservative systems described by the diffusion equation [9]. Here we present the explicit construction of T for a Hamiltonian quantum system, the Friedrichs model, with energy bounded from below (C. Lockhart [3] has constructed a time operator $\hat{t}$ acting on wave functions for the Friedrichs model with no lower bound on the energy).

In Section II we outline a general procedure to construct the time superoperator for systems with a continous spectrum.[1] In Section III we apply this construction to the Friedrichs model. We estimate the average age of the

---

[1] Many of the derivations in the present chapter have been included in Ref. 10. Therefore here we shall omit some details.

excited state in the case of weak coupling. The main result is that the average age coincides with the lifetime of the excited state. In Section IV we estimate the age fluctuations of the excited state. It is found that these fluctuations are large, on the order of the lifetime. In Section V we make a few comments on the entropy superoperator. This operator is formulated in terms of a nonunitary transformation Λ [5] that leads to a nonlocal representation of dynamics with broken time symmetry [11].[2]

II. CONSTRUCTION OF THE TIME SUPEROPERATOR

Let us consider the Hamiltonian

$$H = \sum_\alpha \int_0^\infty dw |\phi_w, \alpha\rangle w \langle \phi_w, \alpha| \qquad (2)$$

where w is the energy and α is a degeneracy index. The energy eigenstates satisfy the usual orthonormality and completeness relations,

$$\sum_\alpha \int_0^\infty dw |\phi_w, \alpha\rangle \langle \phi_w, \alpha| = 1$$

$$\langle \phi_{w'}, \alpha' | \phi_w, \alpha \rangle = \delta(w - w')\delta_{\alpha,\alpha'} \qquad (3)$$

In the Liouville space [10,14] the dyads $|\phi_w, \alpha\rangle\langle\phi_{w'}, \alpha'|$ are eigenstates of L_H with eigenvalues $w - w'$. It is convenient to introduce the variables v and $\bar{w}$,

$$v \equiv w - w', \qquad \bar{w} \equiv \frac{w + w'}{2} \qquad (4)$$

as well as the notation

$$|\Phi_{v, \bar{w}, \alpha, \alpha'}\rangle\rangle \equiv |\phi_w, \alpha\rangle\langle\phi_{w'}, \alpha'| \qquad (5)$$

Then the eigenvalue equation of L_H is written as

$$L_H |\Phi_{v, \bar{w}, \alpha, \alpha'}\rangle\rangle = v|\Phi_{v, \bar{w}, \alpha, \alpha'}\rangle\rangle \qquad (6)$$

---

[2] In recent years an alternative approach to obtain representations of dynamics with broken time symmetry has been developed. This approach is based on the construction of complex spectral representations of L_H both in classic [12,13] and quantum [14] mechanics. The relation between this approach and the approach based on the time superoperator is an interesting question we shall consider in a separate publication (see the comment at the end of Section V.)

The spectrum of ν runs from $-\infty$ to $+\infty$. As the eigenstates of L_H are density operators, we need to define their inner product in order to introduce the linear space structure. The inner product of two operators A and B is defined trough the trace $\langle\langle A|B\rangle\rangle \equiv \text{Tr}(AB)$. Now, similar to the usual position and momentum operators, the solution of the commutation relation (1) is constructed for the time superoperator T by taking the Fourier transform of the states $|\Phi_\nu, \bar{w}, \alpha, \alpha'\rangle\rangle$ over the variable ν. However, some care is necessary because the variable ν cannot vary independently of $\bar{w}$. Indeed, we have $w = \bar{w} + \nu/2 \geq 0$ and $w' = \bar{w} - \nu/2 \geq 0$, so we have $\bar{w} \geq |\nu|/2$. In the $(\nu/2, \bar{w})$ plane the allowed region is the shaded region shown in Fig. 1 (see Ref. 15 for a discussion on the spectrum of L_H). For given $\bar{w}$, the variable ν is restricted between the values $-2\bar{w}$ and $2\bar{w}$, as indicated by path I of Fig. 1. In order to remove this restriction, we choose integration paths such as II in Fig. 1 [3]. Along this path the vertical distance

$$E \equiv \bar{w} - |\nu|/2 \tag{7}$$

to the lower edge of the shaded region remains constant.

Introducing the set $\xi \equiv (E, \alpha, \alpha')$, which consists of all the parameters that are constant along path II, we relabel the eigenstates of L_H as $|\Phi_\nu, \xi\rangle\rangle$. Then we arrive at the expression of the eigenstates of T:

$$|\Phi(t), \xi\rangle\rangle = \int_{-\infty}^{\infty} \frac{d\nu}{\sqrt{2\pi}} e^{-i\nu t} |\Phi_\nu, \xi\rangle\rangle \tag{8}$$

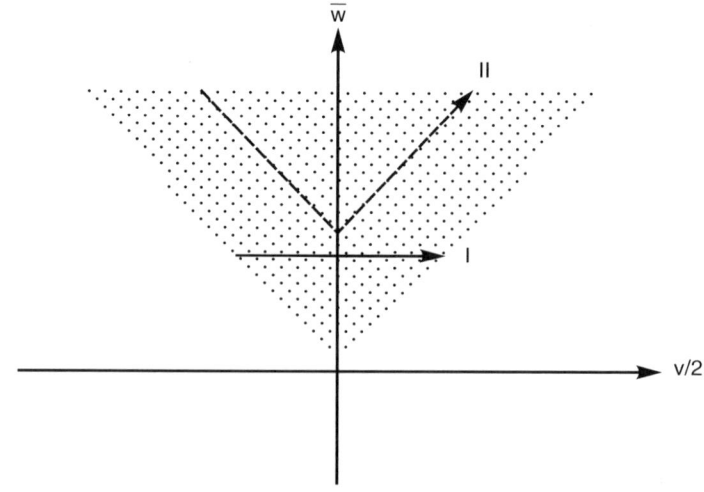

Figure 1. Spectrum of L_H and integration paths for Fourier transform.

Note that we may write the unit superoperator as

$$1 = \sum_{\xi} \int_{-\infty}^{\infty} dv |\Phi_v, \xi\rangle\rangle\langle\langle\Phi_v, \xi| = \sum_{\xi} \int_{-\infty}^{\infty} dt |\Phi(t), \xi\rangle\rangle\langle\langle\Phi(t), \xi| \quad (9)$$

where

$$\sum_{\xi} \equiv \sum_{\alpha,\alpha'} \int_0^{\infty} dE \quad (10)$$

We shall refer to the states (8) as "age eigenstates" [9]. The time superoperator is then

$$T = \int_{-\infty}^{\infty} dt \sum_{\xi} |\Phi(t), \xi\rangle\rangle t \langle\langle\Phi(t), \xi| \quad (11)$$

Similar to the position operator, we may represent the time superoperator in terms of its conjugate variable as $t \Rightarrow i\partial/\partial v$:

$$T = \int_{-\infty}^{\infty} dv \sum_{\xi} |\Phi_v, \xi\rangle\rangle i\frac{\partial}{\partial v} \langle\langle\Phi_v, \xi| \quad (12)$$

We define the expectation value (or average) of T associated with a state ρ as

$$\langle T \rangle_\rho \equiv \frac{\langle\langle\rho|T|\rho\rangle\rangle}{\langle\langle\rho|\rho\rangle\rangle} \quad (13)$$

As shown in Refs. 2 and 10 we have $d\langle T \rangle_{\rho(t)} = dt$; that is, the ordinary time t is the average of T. This allows us to interpret $\langle T \rangle_\rho$ as the "average age" of the state ρ and justifies the names "time superoperator" and "age eigenstate" for T and its eigenstates.[3]

The time superoperator is Hermitian in the Liouville space, and it does not preserve the purity of states [10]. This last property is easily seen in terms of the superoperator $\exp(ivT)$, which is an energy-difference shift operator. Acting on a pure state $|\phi_w, \alpha\rangle\langle\phi_w, \alpha|$, we obtain

$$\exp(ivT)|\phi_w, \alpha\rangle\langle\phi_w, \alpha| = \theta(v)|\phi_{w+v}, \alpha\rangle\langle\phi_w, \alpha| + \theta(-v)|\phi_w, \alpha\rangle\langle\phi_{w-v}, \alpha| \quad (14)$$

where θ is the usual step function. Equation (14) is no longer a pure state.

---

[3] The zero of time plays no special role as we can always shift this point to $t = t_0$ by defining a new operator $T' = T + t_0$. This new operator still satisfies the commutation relation (1) as t_0 is a constant that commutes with L_H. Ages are always measured with respect to the zero of time.

The construction we have shown here is quite general, and it can even be applied to free particles. The interest of our construction, however, is in cases where the dynamics is unstable, as we show next.

III. AVERAGE AGE OF AN EXCITED STATE

Now we apply the results obtained in the previous section to the one-dimensional Friedrichs model [16] with Hamiltionian

$$H = |1\rangle w_1 \langle 1| + \sum_\alpha \int_0^\infty dw |w, \alpha\rangle w \langle w, \alpha|$$
$$+ \lambda \sum_\alpha \int_0^\infty dw f(w)(|w, \alpha\rangle\langle 1| + |1\rangle\langle w, \alpha|) \tag{15}$$

This model describes a bare atom with two discrete states: a ground state and an excited state. The atom interacts with a set of field modes of frequencies w (we use units with $\hbar = 1$ and $c = 1$). The interaction $f(w)$ is assumed to be real. The state $|w, \alpha\rangle$ represents a field mode (or "photon") in the presence of the bare atom in its ground state. The index α is again a degeneracy index. The degeneracy is given by

$$n \equiv \sum_\alpha \tag{16}$$

For example, for a one-dimensional massless scalar field we have $w = |k|$ and $\alpha = \pm 1$, depending on the sign of the momentum k. The state $|1\rangle$ represents the bare atom in its excited state with no photons present. We assume the states $|1\rangle$ and $|w, \alpha\rangle$ form a complete orthonormal basis in the wavefunction space, namely,

$$\sum_\alpha \int_0^\infty dw |w, \alpha\rangle\langle w, \alpha| + |1\rangle\langle 1| = 1 \tag{17}$$
$$\langle 1|w, \alpha\rangle = \langle w, \alpha|1\rangle = 0, \qquad \langle w', \alpha'|w, \alpha\rangle = \delta(w - w')\delta_{\alpha, \alpha'}$$

The Hamiltonian (15) can also be interpreted as a simple model of an unstable particle. In this alternative interpretation the state $|1\rangle$ represents the bare particle and the states $|w, \alpha\rangle$ represent the decay products.

We shall assume that the potential is of the form

$$f(w) = \frac{w^{1/2}}{[1 + (w/w_M)^2]^r} \tag{18}$$

where r is a positive integer and w_M is the ultraviolet cutoff of the interaction with

$$w_M \gg w_1 \tag{19}$$

For the interaction satisfying

$$w_1 > \lambda^2 n \int_0^\infty dw \frac{f^2(w)}{w} \tag{20}$$

the excited state becomes unstable due to resonances [17].

The exact eigenstates of H are known [16] and are given by[4]

$$|\phi_{w,\alpha}\rangle \equiv |w, \alpha\rangle + \frac{\lambda f(w)}{\eta^+(w)} \Bigg[|1\rangle + \sum_{\alpha'} \int_0^\infty dw' \frac{\lambda f(w')}{w - w' + i0} |w', \alpha'\rangle \Bigg] \tag{21}$$

where $\eta^\pm(w) \equiv \eta(w \pm i0)$, with

$$\eta(z) \equiv z - w_1 - n \int_0^\infty dw \frac{\lambda^2 f^2(w)}{z - w} \tag{22}$$

We assume that the function $[\eta^+(w)]^{-1}$ continued to the lower half-plane has one pole (with $\gamma > 0$) at

$$z_1 \equiv \tilde{w}_1 - i\gamma/2 \tag{23}$$

where γ is the decay rate.

The eigenstates (21) form a complete orthonormal set as in Eq. (3). As a result of the resonance instability, the excited state (corresponding to the point spectrum) disappears from the complete set of eigenstates of H [16].

We shall calculate the average $\langle T \rangle_{\rho_1} = \langle\langle \rho_1 | T | \rho_1 \rangle\rangle$ where

$$|\rho_1\rangle\rangle \equiv |1\rangle\langle 1| \tag{24}$$

---

[4] The states (21) correspond to the "in" states of scattering theory—that is, the asymptotic time-evolved free states for $t \to +\infty$. We may obtain "out" states by taking the complex conjugate of (21). This corresponds to the change to $T \Rightarrow -T$ in the time superoperator constructed from these states.

is the bare excited state [we have $\langle\langle \rho_1|\rho_1\rangle\rangle = 1$]. For simplicity we shall consider the case of weak coupling $\lambda \ll 1$ and assume that $\tilde{w}_1 \gg \gamma$. Under these assumptions we may focus on the exponential region of the decay of the excited state, where we can neglect the deviations from exponential decay (note that the average value of T involves the complete "history" of the evolution of ρ_1).

Going back to the original variables w, w' and using (21), we find the exact form of the states $|\Phi_v, \xi\rangle\rangle = |\Phi_v, E, \alpha, \alpha'\rangle\rangle$ as

$$|\Phi_v, E, \alpha, \alpha'\rangle\rangle = |\phi_w, \alpha\rangle\langle\phi_{w'}, \alpha'| = \begin{cases} |\phi_{E+v}, \alpha\rangle\langle\phi_E, \alpha'| & \text{for } v > 0 \\ |\phi_E, \alpha\rangle\langle\phi_{E-v}, \alpha'| & \text{for } v < 0 \end{cases} \quad (25)$$

Defining

$$g(v, E) \equiv \langle\langle \rho_1|\Phi_v, \xi\rangle\rangle = \begin{cases} \lambda^2 f_{E+v} f_E [\eta^+(E+v)\eta^-(E)]^{-1} & \text{for } v > 0 \\ \lambda^2 f_E f_{E-v} [\eta^+(E)\eta^-(E-v)]^{-1} & \text{for } v < 0 \end{cases} \quad (26)$$

we obtain [see Eq. (12)]

$$\langle T \rangle_{\rho_1} = \langle\langle \rho_1|T|\rho_1\rangle\rangle = n^2 \int_{-\infty}^{\infty} dv \int_0^{\infty} dE\, g(v, E) i \frac{\partial}{\partial v} g^*(v, E) \quad (27)$$

A few manipulations lead to [10]

$$\langle T \rangle_{\rho_1} = -i\lambda^4 n^2 \int_0^{\infty} dv \int_0^{\infty} dE\, \frac{f^2(E+v)}{|\eta^+(E+v)|^2} \\ \times \frac{f^2(E)}{|\eta^+(E)|^2} \left(\frac{\eta^-(E+v)'}{\eta^-(E+v)} - \frac{\eta^+(E+v)'}{\eta^+(E+v)} \right) \quad (28)$$

where $\eta^\pm(w)' \equiv \partial \eta^\pm(w)/\partial w$. Using the relation

$$\frac{n}{2\pi i} \frac{\lambda^2 f^2(w)}{|\eta^+(w)|^2} = \left[\frac{1}{\eta^+(w)} - \frac{1}{\eta^-(w)} \right] \quad (29)$$

and the approximations

$$\eta^\pm(w)' = 1 + O(\lambda^2), \quad [\eta^+(w)]^{-1} = [w - z_1]^{-1} + O(\lambda^2) \quad (30)$$

which are valid for $\lambda \ll 1$, we get

$$\langle T \rangle_{\rho_1} \approx \int_0^{\infty} \frac{dE}{2\pi} \frac{\gamma}{(E - \tilde{w}_1)^2 + \gamma^2/4} \int_0^{\infty} \frac{dv}{2\pi} \frac{\gamma^2}{[(E + v - \tilde{w}_1)^2 + \gamma^2/4]^2} \quad (31)$$

Using $\tilde{w}_1 \gg \gamma$ this leads to the average age of the excited state

$$\langle T \rangle_{\rho_1} \approx \frac{1}{\gamma} \tag{32}$$

which in our approximation is the same as the lifetime γ^{-1}.

IV. AGE FLUCTUATIONS

Because the bare excited state is not an eigenstate of T, we expect that its average age will have fluctuations. We define the time fluctuation associated with a state ρ as ΔT_ρ with

$$(\Delta T)_\rho^2 \equiv \langle T^2 \rangle_\rho - (\langle T \rangle_\rho)^2 \tag{33}$$

Due to the commutation relation in Eq. (1), the fluctuations of T and L_H must obey the relation

$$(\Delta T)_\rho (\Delta L_H)_\rho \geq 1/2 \tag{34}$$

(with $\hbar = 1$). For pure states ρ normalized as $\text{Tr}(\rho) = 1$, the fluctuations of the Liouvillian are easily shown to be related to the fluctuations of the Hamiltonian as

$$(\Delta L_H)_\rho^2 = 2(\Delta H)_\rho^2 \tag{35}$$

where

$$(\Delta H)_\rho^2 \equiv \text{Tr}(H^2 \rho) - [\text{Tr}(H\rho)]^2 \tag{36}$$

as usual. In the following we shall calculate the time fluctuation for the bare excited state and verify that it satisfies the relation (34).

The operator T^2 may be written as [c.f. Eq. (12)]

$$T^2 = -\int_{-\infty}^{\infty} dv \sum_\xi |\Phi_v, \xi\rangle\rangle \frac{\partial^2}{\partial v^2} \langle\langle \Phi_v, \xi| \tag{37}$$

Therefore we have [c.f. Eq. (26)]

$$\langle T^2 \rangle_{\rho_1} = -n^2 \int_{-\infty}^{\infty} dv \int_0^{\infty} dE\, g(v, E) \frac{\partial^2}{\partial v^2} g^*(v, E) \tag{38}$$

Integrating by parts, we obtain

$$\langle T^2 \rangle_{\rho_1} = n^2 \int_{-\infty}^{\infty} dv \int_0^{\infty} dE \left| \frac{\partial}{\partial v} g(v, E) \right|^2 \tag{39}$$

Using Eq. (26) and the relations (29) and (30), we obtain

$$\langle T^2 \rangle_{\rho_1} \approx 2 \int_0^{\infty} \frac{dE}{2\pi} \frac{\gamma}{(E - \tilde{w}_1)^2 + \gamma^2/4} \int_0^{\infty} \frac{dv}{2\pi} \frac{\gamma}{(E + v - \tilde{w}_1)^2 + \gamma^2/4}$$
$$\times \left[\left(\frac{f(E+v)'}{f(E+v)} \right)^2 - \frac{f(E+v)'}{f(E+v)} \left(\frac{1}{E+v-z_1} + \text{c.c.} \right) \right.$$
$$\left. + \frac{1}{(E+v-\tilde{w}_1)^2 + \gamma^2/4} \right] \tag{40}$$

where $f(w)' \equiv \partial f(w)/\partial w$. For the potential (18), one can show that the first two terms in brackets in Eq. (40) give contributions on the order of $1/w_1^2$ and $1/(w_1\gamma)$, respectively. These contributions can be neglected as compared with the contribution of the third term, which gives

$$\langle T^2 \rangle_{\rho_1} \approx \frac{2}{\gamma^2} \tag{41}$$

Inserting this result and Eq. (32) in Eq. (33), we obtain

$$\Delta T_{\rho_1} \approx \frac{1}{\gamma} \tag{42}$$

Therefore the age fluctuation of the bare excited state is as large as its average age.

The energy fluctuation of the bare state is given by

$$(\Delta H_{\rho_1})^2 = n \int_0^{\infty} dw \, \lambda^2 f^2(w) \tag{43}$$

Therefore we obtain

$$(\Delta T)_{\rho_1} (\Delta L_H)_{\rho_1} = \frac{1}{\gamma} \sqrt{2n \int_0^{\infty} dw \, \lambda^2 f^2(w)} \tag{44}$$

The square root is of the order of λw_M, while the decay ray γ is of the order $\lambda^2 w_1$. Therefore [c.f. Eq. (19)] we obtain $(\Delta T)_{\rho_1} (\Delta L_H)_{\rho_1} \gg 1/2$. This verifies Eq. (34).

Before closing this section we remark that, in general, higher powers of T (i.e., T^l with $l \geq 3$) may have diverging expectation values, due to the branch-point singularity at $\nu = 0$ (corresponding to the edge of path II in Fig. 1).[5] The branch point appears as a consequence of the positivity of energy. As is well known, this leads to deviations in the exponential decay of the excited state for long times. For long times the decay follows an inverse power-law decay in time as t^{-p} with p positive integer [18]. Therefore, the integral over time in the expectation value of T^l will diverge if $l > p - 2$. In spite of this divergence, one can still define other functions of T such as the energy-difference shift operator $\exp(i\nu T)$.[6]

V. ENTROPY SUPEROPERATOR

We first define a superoperator $\Lambda = F(T)$ as

$$\Lambda = \int_{-\infty}^{\infty} dt \sum_{\xi} |\Phi(t), \xi\rangle\rangle F(t) \langle\langle\Phi(t), \xi| \tag{45}$$

where $F(t)$ is an arbitrary nonincreasing and bounded function of t. This is a nonunitary transformation, first introduced by Misra et al. [5,11] for unstable dynamical systems such as the baker transformation. In our case, Λ acts on density operators as $\tilde{\rho} \equiv \Lambda \rho$. The transformed density operators $\tilde{\rho}$ evolve according to a Markovian equation that breaks time symmetry. Through Λ we define the entropy superoperator as $M = \Lambda^\dagger \Lambda$. This satisfies $M \geq 0$ and

$$\frac{d}{dt} \langle M \rangle_{\rho(t)} \leq 0 \tag{46}$$

The expectation value of M then gives the negative of an entropy function. This expectation value may be put in the form [5,11] $\langle M \rangle_\rho = \text{Tr}(\tilde{\rho}^\dagger \tilde{\rho})$.

As shown in Ref. 11, there is a close relation between entropy and nonlocality. For example, for the baker map, Λ transforms points in phase space into ensembles corresponding to nonlocal distributions $\tilde{\rho}$ [11]. In quantum mechanics, Λ has an analogous properties. First of all (similar to T) it does not preserve the purity of states. This is analogous to the nonpreservation of trajectories in classic mechanics. Second, the transformation Λ introduces delocalization of states in the space representation. To see this we take as an example $F(t) = \theta(-t)$. Associated with the photons we introduce the position

---

[5] We thank S. Tasaki for pointing this out.
[6] The situation is quite similar to what happens with the integral $\int_{-\infty}^{\infty} dx\, f(x)(x^2 + x_0^2)^{-2}$. This diverges for $f(x) = x^{2l}$ with $l > 1$. However, for $f(x) = \exp(ikx)$ the integral is well-defined.

kets $|x\rangle$, which are related to the momentum states as $\langle x|w, \alpha\rangle = (2\pi)^{-1/2}$ $\exp(i\alpha wx)$, for $k = \alpha|w|$ and $\alpha = \pm$. The density operator corresponding to the position kets is $\rho_x = |x\rangle\langle x|$. Because in the bare state ρ_1 there are no photons, we have $\langle x|\rho_1|x\rangle = \langle\langle\rho_x|\rho_1\rangle\rangle = 0$. On the other hand, for the transformed states $\tilde\rho_1 = \Lambda\rho_1$ and $\tilde\rho_x = \Lambda\rho_x$ we obtain a nonvanishing overlap [10]

$$\langle\langle\tilde\rho_x|\tilde\rho_1\rangle\rangle = \langle\langle\rho_x|M|\rho_1\rangle\rangle \sim \gamma e^{-\gamma|x|} \tag{47}$$

This demonstrates the delocalization induced by the Λ transformation and the entropy operator.

We remark that the choice of the decreasing function $F(t)$ is not unique. The main point is that the existence of the time superoperator implies the existence of an entropy superoperator. The function $F(t)$ should be linked to the intrinsic decay rates of the system (i.e., to γ). One possibility to achieve this is to relate the construction we have presented here with the construction based on the complex spectral representation of the Liouville operator [14, 16, 19, 20]. This will be considered elsewhere.

VI. CONCLUDING REMARKS

Through the time superoperator we have introduced the concepts of age, age fluctuation, and entropy for the Friedrichs model describing a single excited atom or unstable particle.

There is a close relation between (a) the existence of instabilities and resonances and (b) the existence of the time superoperator. This can be seen explicitly in the Friedrichs model. In this model one can obtain a stable configuration when the inequality (20) is not satisfied. In this case there is no resonance between the discrete state and the continuous states. The discrete state does not decay (it has an infinite lifetime as we have $\gamma = 0$). In this case one can show that the time operator cannot be constructed because there appears an isolated point in the spectrum of H. This isolated point spoils the possibility of obtaining a complete set of age eigenstates by Fourier transformation of the continuous states [c.f. Eq. (9)].

The time superoperator leads to a formulation of quantum unstable dynamics where time and ensembles play a fundamental role [2]. In the future we hope to extend this formulation to relativistic field theories.

Acknowledgments

We thank Professor I. Prigogine for suggesting this work and for his helpful comments, as well as for his continuous support and encouragement. We thank I. Antoniou, M. Courbage, D. Driebe, B. Misra, E. Qubain, A. Shaji, E. C. G. Sudarshan, and S. Tasaki for their helpful suggestions. We acknowledge the European Community ESPRIT Project 28890 NTGONGS, U.S. Department of Energy Grant No. DE-FG03-94ER14465, the Robert A. Welch Foundation Grant No. F-0365, and the Loterie Nationale and the Communauté Francaise of Belgium for support of this work.

References

1. M. Jammer, *The Philosophy of Quantum Mechanics*, Wiley Interscience, New York, 1974, p. 141.
2. I. Prigogine, *From Being to Becoming*, Freeman, New York, 1980.
3. C. Lockhart, Thesis, University of Texas, Austin, 1981.
4. B. Misra, *Proc. Natl. Acad. Sci. U.S.A.* **75**, 1627 (1978).
5. B. Misra, I. Prigogine, and M. Courbage, *Physica A* **98**, 1 (1979); *Proc. Natl. Acad. Sci. U.S.A.* **80**, 2412 (1983).
6. I. Antoniou, Thesis, Free University of Brussels, 1988.
7. I. Antoniou, V. A. Sadovnichii, and S. A. Shkarin, *Physica A*, **269**, 299 (1999).
8. B. Misra, *Found. Physics* **25**, 1087 (1995).
9. I. Antoniou, I. Prigogine, V. A. Sadovnichii, and S. A. Shkarin, *Chaos, Solitons and Fractals* **11**, 465 (2000).
10. G. Ordonez, T. Petrosky, E. Karpov, and I. Prigogine, *Chaos Solitons Fractals*, **12**, 2591 (2001).
11. B. Misra and I. Prigogine, *Lett. Math. Phys.* **7**, 421 (1983).
12. H. Hasegawa and D. Driebe, *Phys. Rev. E*, **50**, 1781 (1994); I. Antoniou and S. Tasaki, *Physica A* **190**, 202 (1992).
13. T. Petrosky and I. Prigogine, *Chaos Solitons Fractals* **7**, 441 (1996).
14. T. Petrosky and I. Prigogine, *Adv. Chem. Phys.* **99**, 1 (1997).
15. E. C. G. Sudarshan, *Phys. Rev. A* **46**, 37 (1992).
16. T. Petrosky, I. Prigogine, and S. Tasaki, *Physica A* **173**, 175 (1991).
17. A. K. Likhoded and G. P. Pronko, *Int. J. Theor. Phys.* **36**, 2335 (1997).
18. L. A. Khalfin, *Zh. Eksp. Theor. Fiz.* **33**, 1371 (1957); *Sov. Phys. JETP* **6**, 1053 (1958).
19. I. Antoniou and S. Tasaki, *Int. J. Quant. Chem.* **46**, 425 (1993).
20. G. Ordonez, T. Petrosky, and I. Prigogine, *Phys. Rev. A.* **63**, 052106 (2001).

MICROPHYSICAL IRREVERSIBILITY AND TIME ASYMMETRIC QUANTUM MECHANICS*

A. BOHM

Department of Physics, University of Texas, Austin, Texas, U.S.A.

CONTENTS

I. Resonances and Decay
   A. Introduction
   B. From the Golden Rule to the Gamow Kets
II. Probability and Time Asymmetry
   A. Introduction
   B. From Gamow Kets to Microphysical Irreversibility
Acknowledgments
References

I. RESONANCES AND DECAY

A. Introduction

This is the first time that I find in one room together many people from whom I have learned and who got me interested in irreversibility and who provided advice and criticism: Ilya Prigogine, Manolo Gadella, Ioannis Antoniou, and Nico van Kampen. And there are of course others who are not here. To express my appreciation and pay my respect to them, I shall include some personal remarks after I have introduced the subject.

But the real heroine in this story is the mathematics of the Rigged Hilbert Space, which led to an idea so foreign to my upbringing that I could have never dreamt of it.

---

*Proceedings of the XXI Solvay Conference, 1998.

Dynamical Systems and Irreversibility: A Special Volume of Advances in Chemical Physics, Volume 122, Edited by Ioannis Antoniou. Series Editors I. Prigogine and Stuart A. Rice. ISBN 0-471-22291-7. © 2002 John Wiley & Sons, Inc.

B. From the Golden Rule to the Gamow Kets

For me the story of (quantum mechanical) irreversibility started with the Golden Rule of Dirac:

$$\dot{\mathscr{P}}(0) = \frac{2\pi}{\hbar} \int dE \sum_{\substack{\text{all } b \\ b \neq b^D}} |\langle b, E|V|f^D\rangle|^2 \delta(E - E_D) \tag{1}$$

Here $\dot{\mathscr{P}}(0)$ is the initial decay rate or transition rate for the decaying state $W^D = |f^D\rangle\langle f^D|$ with energy E_D, which decays into the decay products described by the projection operator:

$$\Lambda = \sum_{\substack{\text{all } b \\ b \neq b^D}} \int_0^\infty dE |E, b\rangle\langle E, b|, \qquad \text{where} \quad H_0|E, b\rangle = E|E, b\rangle \tag{2}$$

V is the interaction Hamiltonian for the decay, and the total Hamiltonian is $H = H_0 + V$. Example: f^D is a K_S^0 decaying into the channels $\pi^+\pi^-$, $\pi^0\pi^0$, $\pi^+\pi^-\gamma$, and so on, which we label by the channel quantum number η. E is the energy and $b = b_1 b_2 \ldots b_n$ stands for all other quantum numbers (e.g., directions) of all asymptotic decay products. If one chooses in place of $\Lambda = \sum_\eta \Lambda_\eta$ only the Λ_η of the decay channel, η (e.g., $\eta = \pi^+\pi^-$), then the sum over all b in (1) and (2) is replaced by the sum over the b_η for the ηth decay channel only. The rate in (1) will then become the partial initial decay rate $\dot{\mathscr{P}}_\eta(0)$ and $\dot{\mathscr{P}}(0) = \sum_\eta \dot{\mathscr{P}}_\eta(0)$.

According to the fundamental postulates of quantum mechanics, the transition or decay probability at the time t—that is, the probability to find the observable Λ in the state

$$W^D(t) = e^{-iHt} W^D e^{iHt} \quad \text{or} \quad f^D(t) = e^{-iHt} f^D \tag{3}$$

at the time t—is

$$\mathscr{P}(t) = \text{Tr}(\Lambda W^D(t)) = \text{Tr}(e^{iHt} \Lambda e^{-iHt} W^D) = \langle f^D(t)|\Lambda|f^D(t)\rangle \tag{4}$$

The decay rate (transition probability per unit time) at t should then be given by the time derivative $\dot{\mathscr{P}}(t) = d\mathscr{P}(t)/dt$ of $\mathscr{P}(t)$, and the right-hand side of (1) should be obtained by taking the time derivative of the right-hand side of (4) and then setting $t = 0$.

This is how I wanted to derive the Golden Rule (1) from the fundamental quantum mechanical probabilities (4) for my quantum mechanics class in the

1970s. The mathematical restrictions of the Hilbert space (HS) were no considerations, because Dirac kets

$$H|E, b^{\mp}\rangle = E|E, b^{\mp}\rangle, \qquad e^{iHt}|E, b^{\mp}\rangle = e^{iEt}|E, b^{\mp}\rangle \qquad (5)$$

and the Lippmann–Schwinger equation

$$|E, b^{\mp}\rangle = |E, b\rangle + \frac{1}{E - H \mp i\epsilon} V|E, b\rangle \qquad (6)$$

were freely used.

It turned out that (1) cannot be derived from (4), certainly not if f^D is an eigenvector of H_0, $H_0 f^D = E_D f^D$ or if f^D is an eigenvector of H. Also, taking any arbitrary element of the HS for f^D or more complicated mixed states W^D of the HS does not allow us to derive something like (1) from 4.[1] The reason is that Dirac's Golden Rule [1] (and also the method of Weisskopf and Wigner [2] or Lee–Oehme–Yang's effective theory with a two-dimensional complex Hamiltonian matrix [3]) are approximate methods. And "there does not exist...a rigorous theory to which these methods can be considered as approximations" [4].

In order to derive a Golden Rule, an eigenvector $\psi^G = |E_R - i\Gamma/2\rangle \sqrt{2\pi\Gamma}$ of the (self-adjoint, semibounded) Hamiltonian H with complex eigenvalue $(E_R - i\Gamma/2)$, that is,

$$H|E_R - i\Gamma/2^-\rangle = (E_R - i\Gamma/2)|E_R - i\Gamma/2^-\rangle, \qquad (7)$$

and the exponential time evolution

$$\psi^G(t) = e^{-iHt}\psi^G = e^{-iE_R t} e^{-\frac{\Gamma}{2}t}\psi^G \qquad (8)$$

had to be postulated. Such vectors do not exist in the HS, but this was not a big issue, because Dirac-Lippmann-Schwinger kets (5) and (6) are not in HS.

In the meanwhile, both the well-accepted Dirac kets (5) [7] and the much less popular Gamow kets (7) [8] have been given a mathematically precise meaning as functionals of a rigged Hilbert space (RHS):

$$\Phi_+ \subset \mathcal{H} \subset \Phi_+^\times \qquad (9)$$

---

[1] At that time the theorem [5]—which states that the probabilities like (4) are identical to zero, under the ususal assumptions for decay probabilities and for the Hamiltonian H, if $f^D \in$ HS and Λ is a positive operator in HS—was not available. But it was pretty obvious that a HS vector f^D would not do the job. However, it was well known that the exponential law could not hold for the survival probability $(1 - \mathcal{P}(t))$ if f^D was in HS [6].

The eigenvalue equation (5) as well as (7) and (8) are understood as a generalized eigenvalue equation in RHSs (of Hardy class).

A generalized eigenvector of an operator A is defined to be that $F \in \Phi_+^\times$ which fulfills

$$\langle A\phi|F\rangle \equiv \langle \phi|A^\times F\rangle = \omega\langle\phi|F\rangle \qquad \text{for all} \quad \phi \in \Phi_+ \tag{10}$$

where the number ω is called the generalized eigenvalue. The first equality in (10) defines (uniquely) the conjugate operator $A^\times$ of any *continuous* operator A in the space Φ_+. Equation (10) is also written as

$$A^\times |F\rangle = \omega |F\rangle, \qquad |F\rangle \in \Phi_+^\times \tag{11}$$

Therefore, to be precise, in (5) and (7) the operator $H^\times$ should have been used, whereas Dirac just wrote H. The conjugate operator for the unitary operator $U(t) = e^{iHt}$ in Φ_+ is the operator in $\Phi_+^\times$:

$$U_+^\times(t) = \left(e^{iHt}\right)_+^\times = e_+^{-iH^\times t} \tag{12}$$

where the subscript $+$ is the label of the space Φ_+ (Hardy class of upper half-plane).

Thus the Rigged Hilbert Space (9) is already needed for the Dirac kets (5) and the new Gamow kets (7) do not require a new mathematical theory.[2]

A resonance and decaying state is associated with an S-matrix pole at the position $z_R = E_R - i\Gamma/2$ on the second sheet of the analytically continued jth partial S-matrix element (j = angular momentum of the decay products or spin of the resonance), and has a Breit–Wigner energy distribution. The decaying state vector f^D in (1), though a problematic notion in the HS theory, should therefore have a Breit-Wigner energy wave function:

$$f^D = \frac{1}{i}\int_0^\infty dE |E, b^-\rangle \frac{\sqrt{\frac{\Gamma}{2\pi}}}{E_R - i\frac{\Gamma}{2} - E} \tag{13}$$

Using (7), (8), and $\psi^G = f^D$ of (13) as postulates, one can obtain in a heuristic way something like (1) from the fundamental probabilities (14). However, postulates cannot be made arbitrarily; instead, they have to be consistent with each other. But there is no way that one can derive (7) for the vector f^D of (13), not to speak of justifying (8).

---

[2] Except that the Dirac kets can be defined on a larger dense subspace Φ of $\mathcal{H}$ (Schwartz space) and the Gamow kets only on the Hardy space Φ_+ because these require continuation to negative and complex energies.

However, one can modify (13) a little bit and postulate a vector ψ^G:

$$\psi^G = \frac{1}{i}\int_{-\infty_{II}}^{+\infty} dE|E,b^-\rangle \frac{\sqrt{\frac{\Gamma}{2\pi}}}{E_R - i\frac{\Gamma}{2} - E} \qquad (14)$$

(where $-\infty_{II}$ means integration along the negative semiaxis in the second sheet[3]). For practical purposes, (13) and (14) differ very little when $\Gamma/2E_R$ is small. But for ψ^G of (14) one can prove the following:

1. ψ^G, if considered as functional over the Hardy class space Φ_+, can be derived as pole term at $z_R = E_R - i\Gamma/2$ in the lower half-plane of the second sheet of the analytically continued S-matrix element.
2. Equation (7) is fulfilled in the sense of (10) or (11), as a generalized eigenvalue equation with $\psi^G \in \Phi_+^\times$.

In order to obtain statement 1, the Hardy class property of the spaces Φ_+ is needed. In order to prove statement 2 and a similar equation for all powers H^n of H, the Schwartz space property is needed. Therefore we chose for Φ_+ the space of "very well behaved" vectors, which means the following:

1. $\psi^- \in \Phi_+$ is well-behaved (i.e., the energy wavefunction $\langle ^-E|\psi^-\rangle \in$ Schwartz space).
2. $\psi^- \in \Phi_+$ has some analyticity properties (precisely $\langle ^-E|\psi^-\rangle$ is a Hardy class function analytic in the upper half-plane).[3]

Gadella has shown that, with this property of Φ_+, the triplet of spaces (9) is indeed an RHS [9], which is needed in order to justify (5) and the Dirac basis vector expansion [c.f. (26) below (Dirac bra-ket formalism)].

For this RHS (9) of Schwartz and Hardy class, we now have to derive (8) as a generalized eigenvalue equation [in the sense of (10) with $\phi = \psi^-$, $F = \psi^G$, $A = U(t) = e^{iHt}$] for the vector ψ^G given by (14). The surprising, totally unforeseen and unintended result was that for $\psi^G \in \Phi_+^\times$ (Hardy class) one can show that (8) cannot hold for $t < 0$, but one can prove it as a generalized eigenvalue equation for $t \geq 0$:

$$e_+^{-iH^\times t}\psi^G = e^{-iE_R t}e^{-\frac{\Gamma}{2}t}\psi^G \qquad \text{for} \quad t \geq 0 \text{ only} \qquad (15)$$

This means that the time evolution operators (12) on the space $\Phi_+^\times$ form only a semigroup, not a reversible unitary group like the $U(t)$ in the Hilbert space $\mathcal{H}$.

---

[3] For the theory of resonances the first sheet of the S-matrix is irrelevant; when we talk of analytic functions we will always mean the second (or higher) sheet of the S-matrix.

With equations (5), (7), (14), and (15) mathematically well established, one can obtain the Golden Rule from the fundamental postulate (4) of quantum mechanics. For the decaying state W^D, one takes the Gamow state

$$W^G = \frac{1}{f^2} |\psi^G\rangle\langle\psi^G| \qquad (f \text{ is a normalization factor}) \qquad (16)$$

and derives for

$$\mathscr{P}(t) = \text{Tr}(\Lambda W^G(t)) = \langle\psi^G(t)|\Lambda|\psi^G(t)\rangle \qquad (4)$$

using (15) and the Lippmann–Schwinger equation (6), with some mathematical qualifications, the result [10]:

$$\mathscr{P}(t) = 1 - e^{-\Gamma t} \int_0^\infty dE \sum_{b \neq b^D} |\langle E, b|V|\psi^G\rangle|^2 \frac{1}{(E - E_R)^2 + (\Gamma/2)^2} \, ; \qquad t \geq 0 \qquad (17)$$

This is the probability for the transition of the decaying state W^G into all mixtures of decay products Λ of (2). We have normalized[4] W^G such that $\mathscr{P}(\infty) = 1$ (at $t \to \infty$ the probability for the decay products is certainty).

Taking the time derivative of (17), one obtains the decay rate

$$\dot{\mathscr{P}}(t) = e^{-\Gamma t} 2\pi \int_0^\infty dE \sum_{b \neq b^D} |\langle E, b|V|\psi^G\rangle|^2 \frac{\Gamma/2\pi}{(E - E_R)^2 + \Gamma/2^2} \, ; \qquad t \geq 0 \qquad (18)$$

The time directedness of the decay (*increase* of the probability for the decay products), $t \geq 0$, is a consequence of the semigroup property (15). The initial decay rate is then obtained from (18) as

$$\dot{\mathscr{P}}(0) = 2\pi \int_0^\infty dE \sum_{b \neq b^D} |\langle E, b|V|\psi^G\rangle|^2 \frac{\Gamma/2\pi}{(E - E_R)^2 + (\Gamma/2)^2} \qquad (19)$$

We will call formulas (18), (19), and also (17) the *exact* Golden Rule.

Comparing (19) with (17) for $\mathscr{P}(0) = 0$ (at $t = 0$ the decay is to begin and the probability for the decay products is zero), one finds

$$\dot{\mathscr{P}}(0) = \Gamma \qquad (= \Gamma/\hbar) \qquad (20)$$

---

[4] The kets $|E_R - i\Gamma/2\rangle$ inherit their normalization from the δ-function normalization $\langle{}^-bE'|Eb^-\rangle = \delta(E' - E)$ of the Dirac kets; the other factors in (14), including the phase factor, are arbitrary. Then f in (16) is chosen such that $\mathscr{P}(\infty)$ is normalized to unity.

From the exponential time dependence in (17) or (18), we obtain that the lifetime τ_R [which is defined by the exponential law as the time during which the rate went down to $\frac{1}{e}$ of its initial value, and therefore precisely defined only if the exponential law (1.18) holds] is given by

$$\tau_R = \frac{1}{\Gamma} \quad (= \hbar/\Gamma) \tag{21}$$

The results (20) and (21) mean that the imaginary part of the complex energy in (7), which is also the imaginary part of the S-matrix pole position and the width of the Breit–Wigner energy distribution (14) is equal to the initial rate $\dot{\mathcal{P}}(0)$ of the decay probability (4) and also equal to the inverse lifetime. All the equations (17), (18), (19), (20), (21) are exact, [although (19) is not of much practical use for the calculation of $\dot{\mathcal{P}}(0) = \Gamma$ because Γ appears on both sides of the equations].

To obtain the very useful Golden Rule of Dirac, one makes the following (Born) approximation:

$$\langle b, E|V|\psi^D\rangle \approx \langle b, E|V|f^D\rangle \tag{22a}$$

$$\frac{\Gamma}{2E_R} \to 0 \tag{22b}$$

$$E_R \approx E_D \tag{22c}$$

The Breit–Wigner energy distribution (natural line shape) has the property

$$\lim_{\frac{\Gamma}{2E_R} \to 0} \frac{\Gamma/2\pi}{(E-E_R)^2 + (\Gamma/2)^2} = \delta(E - E_R) \tag{22d}$$

Using (22), one obtains from (19) the initial decay rate in this Born approximation:

$$\dot{\mathcal{P}}(0) = \frac{2\pi}{\hbar}\int dE \sum_b |\langle b, E|V|f^D\rangle|^2 \delta(E - E_D) \tag{1}$$

which is the Dirac Golden Rule (1).

Summarizing, to satisfy our desire to "derive" Dirac's Golden Rule (1), we had to postulate the existence of a new generalized vector ψ^G with the properties (7) and (8). These Gamow vectors come in pairs. To every Gamow vector $\psi^G = |E_R - i\Gamma/2^-\rangle\sqrt{2\pi\Gamma} \in \Phi_+^\times$, there is also another Gamow vector $\tilde{\psi}^G = |E_R + i\Gamma/2^+\rangle\sqrt{2\pi\Gamma}$

$$H^\times \tilde{\psi}^G = (E_R + i\Gamma/2)\tilde{\psi}^G, \qquad |E_R + i\Gamma/2^+\rangle \in \Phi_-^\times \tag{23}$$

in another RHS

$$\Phi_- \subset \mathcal{H} \subset \Phi_-^\times \tag{24}$$

with the same HS $\mathcal{H}$ but with a space Φ_- that is Hardy class in the lower half-plane.

There can be several Gamow vectors of given angular momentum j in a particular quantum physical system $\psi_i^G = |E_R - i\Gamma_i/2\rangle$, $i = 1, 2, \cdots, N$ corresponding to N poles of the jth partial S-matrix. The Gamow vectors have the following features, some of which we have not yet mentioned above:

1. They are derived as functionals of the resonance pole term at $z_R = E_R - i\Gamma/2$ (and at $z_R^* = E_R + i\Gamma/2$) in the second sheet of the analytically continued S-matrix [11].

2. They are given by (14) and have a Breit–Wigner energy distribution $|\langle ^-E|\psi^G\rangle|^2 = \frac{\Gamma}{2\pi}\frac{1}{(E-E_R)^2+(\Gamma/2)^2}$ which extends to negative energy values on the second sheet indicated in the representation (14) by $-\infty_{II}$ [11].

3. The decay probability $\mathscr{P}(t) = \text{Tr}(\Lambda|\psi^G\rangle\langle\psi^G|)$ of $\psi^G(t)$, $t \geq 0$, into the final non-interacting decay products described by Λ can be calculated as a function of time, and from this the decay rate $\dot{\mathscr{P}}(t) = \frac{d\mathscr{P}(t)}{dt}$ is obtained by differentiation [10,12] This leads to an exact Golden Rule (with the natural line shape given by the Breit–Wigner) and the exponential decay law

$$\dot{\mathscr{P}}_\eta(t) = e^{-\Gamma t}\Gamma_{\Lambda_\eta}, \qquad t \geq 0 \tag{25}$$

where Γ_{Λ_η} is the partial width (or partial initial decay rate) for the decay products Λ_η (Γ_{Λ_η} = branching ratio × Γ). In the Born approximation ($\psi^G \to f^D$ = an eigenvector of $H_0 = H - V$; $\Gamma/E_R \to 0$; $E_R \to E_D$ = a discrete eigenvalue of H_0 in its continuous spectrum) this exact Golden Rule goes into Fermi's Golden Rule No. 2 of Dirac.

4. The Gamow vectors ψ_i^G are members of a "complex" basis vector expansion [11]. In place of the well-known Dirac basis system expansion (Nuclear Spectral Theorem of the RHS) given by

$$\phi^+ = \sum_n |E_n)(E_n|\phi^+) + \int_0^{+\infty} dE|E^+\rangle\langle^+E|\phi^+\rangle \tag{26}$$

(where the discrete sum is over bound states, which we henceforth ignore), every state vector $\phi^+ \in \Phi_-$ can also be expanded as

$$\phi^+ = -\sum_{i=1}^N |\psi_i^G\rangle\langle\psi_i^G|\phi^+\rangle + \int_0^{-\infty_{II}} dE|E^+\rangle\langle^+E|\phi^+\rangle \tag{27}$$

where $-\infty_{II}$ indicates that the integration along the negative real axis (or other contour) is in the second Riemann sheet of the S-matrix).[3] N is the number of resonances in the system (partial wave), each occurring at the pole position $z_{R_i} = E_{R_i} - i\Gamma_i/2$. This allows us to mathematically isolate the exponentially decaying states ψ_i^G.

The "complex" basis vector expansion (27) is rigorous. The Weisskopf–Wigner approximate methods are tantamount to omitting the background integral in (27), that is,

$$\phi^+ \stackrel{W-W}{=} \sum_{i=1}^{N} |\psi_i^G\rangle c_i, \qquad c_i = -\langle \psi_i^G | \phi^+ \rangle \tag{28}$$

For instance, for the $K_L^0 - K_S^0$ meson system with $N = 2$,

$$\phi^+ = \psi_S^G b_S + \psi_L^G b_L \tag{29}$$

For the case of a single decaying state, $N = 1$ (27) becomes

$$\phi^+ = -|\psi^G\rangle\langle\psi^G|\phi^+\rangle + \int_0^{-\infty} dE |E^+\rangle\langle {}^+E|\phi^+\rangle \tag{30}$$

and a prepared state $\phi^+ \in \Phi_-$ is only approximately represented by an exponentially decaying Gamow vector $\psi^G \in \Phi_+^\times$:

$$\phi^+ \stackrel{W-W}{=} \psi^G c_i \tag{31}$$

5. The time evolution of every state prepared by a macroscopic apparatus is obtained from (30) as

$$\phi^+(t) = e^{-iHt}\phi^+ = -e^{-iE_R t} e^{-\Gamma/2t} |\psi^G\rangle\langle\psi^G|\phi^+\rangle$$

$$+ \int_0^{-\infty} dE\, e^{-iEt} |E^+\rangle\langle {}^+E|\phi^+\rangle \tag{32}$$

In addition to the exponential time dependence, it has the time dependence given by the "background integral," which is nonexponential. Theoretically, such a background term is always present in the prepared state vector $\phi^+ \in \Phi_- \subset \mathcal{H}$, but it could be arbitrary small.

The properties (7), (14), (15), (25), (26), (27), and (32) are not independently postulated conditions for the Gamow vectors but are, instead, derived from each other in the mathematical theory of the RHS. These are properties that one would require of a vector which is to represent the "state" of an unstable particle or of a resonance. In fact the Gamow vector combines features which one has observed as different aspects of the phenomena associated with resonance scattering and decay and which were in the past only vaguely and approximately related to each other in a collision theory [13][5] based on Weisskopf–Wigner's approximate methods [2].

For instance, the exponential decay law (25) and (17), which allows the definition of a lifetime τ_R, has been observed and confirmed with high accuracy for numerous radioactive decays and other decay phenomena with $\hbar/\tau_R E_R$ (or $\hbar/\tau_R M_R$ in the relativistic case) $\approx 10^{-14}$ and less [14]. To describe this phenomenon, one needed a vector with the property (7) and (8).[6] Because such vectors were not available in the standard theory, one postulated a finite dimensional complex Hamiltonian matrix and used its eigenvectors [3,15]. However, it remained unclear how this finite matrix could be a submatrix of the infinite-dimensional self-adjoint Hamiltonian. Now (27) explains this as the approximation (28).

The converse problem of the mathematically derived *deviations* from the exponential law for any vector [e.g., also f^D of (13)] in the Hilbert space evolving with a self-adjoint semibounded Hamiltonian $f^D(t) = e^{-iHt}f^D$ [6] is also overcome by ψ^G because $\psi^G \in \Phi_+^\times$ and ψ^G is not in $\mathcal{H}$, thus allowing the exponential law (8). This exponential law (8) for the vector leads to the exponential law (17) and (25) for the decay rates. This, in turn, permits us to define a lifetime τ_R. From (17) or (25) it follows that $\tau_R = \hbar/\Gamma$, where Γ is the imaginary part of the energy eigenvalue of (8), which, in turn, is the same as the Γ of the Breit–Wigner energy distribution in (14).

However, in spite of the exponential law for the decaying state per se, deviations from the exponential law, as may have been observed in Ref. 16, (c.f. Fig. 1), can be explained in our theory if one interprets $\phi^+ \in \Phi_-$ as the state prepared by the (macroscopic) experimental apparatus. Due to the background term in (32), there are deviations from the exponential law for a prepared state. And according to (32), these deviations change with the energy distribution $\langle ^+E|\phi^+\rangle$—that is, with the experimental conditions of state preparation. The nonexponential background term can become substantial. But the time evolution of the unstable particle ψ^G will always be exponential with the same lifetime $\tau_R = \hbar/\Gamma$ because this is according to (32) independent of experimental changes in state preparation. An unstable particle is something defined by its mass (or E_R) and its lifetime and independent of the experimental preparations.[7]

In scattering experiments, one observes a different aspect of the resonance: unstable particle phenomenon for a different range of the resonance parameter Γ/E_R (or Γ/M_R) $> 10^{-7}$ (e.g., Z boson, hadron resonances). One sees a bump

---

[5] Using Dirac kets and the Lippmann–Schwinger equation but not Gamow kets [13].

[6] For $N = 2$ [3] or $N =$ finite [15], one needs the superpositions (29) or (28), respectively.

[7] If more than one unstable particle is involved as in (29), the rate $\dot{\mathcal{P}}(t)$ may not be exponential as in (18) due to the interference of the two (or more) Gamow states.

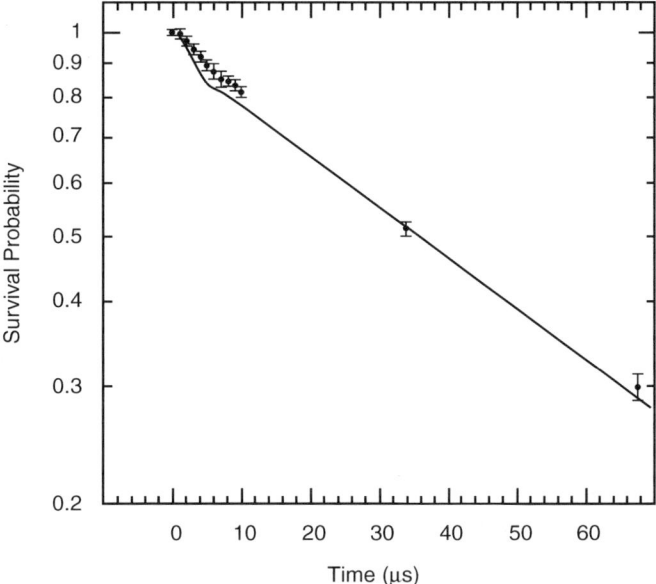

Figure 1. The logarithm of the survival probability $(1 - \mathscr{P}(t))$ as a function of time shows some deviations from linearity. (From Ref. 16, with permission of the author.)

in the cross section which one analyzes in terms of a Breit–Wigner amplitude B.W. and a background term B (or a resonant and a background phase shift). For a resonance in this range of the parameter Γ/E_R, one fits the cross-section data to $|B + \text{B.W.}|^2$ and determines E_R as the position of the maximum and Γ as the width of the Breit–Wigner. Because of the ever-present background B, which is related to the second term in (30), and other corrections, this is not without ambiguities. It is certainly impossible to distinguish experimentally between an "ideal" Breit–Wigner that extends over E from $-\infty$ to $+\infty$ as in (14) and a Breit–Wigner of (13) which is zero for "unphysical" values $E < 0$.

The width Γ and the lifetime τ_R have a completely different origin. Γ is the width of a Breit–Wigner amplitude, which together with the background give the scattering amplitude (or S-matrix element). For the S-matrix definition of a resonance, one does not need a vector; but if one wants to associate with the resonance a "resonance state vector," f^D of (13) would do fine.[8] τ_R is the lifetime measured by a fit of the experimental counting rate $(\dot{N}_\eta(t)/N) \approx \dot{\mathscr{P}}_\eta(t)/\Gamma$ to the exponential e^{-t/τ_R} of (25) $(\log(\dot{N}(t)/N) = -t/\tau_R)$. Thus

---

[8] f^D is in the Hilbert space $\mathscr{H}$, but it is not in the domain of the Hamiltonian $H(\|Hf^D\| \to \infty)$; so it could, in particular, not be an energy eigenvector in the Hilbert space sense.

observationally the width Γ and the inverse lifetime $1/\tau_R$ are different quantities: One comes from the Breit–Wigner distribution for the energy, whereas the other comes from the (linear) time dependence (of the logarithm) of the counting rate. These observationally different quantities Γ and $1/\tau_R$ become the same only through the Gamow vector (14). And it is important that ψ^G is a generalized vector $\psi^G \in \Phi_+^\times$ (14) and not the Hilbert space vector $f^D \in \mathscr{H}$ (13). Changing the integration from $0 \leq E < \infty$ to $-\infty_{II} < E < +\infty$ is important [17]—mathematically because it changes a Hilbert space vector into a Rigged Hilbert Space vector, and physically because it makes the observationally different quantities Γ and $1/\tau_R$ the same ($\Gamma = \hbar/\tau_R$), not only approximately and vaguely, but precisely.

For this, one needs the Rigged Hilbert Space quantum mechanics. Gamow vectors ψ^G have all the properties that one heuristically wanted for unstable particles and resonances. They are the central ingredients of a mathematical theory of which the Weisskopf–Wigner approximate method, the effective theories with finite complex Hamiltonian (e.g., the Lee–Oehme–Yang theory [3] and many in nuclear physics [15]), and Dirac's Golden Rule are approximations. But, in addition, they have an unwanted and mostly undesired feature, the semigroup time evolution (15)—which is in contrast to the reversible unitary group evolution of the Hilbert space and expresses something like irreversibility on the microphysical level. Without the Gamow vector and without the RHS there is no rigorous theory in which (18) can be derived and for which (20), (21), and (23) hold. But, in welcoming these results, one also has to contend with the semigroup time evolution (15). We want to discuss this feature next, and we shall see that it is, after all, not as bad as one may have thought.

II. PROBABILITY AND TIME ASYMMETRY

A. Introduction

Because I did not know what to do with the puzzling mathematical result (15) that I obtained when I asked the question about the relationship between the quantum mechanical probabilities (4) and the decay rate (1), I showed my paper [11] to Ilya Prigogine. But I could not get him interested in it; and because I also had a lot of difficulty in getting the paper published, I kept working on my other subject using unitary group representations, for which not reversing time (or any other parameter) is abominable. In the meanwhile, Gadella, working by himself, established the mathematics of the Rigged Hilbert Space of Hardy class functions. About 10 years later, when also Gadella had given up on this subject, I happened to hear a talk at a conference by Ioannis Antoniou. To my surprise, he had not only understood the idea of quantum mechanics in Rigged Hilbert Space, but had also learned the mathematics and was using it for the intrinsic irreversibility of the Brussels group. Antoniou's talk convinced me to return to the subject of resonances, decaying states, and their time evolution, and I also persuaded my Rigged Hilbert Space collaborator Gadella to rejoin me. In order to study irreversibility and to learn about the arrows of time, I started my visits to Dr. Prigogine's office and in these sessions he got me really sold on this subject with such statements as "all irreversibility comes from the time evolution of resonances."

At that time Nico van Kampen was visiting Texas and I wanted to get a second opinion. I remember that we talked for some time on the phone and we did not seem to understand each other, but then he must have finally caught on, because he remarked something like "O there are people who say that there is irreversibility on the microphysical level, but that is heresy."

I liked his choice of words and quoted van Kampen in my talk at a subsequent conference so that the audience would not mistake this semigroup irreversibility for the conventional irreversibility in quantum statistical mechanics due to external influences of a reservoir or measurement apparatus upon an open quantum system. But then this came back to me in a referee's report in which the referee—who must have been in the audience—rejected the paper saying among others that "the authors themselves call this heresy." When I showed this report to Ilya Prigogine, his reaction was something like "I am proud to be a superheretic." In the meanwhile I have learned that the semigroup evolution (15) is not that heretic after all.

The irreversibility of (15) is the time asymmetry of the solutions of a time-symmetric dynamical equation with time-asymmetric boundary conditions. In classic physics, there are several well-known examples of time-asymmetric boundary condition for time-symmetric dynamical equations (the radiation arrow of time, the cosmological arrow of time). It is the peculiarity of the Hilbert space[9] which disallows time-asymmetric boundary conditions for solutions of the time-symmetric Schrödinger equation. But time-asymmetric, specifically purely outgoing boundary conditions were suggested a long time ago [18].

B. From Gamow Kets to Microphysical Irreversibility

Time-asymmetric quantum theory is mathematically described by the appropriate choice of spaces for the solutions of the usual time-symmetric Schrödinger equation with asymmetric boundary conditions. This is done with the aid of the two Rigged Hilbert Spaces (9) and (24).[10] The questions then are, What is the physical interpretation of the two rigged Hilbert spaces, are there

---

[9] In HS the exponential e^{iHt} is not defined by the converging exponential series but by the Stone–von Neumann calculus.

[10] And also by the Lippmann–Schwinger (integral) equation. Vaguely, the Gamow kets $|E_R \mp i\Gamma/2^{\mp}\rangle$ of (7) and (23) are just an analytic continuation of the Lippmann–Schwinger–Dirac kets (6).

fundamental reasons for boundary conditions that lead to the two rigged Hilbert spaces, and what is the basis and the physical evidence for quantum mechanical irreversibility of the kind described by (15)?

In quantum physics one measures probabilities (4). In experiments, the state W [or the pure (idealized) state ϕ] is prepared by a preparation apparatus, and the observable Λ (or the idealized observable $|\psi\rangle\langle\psi|$) is registered by a registration apparatus (e.g., a detector). For instance, ϕ can be the in-states ϕ^+ of a scattering experiment, and the observables ψ can be the detected out-states ψ^- of a scattering experiment.

The measured (or registered) quantities are ratios of (usually) large numbers $N_\eta(t)/N$, the detector counts. They are interpreted as probabilities (e.g., as the probability to measure the observable Λ_η in the state W at the time t) and represented in the theory according to (4) by $\mathscr{P}_W(\Lambda_\eta(t)) = \text{Tr}(\Lambda_\eta W(t))$:

$$N_\eta(t)/N \approx \mathscr{P}(t) \equiv \text{Tr}(\Lambda_\eta(t)W(0)) = \text{Tr}(\Lambda_\eta(0)W(t)) = |\langle\psi^-(t)|\phi^+\rangle|^2 \quad (33)$$

In Hilbert space quantum mechanics the theoretical probabilities can be calculated at any time using the unitary reversible group evolution

$$W(t) = e^{-iHt}W(0)e^{iHt}, \qquad \phi(t) = U^\dagger(t)\phi_0 = e^{-iHt}\phi(0)$$
$$\text{where} \quad -\infty < t < \infty \quad (34a)$$

or, in the Heisenberg picture,

$$\Lambda(t) = e^{iHt}\Lambda_0 e^{-iHt}, \qquad \text{where} \quad -\infty < t < \infty \quad (34b)$$

The probabilities $\mathscr{P}_\eta(t) = N_\eta(t)/N$ cannot be observed at any arbitrary positive or negative time t. The reason is the following:

A state needs to be prepared before an observable can be measured, or registered in it.

We call this truism the preparation $\Rightarrow$ registration arrow of time [19]; it is an expression of causality. Let $t_0 (= 0)$ be the time at which the state has been prepared. Then, $\mathscr{P}(t)$ is measured as the ratio of detector counts:

$$\mathscr{P}_\eta^{\text{exp}}(t) \approx \frac{N_\eta(t)}{N} \qquad \text{for} \quad t > t_0 = 0 \quad (35)$$

If there are some detector counts before $t = t_0$, they are discounted as noise because the experimental probabilities

$$\text{can } not \text{ fulfill} \quad \mathscr{P}^{\text{exp}}(t) \not\approx 0, \qquad \text{for} \quad t < t_0 = 0 \quad (36)$$

Though in the Hilbert space theory $\mathscr{P}_W(\Lambda(t)) = \mathscr{P}_{W(t)}(\Lambda)$ can be calculated at positive or negative values of $t - t_0$ using the unitary group evolution (34a) or

(34b), an experimental meaning can be given to $\mathscr{P}(t)$ only for $t > t_0$. The preparation $\Rightarrow$ registration arrow of time clearly sets states apart from observables. In standard quantum mechanics the same set of vectors $\mathscr{H}$ (usually the whole HS) is used for states $W = |\phi\rangle\langle\phi|$ as well as observables $\Lambda = |\psi\rangle\langle\psi|$ (and the $|\psi\rangle$ are usually called "eigenstates" even if they are observables). In scattering theory the vectors controlled by the preparation apparatus ϕ^{in}, ϕ^+ are called in-states and the vectors controlled by the registration apparatus (e.g., detector) are called out-states ψ^{out}, ψ^-, though $|\psi^{out}\rangle\langle\psi^{out}|$ is really an observable defined by the detector [c.f. Eq. (38) below]. Thus in scattering theory, one at least introduces the set of (pure) states $\{\phi^+\}$ and the set of observed "properties" $\{\psi^-\}$ separately, though soon one makes the hypothesis $\{\phi^+\} = \{\psi^-\}(= \mathscr{H})$, in exceptional cases realizing that this is not quite in agreement with our intuition of causality [20]. Based on the preparation $\Rightarrow$ registration arrow of time, we will distinguish meticulously between states and observables by choosing for the in-state vectors the space Φ_- of (24), $\{\phi^+\} \equiv \Phi_-$ and choosing for the out-observable vectors the space Φ_+ of (9), $\{\psi^-\} \equiv \Phi_+$. Thus we make the following hypothesis: Each species of quantum physical systems has a pair of Rigged Hilbert Spaces: $\phi^+ \in \Phi_- \subset \mathscr{H} \subset \Phi_-^\times$ defined by the preparation apparatus and $\psi^- \in \Phi_+ \subset \mathscr{H} \subset \Phi_+^\times$ defined by the registration apparatus. Both have the same $\mathscr{H}$. One can then see [21] that the Hardy class properties of Φ_+ and Φ_- are such that prepared states $\phi^+ \in \Phi_-$ must be there before observables $|\psi^-\rangle\langle\psi^-| \in \Phi_+$ can be measured in them.

The vectors ψ^G with the semigroup time evolution (15) describe quantum physical states in nature that evolve only into the positive direction of time, $t > t_0$. Resonances and unstable particles have this property, though the irreversible character of quantum mechanical decay has rarely been mentioned in the literature. There are, however, exceptions [22].

We shall now demonstrate the time asymmetry of the quantum mechanical decay process and explain the meaning of the time t_0, using the neutral kaon state K_S^0 as an example. Its time evolution is entirely due to the Hamiltonian of the neutral kaon system and free of external influences.[11]

The process (idealized, because in the real experiment one does not use a π^- but a proton beam) that prepares the neutral kaon state is

$$\pi^- p \Rightarrow \Lambda K^0, \quad K^0 \Rightarrow \pi^+ \pi^- \qquad (37)$$

K^0 is strongly produced with a time scale of 10^{-23} sec and it decays weakly, with a time scale of 10^{-10} sec, which is roughly equal to the lifetime τ_{K_S} of K_S^0. Thus t_0, the time at which the preparation of the K^0 state W^{K^0} is completed and the

---
[11]Because here we are only interested in the fundamental concepts of decay, we discuss a simplified K^0 system for which the K_L^0 as well as the CP violation is ignored [23].

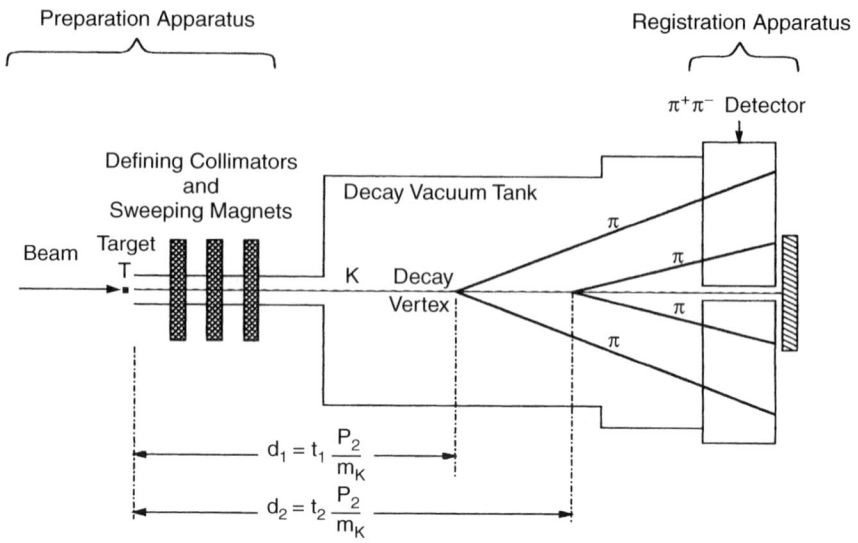

Figure 2. Schematic diagram of the neutral K-meson decay experiment.

registration can begin, is very well-defined (within $10^{-13}\tau_K$). A schematic diagram of a real experiment [24] is shown in Fig. 2. The state W^{K^0} is created instantly (at $t_0 \pm 10^{-13}\tau_K$) at the baryon target T (and the baryon B is excited from the ground state (proton) into the Λ state, with which we are no further concerned). We imagine that a single particle K^0 is moving into the forward beam direction, because somewhere at a distance, say at d_2 from T, we "see" a decay vertex for $\pi^+\pi^-$; that is, a detector (registration apparatus) has been built such that it counts $\pi^+\pi^-$ pairs that are coming from the position d_2. The observable registered by the detector is the projection operator for the $\pi^+\pi^-$ decay channel ($\eta = \pi^+\pi^-$)

$$\Lambda_\eta(t_2) = |\pi^+\pi^-, t_2\rangle\langle\pi^+\pi^-, t_2| = |\psi^{\text{out}}(t_2)\rangle\langle\psi^{\text{out}}(t_2)| \tag{38}$$

for those $\pi^+\pi^-$ which originate from the fairly well-specified location d_2. From the position (in the lab frame) d_2^{lab}, the momentum p of the K^0 (= the z component of the momentum of the $\pi^+\pi^-$ system), and the mass m_K of K^0, one obtains the time t_2^{rest} (in the K^0 rest frame) that the K^0 has taken to move from T to d_2^{lab}. This is given by the simple formula of relativity $d_2^{\text{lab}} = t_2^{\text{rest}}\frac{p}{m_K}$, which we write $d_2 = t_2\frac{p}{m_K}$.

We do not have to focus at only one location d_2 but can count decay vertices at any distance d. The detector (described by the projection operator $\Lambda_\eta(t) \equiv |\pi^+\pi^-, t\rangle\langle\pi^+\pi^-, t|$) counts the $\pi^+\pi^-$ decays at different times

$t = t_1, t_2, t_3, \ldots$ (in the rest frame of the K^0), and these correspond to the distances from the target $d_1 = pt_1/m_K$, $d_2 = pt_2/m_K, \ldots$ (in the lab frame).

One "sees" the decay vertex d_i for each single decay and imagines a single decaying K^0 microsystem that had been created on the target T at time t_0 ($= 0$) and then traveled for a time t_i until it decayed at the vertex d_i. We give the following interpretation to these observations: A single microphysical decaying system K^0 described by W^{K^0} has been produced by a macroscopic preparation apparatus and a quantum scattering process, at a time $t = 0$. Each count of the detector is the result of the decay of such a single microsystem. This particular microsystem has lived for a time t_i—the time that it took the decaying system to travel from the scattering center T to the decay vertex d_i. The whole $\pi^+\pi^-$ detector registers the counting rate $\Delta N_\eta(t_i)/\Delta t \approx N\mathscr{P}(t)$ as a function of d_i—that is, of $t_i = \frac{m_k}{p} d_i$, for $\cdots t_i > \cdots t_2 > t_1 > t_0 = 0$. ($N$ is the total number of counts.)

Figure 3 shows the counting rate $\Delta N_\eta(t_i)/\Delta t$ plotted as a function of time t (in the K^0 rest frame). (It is normalized to $\mathscr{P}(0) \doteq 1$, i.e., to $\frac{\Delta N_\eta(0)}{\Delta t}\frac{1}{N} \doteq 1$.)

No $\pi^+\pi^-$ are registered for $t < t_0 = 0$—that is, clicks of the counter for $\pi^+\pi^-$ that would point to a decay vertex at the position d_{-1} in front of the target T are not obtained (if there were any, they would be discarded as noise). One finds for the counting rate

$$\frac{\Delta N_\eta(t_i)}{\Delta t} \approx 0, \qquad t < t_0 = 0 \tag{39}$$

in agreement to the preparation $\Rightarrow$ registration arrow of time (36). This is so obvious that one usually does not mention it.

For $t > 0$ one expects the following from the prediction (18) or (25):

$$\frac{1}{N}\frac{\Delta N(t_i)}{\Delta t} \approx e^{-\Gamma t}, \qquad t > 0 = t_0 \tag{40}$$

From Fig. 3 one can see that this is indeed fulfilled to a high degree of accuracy.

More importantly, we see that the state $W^{K^0}(t) = |\psi_S^G(t)\rangle\langle\psi_S^G(t)|$ or the Gamow vector $\psi_S^G(t)$ of (15) (with $E_R = M_S$ and $\Gamma = \Gamma_S$) describes the state of an ensemble of single microsystems K^0 created at an "ensemble" of times t_0, all of which are chosen to be identical to the initial time $t = 0$ for the (mathematical) semigroup. The decay probabilities are the statistical probabilities for this ensemble of individual K^0 systems, and t in W^{K^0} is the time in the "life" of each single decaying K^0 system that had started at $t = 0$. It is not the time in the experimentalists life in the laboratory or the time of a "wave packet" of K^0's. The mathematical semigroup time $t = 0$ and the physical time t_0 at which the preparation of the state is completed can be thought of as

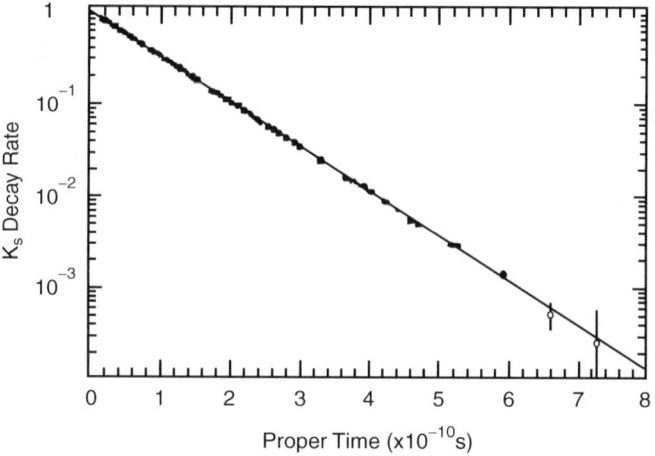

Figure 3. K_S decay rate versus proper time. (From Ref. 24, with permission.)

the ensemble of times t_0 at which each single quasistable particle described by the state ψ^G has been created. If this time of preparation t_0 can be accurately identified (e.g., for K^0 within $10^{-13}\tau_K$), then the decaying Gamow state can be experimentally isolated as a quasistationary microphysical system. The experimental time ordering (35) for the observed probabilities are the same as the time ordering of (17) calculated for the probabilities $\mathscr{P}(t) = \text{Tr}(\Lambda_\eta |\psi^G(t)\rangle\langle\psi^G(t)|)$ from the semigroup evolution (15) of ψ^G.

The semigroup time evolution of Gamow kets describes the irreversible character of microphysical decay. The time asymmetric boundary condition of the RHSs, the preparation $\Rightarrow$ registration arrow of time, and the time ordering of the quantum mechanical probabilities are manifestations of a fundamental time asymmetry in quantum physics.

Acknowledgments

Discussions with many people have contributed to the final version of this chapter. I particularly wish to acknowledge conversations with Nico van Kampen on irreversibility in quantum decay processes. I am very grateful to Mark Raizen, who provided one of the figures, read the manuscript, and explained his experiment to me. Financial support from the Solvay Foundation and in particular from the Welch Foundation is gratefully acknowledged.

References

1. E. Fermi, *Nuclear Physics*, Chicago University Press, Chicago, 1950, Chapter VIII; P. A. M. Dirac, *Proc. R. Soc. A* **114**, 243 (1927); *Principles of Quantum Mechanics*, 4th edition, Chapters VII and VIII, Oxford, Claredon Press, 1958.

2. V. Weisskopf and E. P. Wigner, *Z. Phys.* **63**, 54 (1930); **65**, 18 (1930).
3. T. D. Lee, R. Oehme, and C. N. Yang, *Phys. Rev.* **106**, 340 (1957).
4. M. Levy, *Nuovo Cimento* **13**, 115 (1959).
5. G. C. Hegerfeldt, *Phys. Rev. Lett.* **72**, 596 (1994).
6. L. A. Khalfin, *JETP* **33**, 1371 (1957); *JETP* **6**, 1053 (1958); L. Fonda, G. C. Ghirardi, and A. Rimini, *Repts. Prog. Phys.* **41**, 587 (1978), and references therein.
7. E. Roberts, *J. Math. Phys.* **7**, 1097 (1966); A. Bohm, *Boulder Lectures in Theoretical Physics 1966*, Vol. 9A, Gordon and Breach, New York, 1967; J. P. Antoine, *J. Math. Phys.* **10**, 53 (1969); **10**, 2276 (1969); O. Melsheimer, *J. Math. Phys.* **15**, 902, 917 (1974).
8. A. Bohm, *Springer Lecture Notes in Physics* **94**, 245 (1978); *Lett. Math. Phys.* **3**, 455 (1979).
9. M. Gadella, *J. Math. Phys.* **24**, 1462 (1983).
10. A. Bohm and N. L. Harshmann, in *Irreversibility and Causality*, A. Bohm, H. D. Doebner, and P. Kielanowski, eds., Springer, Berlin, 1998, p. 225, Section 7.4.
11. A. Bohm, *J. Math. Phys.* **22**, 2813 (1981); A. Bohm, S. Maxson, M. Loewe, and M. Gadella, *Physica A* **236**, 485 (1997).
12. A. Bohm, *Quantum Mechanics*, 1st edition, Springer, New York, 1979, Chapter XXI; 3rd edition, 2nd revised printing, 1994, Chapter XXI.
13. M. L. Goldberger and K. M. Watson, *Collision Theory*, Wiley, New York, 1964, Chapter 8.
14. K. L. Gibbons et al., *Phys. Rev. D* **55**, 6625 (1997); V. L. Fitch et al., *Phys. Rev. B* **140**, 1088 (1965); N. N. Nikolaev, *Sov. Phys. Usp.* **11**, 522 (1968) and references therein; E. B. Norman, *Phys. Rev. Lett.* **60**, 2246 (1988).
15. L. S. Ferreira, Resonances in *Lecture Notes in Physics*, Vol. 325, E. Brändas, ed., Springer, Berlin, 1989, p. 201; C. Mahaux, Resonances, Models and Phenomena, in *Lecture Notes in Physics*, Vol. 211, Springer, Berlin, 1984, p. 139; O. I. Tolstikhin, V. N. Ostrovsky, and H. Nakamura, *Phys. Rev.* **58**, 2077 (1998); V. I. Kukulin, V. M. Krasnopolsky, and J. Horacek, *Theory of Resonances*, Kluwer Academic Publishers, Dordrecht, 1989.
16. S. R. Wilkinson, C. F. Bharucha, M. C. Fischer, K. W. Madison, P. R. Morrow, Q. Niu, B. Sundaram, and M. Raizen, *Nature* **387**, 575 (1997); c.f. also C. F. Barucha et al., *Phys. Rev. A* **55**, R857 (1997).
17. This changing of the integration boundary, intentionally or inadvertently, has played some important role in the past. E. Fermi, *Rev. Mod. Phys.* **4**, 87 (1932); M. I. Shirokov, *Sov. J. Nucl. Phys.* **4**, 774 (1967).
18. A. F. J. Siegert, *Phys. Rev.* **56**, 750 (1939); R. Peierls, *Proceedings of the 1954 Glasgow Conference on Nuclear and Meson Physics*, E. M. Bellamy et al., eds., Pergamon Press, New York, 1955; G. Garcia-Calderon and R. Peierls, *Nucl. Phys. A* **265**, 443 (1976); A. Mondragon and E. Hernandez, *Ann. Phys.* **48**, 503 (1991).
19. G. Ludwig, *Foundations of Quantum Mechanics*, Vol. I, Springer-Verlag, Berlin, 1983; and Vol. II, 1985; where this arrow of time was recognized for the preparation and registration apparatuses, but not incorporated into the quantum theory of states and observables for which the Hilbert space was used.
20. R. G. Newton, *Scattering Theory of Waves and Particles*, McGraw-Hill, 1966, Chapter 7; in particular, p. 188.
21. A. Bohm, I. Antoniou, and P. Kielanowski. *J. Math. Phys.* **36**, 2593 (1995).
22. C. Cohen-Tannoudji et al., *Quantum Mechanics*, Vol. II, Wiley, New York, 1977, pp. 1345, 1353–1354; T. D. Lee, in *Particle Physics and Introduction to Field Theory*, Harwood Academic, New York, 1981, Chapter 13.

23. For a detailed analysis of the K_L–K_S system in the framework of time-asymmetric quantum mechanics with and without CP violation, see A. Bohm, *Proceedings, Workshop on Cosmology and Time Asymmetry*, Peyresq, France, 1996; *Int. J. Theor. Phys.* **36**, 2239 (1997).
24. This is the simplified schematic diagram of several generations of experiments measuring CP violation in the neutral kaon system; J. Christenson, J. Cronin, V. Fitch, and R. Turley, *Phys. Rev. Lett.*, **13**, 138 (1964); K. Kleinknecht, in *CP Violation*, C. Jarskog, ed., World Scientific, Singapore, 1989, p. 41, and references therein; NA**31**, G. D. Barr et al., *Phys. Lett.* **B317**, 233 (1993); E731; L. K. Gibbons et al., *Phys. Rev. D* **55**, 6625 (1997).

POSSIBLE ORIGINS OF QUANTUM FLUCTUATION GIVEN BY ALTERNATIVE QUANTIZATION RULES

MIKIO NAMIKI

Department of Physics, Waseda University, Tokyo, Japan

CONTENTS

I. Introduction
II. Canonical Quantization and Bohm's Theory
III. Stochastic and Microcanonical Quantizations
   A. SQ in a $(D+1)$-Dimensional Space–Time World
   B. MCQ in a $(D+1)$-Dimensional Space–Time World
IV. Conclusion
References

I. INTRODUCTION

As is well known, the present formalism of physics contains Planck constant $\hbar$, light speed c, and gravitation constant G, as given universal constants, but never ask their origins. Besides, we have no theory to give the elementary charge e (of an electron) in terms of other fundamental parameters. They must be unsatisfactory from the fundamental point of view [1].

In this chapter we discuss possible origins of the quantum fluctuation, within the framework of new quantization schemes such as *stochastic* and *microcanonical* quantizations, to give quantum mechanics in the infinite limit of an additionally introduced time (i.e., *fictitious time*) different from the ordinary one, assuming that $G = 0$. This work also leads us to another view that "a D-dimensional quantum system is equivalent to a $(D+1)$-dimensional classic system."

Dynamical Systems and Irreversibility: A Special Volume of Advances in Chemical Physics, Volume 122, Edited by Ioannis Antoniou. Series Editors I. Prigogine and Stuart A. Rice. ISBN 0-471-22291-7. © 2002 John Wiley & Sons, Inc.

For this purpose, it would be better to start this chapter from the following formula:

$$\hbar \sim \frac{mc^2 \tau}{\sqrt{N}} \qquad (1)$$

which was given by Hayakawa and Tanaka [2] under the following assumptions:

1. The universe is composed of N ($\gg 1$) fundamental particles, with mass m and lifetime τ.
2. The universe is an open system put in a bigger space–time world to give $\Delta N/N \sim 1/\sqrt{N}$.
3. The quantum mechanical energy–time uncertainty relation still holds: We remark that, at least, this prepares a consistency check of modern physics theory and the structure of the universe or, at most, a doorway to deeper dynamics beyond quantum mechanics.

The original authors found a nice order of magnitude by identifying the fundamental particle with *neutrino* 40 years ago [2]. Nevertheless, everything (in particular, neutrino mass $\neq 0$) should be improved by recent observations, and the above three assumptions should be examined by new information of cosmology and particle physics. However, we can hardly go further beyond quantum mechanics unless we have a deeper dynamics.

In this chapter we examine a few quantization schemes (such as *stochastic* and *microcanonical* quantizations) as a preliminary step to future dynamics, as to whether they can suggest *possible origins* of the quantum fluctuation or of Planck constant, and also show another view that a D-dimensional quantum system is equivalent to a $(D + 1)$-dimensional classic system. Remember that we keep $\hbar$ as a *given* universal constant and assume $G \sim 0$.

One may expect to find a doorway to new future dynamics beyond quantum mechanics, in recent cosmic ray observations—anomalies of high-energy particle distributions, the nonvanishing neutrino mass, and so on.

II. CANONICAL QUANTIZATION AND BOHM'S THEORY

As is well known, the *canonical* quantization brings quantum mechanics into an operator theory, in which every dynamical quantity is represented by an operator on a Hilbert space. Its operator nature is specified by canonical commutation relations. Note that the reasoning that $\hbar$ is a measure of the classic random motion can hardly be derived from the canonical quantum mechanics itself. On the other hand, we know that the path-integral method (which is equivalent to the conventional canonical quantum mechanics) is a sort of c-number quantization, but never asks a deeper origin of the quantum fluctuation. In order to avoid the mathematically ill-posed nature of the original path-integral method, we are sometimes recommended to use a Euclid measure $\exp(-S/\hbar)$ (S being the Euclid action) introduced by means of transformation $x_0 \to -ix_0$ (x_0 standing for the real time). Remember that the final goal of the stochastic or microcanonical quantizations is to give quantum mechanics of this type. These theories have introduced $\hbar$ as a given measure of the *quantum fluctuation*, but never ask its deeper origins.

As far as we know, Bohm's theory of quantum mechanics is a nice example to discuss a possible origin of the quantum fluctuation [3]. For the sake of simplicity, let us discuss Schrödinger equation in the one-dimensional nonrelativistic case. He reformulated the Schrödinger equation in terms of R and S defined by

$$\psi = \sqrt{R} \exp\left(\frac{iS}{\hbar}\right), \qquad R = |\psi|^2, \qquad \mathbf{p} = \nabla S \tag{2}$$

as a "Newtonian equation," that is,

$$\frac{d\mathbf{p}}{dx_0} = -\nabla(V + V_Q) \tag{3}$$

with the quantum mechanical potential

$$V_Q \equiv -\frac{\hbar^2}{2m} \frac{\nabla^2 |\psi|}{|\psi|} \tag{4}$$

Bohm planned to eliminate quantum mechanics within this fomulation: A quantum mechanical particle (originally obeying classic Newtonian equation of motion) moves in unknown *ether* that makes random fluctuating forces, $\mathbf{f}_Q = -\nabla V_Q$ [3].

This line of thought, if turned inversely, is considered to be a sort of quantization method. Actually, Nelson formulated his way of stochastic quantization analogously to this idea [3].

According to Bohm–Nelson's idea, the quantum fluctuation is rooted in the classic random process in the *real* space–time world, where $\hbar$ is introduced as a given measure. As is well known, every quantum property must come from the classic fluctuating forces. In order to reproduce quantum mechanics, therefore, we have to ascribe all strange properties, such as the *nonlocal long-distance correlations*, to the *classic* random forces, for example, $\mathbf{f}_Q = -\nabla V_Q$ in the Bohm's theory. *Mathematically* it may be accepted, but *physically* not.

In order to avoid this kind of difficulty, we have only to introduce the *stochastic* or *microcanonical* quantization method, which quantizes D-dimensional quantum systems through *classic* motions in the $(D+1)$-dimensional space–time world [4,5]. To do this, we have to introduce an *additional* dependence of dynamical quantities on a *fictitious* time (say, t) but not only on the *ordinary* one (say, x^0).

III. STOCHASTIC AND MICROCANONICAL QUANTIZATIONS

We briefly describe a basic nature of these new quantization schemes, such as *stochastic* and *microcanonical* quantizations, by introducing the above additional dependences of dynamical quantities on the *fictitious* time t.

1. The former (i.e., *SQ*) is based on the idea that a D-dimensional quantum system is equivalent to a hypothetical $(D+1)$-dimensional *classic* stochastic process [4]. We assume that the Planck constant $\hbar$ is a measure of magnitude of the fluctuation to generate the stochastic process, but never ask its deeper origins. This idea was not quite new, because Suzuki invented a new calculation method for quantum spin systems by making use of Trotter's formula [6].
2. The latter (i.e., *MCQ*) is based on the idea that a D-dimensional quantum system is equivalent to a $(D+1)$-dimensional *classic deterministic* motion [5]. Note that the Planck constant is defined with a statistical average of $(D+1)$-dimensional kinetic energy over fictitious-time initial values. One might expect that a sort of *chaotic* behavior in this case would be regarded as the *quantum fluctuation*. Recently, however, we have shown that the average over the initial values is also essentially important (in the harmonic oscillator case) [7].

Someone may be interested in the second, because we would expect that a sort of chaotic behavior of the surrounding classic systems would provoke the quantum fluctuation. We could also suppose that such a classic system should usually provoke any fluctuation by means of a huge number of constituents. In an opposite way to this conventional idea, however, we have to mention a possibility that an appropriate classic system with a few degrees of freedom can provoke dephasing (or irreversibility) to give the wavefunction collapse in quantum measurements [8]. Also note that this kind of idea cannot quantize any linear system without chaotic behavior—for example, the harmonic oscillator system. For this reason, we do not expect any direct connection between the quantum fluctuation and such a chaotic behavior in any quantization scheme.

A. *SQ* in a $(D+1)$-Dimensional Space–Time World

The *SQ* is formulated in terms of the following basic equation and the statistical properties:

$$\frac{\partial q(x,t)}{\partial t} = -\left.\frac{\delta S}{\delta q}\right|_{q=q(x,t)} + \eta(x,t) \tag{5}$$

$$\langle \eta(x,t) \rangle = 0, \qquad \langle \eta(x,t)\eta(x',t') \rangle = 2\hbar\delta(t-t')\delta(x-x') \tag{6}$$

(see Ref. 4). It is true that the SQ is completed by (5) and (6); but in order to discuss a possible relation to the conventional quantum mechanics, it is more convenient to replace the Langevin method with the Fokker–Planck method, by which we obtain

$$P[q,t] \xrightarrow{t \to \infty} N \exp(-S[q]/\hbar) \tag{7}$$

The last equation tells us that the SQ is equivalent to quantum mechanics of the Feynman type, as mentioned before [4]. Also remember that $\hbar$ is a given universal constant.

In the case of a one-dimensional harmonic oscillator [7], we describe the SQ in terms of $q_n(t)$, a Fourier component of $q(x,t)$ on $\{u_n(x) = (1/\sqrt{X})\exp(k_n x)\}$ with $k_n = 2\pi n/X$ for period X. Thus we are led to the basic Langevin equation and the statistical property

$$\frac{dq_n(t)}{dt} = -m\Omega_n^2 q_n + \eta_n(t), \qquad \Omega_n^2 = \omega^2 + k_n^2 \tag{8}$$

$$\langle \eta_n(t) \rangle = 0, \qquad \langle \eta_n(t)\eta_{n'}^*(t') \rangle = 2\hbar \delta_{nn'} \delta(t-t') \tag{9}$$

whose solution is given by

$$q_n(t) = q_n^{(0)} e^{-m\Omega^2 t} + \int_0^t dt' e^{-m\Omega_n^2(t-t')} \eta_n(t') \tag{10}$$

The last equation tells us that $q_n(t)$ should have a definite fluctuation, irrespective of fictitious-time initial values, for sufficiently large t, so that we can safely neglect these fictitious-time initial values. Note that we can obtain the continuous limit of the discretized Fourier representation through

$$\frac{1}{X} \sum_n \frac{1}{\Omega_n^2} G(k_n^2) \to \frac{1}{2\pi} \int_{-\infty}^{\infty} \frac{G(k^2)}{\omega^2 + k^2} dk \tag{11}$$

Apart from the fictitious initial values, we obtain

$$(\Delta q)_t^2 \equiv \langle q_t^2 \rangle = \frac{\hbar}{2m\omega} I(t) \tag{12}$$

$$(\Delta p)_t^2 = \left\langle \left(m\frac{dq}{dx_0}\right)^2 \right\rangle = \hbar m \delta(\epsilon) - \frac{\hbar m \omega}{2} I(t) \tag{13}$$

with a small but finite ϵ and

$$I(t) \equiv \frac{2}{\sqrt{\pi}} \int_0^{\sqrt{2m\omega^2 t}} e^{-z^2} dz \tag{14}$$

We easily understand that the first term of (13) vanishes for finite ϵ, and the negative value of $(\Delta p)_t^2$ comes from the use of Euclidean measure. Here we should notice that the neglect of $\delta(\epsilon)$ is also obtained by replacing k^2/Ω_n^2 with $-\omega^2/\Omega_n$. An additional ansatz is that the k integral in (11) can be evaluated only from the pole, $k = i\omega$, and written as

$$\frac{1}{X}\sum_n \frac{1}{\Omega_n^2} G(k_n^2) \to \frac{1}{2\pi}\int_{-\infty}^{\infty} \frac{G(k^2)}{\omega^2 + k^2}\, dk = \frac{G(k=i\omega)}{2\omega} \qquad (15)$$

in the continuous limit [7]. Remember that the same replacement rule can also work in the *MCQ*.

Finally we obtain the following formula for the uncertainty products:

$$\Delta(t) \equiv (\Delta q)_t^2 |(\Delta p)_t^2| = \frac{\hbar^2}{4}\frac{2}{\sqrt{\pi}}\int_0^{\sqrt{2m\omega^2 t}} e^{-z^2}\, dz \qquad (16)$$

which implies that for increasing t, $\Delta(t)$ starts from an unlikely value, say zero, and reaches the quantum mechanical one $\hbar^2/4$ (see Fig. 1). This is a possible model to generate the quantum fluctuation by means of fictitious-time evolution.

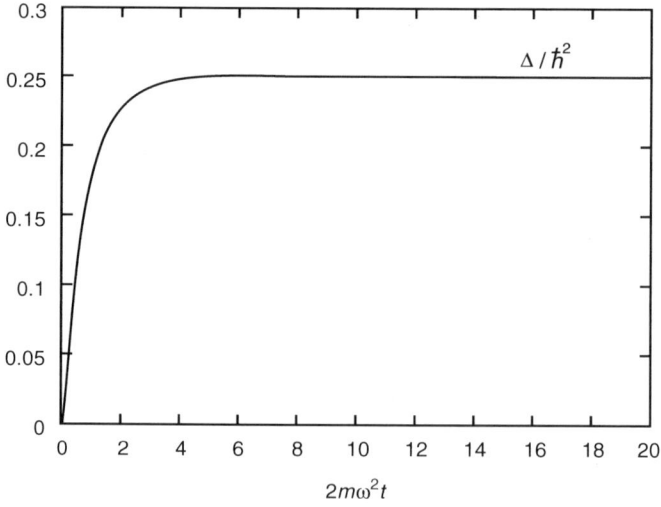

Figure 1. The *SQ*-temporal change of the uncertainty product $\Delta \equiv (\Delta q)_{x,t}^2 (\Delta p)_{x,t}^2$, starting from 0, for increasing t.

B. *MCQ* in a $(D+1)$-Dimensional Space–Time World

The *MCQ* is formulated in terms of the following basic equation:

$$\mathcal{H} \equiv K[\pi(t)] + S[q(t)], \qquad K \equiv \sum_n \frac{1}{2m} \pi_n^*(t)\pi_n(t) \qquad (17)$$

$$\frac{dq_n(t)}{dt} = \{q_n, \mathcal{H}\} = \frac{\pi_n(t)}{m}, \qquad \frac{d\pi_n(t)}{dt} = \{\pi_n, \mathcal{H}\} = -\frac{\partial S[q]}{\partial q_n} \qquad (18)$$

which are to be supplemented by the quantization rule

$$\hbar = \lim_{t \to \infty} 2\langle K[\pi(t)]\rangle \qquad (19)$$

(see Ref. 6). $\langle \cdots \rangle$ stands for the statistical average over fictitious time initial values, $q_n^{(0)}$ and $p_n^{(0)}$.

In the case of one-dimensional harmonic oscillator, we have the effective "Hamiltonian" and "potential" as follows:

$$H_{\text{MCQ}} = K[\pi(t)] + S_E \qquad (20)$$

$$K[\pi(t)] = \sum_n \frac{1}{2m}\pi_n^*(t)\pi_n(t), \qquad S = \sum_n q_n^*(t)\frac{1}{2}m\Omega_n^2 q_n(t) \qquad (21)$$

for the $(D+1)$-dimensional classic motion. This basic equation yields

$$q_n(t) = q_n^{(0)}\cos\Omega_n t + \frac{1}{m\Omega_n}\pi_n^{(0)}\sin\Omega_n t \qquad \text{(exact solution)} \qquad (22)$$

$$\pi_n(t) = \pi_n^{(0)}\cos\Omega_n t - m\Omega_n q_n^{(0)}\cos\Omega_n t \qquad \text{(exact solution)} \qquad (23)$$

$$\hbar = \sum_n \langle \mathcal{H}_n^{(0)}\rangle \qquad \text{(quantization)} \qquad (24)$$

where $q_n^{(0)}$ and $p_n^{(0)}$ stand for the fictitious-time initial values, and $H_n^{(0)} = (1/2m)\pi_n^{(0)*}\pi_n^{(0)} + (m\Omega_n^2/2)q_n^{(0)*}q_n^{(0)}$. Furthermore, we introduce an additional ansatz that those which destroy the fluctuation virial theorem or are vanishingly small in the infinite X be neglected.

For $\langle q_n^{(0)}\rangle = 0$, $\langle \pi_n^{(0)}\rangle = 0$, $\langle q_n^{(0)*}q_{n'}^{(0)}\rangle \propto \langle \pi_n^{(0)*}\pi_{n'}^{(0)}\rangle \propto \delta_{nn'}$, $\alpha_n \equiv \langle q_n^{(0)*}q_n^{(0)}\rangle$, and $\beta_n \equiv \langle \pi_n^{(0)*}\pi_n^{(0)}\rangle$, we obtain

$$(\Delta q)_{x,t}^2 = \frac{1}{X}\sum_n\left[\alpha_n(1+\cos 2\Omega_n t) + \frac{\beta_n}{m^2\Omega_n^2}(1-\cos 2\Omega_n t)\right] \qquad (25)$$

$$(\Delta p)_{x,t}^2 = -(m\omega)^2(\Delta q)_{x,t}^2 \qquad (26)$$

Here we have used the same replacement (as in the SQ case) and the above ansatz. Noting that $\cos^2 2\Omega_n t \xrightarrow{t\to\infty} \frac{1}{2}$ and $\cos(\text{or }\sin)2\Omega_n t \xrightarrow{t\to\infty} 0$, we know that

$$(\Delta q)^2_{x,t} = (\Delta Q)^2 + \frac{1}{2X}\sum_n \left(\alpha_n - \frac{1}{m^2\Omega_n^2}\beta_n\right)\cos 2\Omega_n t \tag{27}$$

$$(\Delta p)^2_{x,t} = (\Delta P)^2 - (m\omega)^2[(\Delta q)^2_{x,t} - (\Delta Q)^2] \tag{28}$$

where $(\Delta Q)^2 \equiv \lim_{t\to\infty}(\Delta q)^2_{x,t} = (\hbar/2m\omega)$, $(\Delta P)^2 \equiv \lim_{t\to\infty}(\Delta p)^2_{x,t} = (\hbar m\omega/2)$, and $\Delta \equiv (\Delta Q)^2|(\Delta P)^2| = (\hbar^2/4)$.

For $\alpha_n = 0$, we obtain

$$(\Delta q)^2_{x,t=0} = 0, \qquad (\Delta p)^2_{x,t=0} = 0 \tag{29}$$

For details, see Ref. 7. The uncertainty product starts from an unlikely value, zero, and reaches the quantum mechanical one, $\hbar^2/4$. The situation is similar to the SQ case (Fig. 1), so that we can also regard this process as another model to generate the quantum fluctuation in fictitious-time evolution.

IV. CONCLUSION

We have suggested possible origins or mechanisms of the quantum fluctuation, by means of *stochastic* and *microcanonical* quantization methods. We have also showed that a D-dimensional quantum system is equivalent to a $(D+1)$-dimensional classic one, in both quantization schemes.

Almost all theories can only explain the situation that $\hbar$ is a measure of the fluctuation, but never ask its deeper origins. Only the Hayakawa–Tanaka formula might suggest a possible reasoning of $\hbar$ within a deeper (cosmological and/or particle-physical) framework, even though we have some questions. At least this theory might suggest that *neutrino* would become one of the most important particles in the twenty-first century.

In addition to the quantum fluctuation mentioned above, we left many other interesting problems unanswered. For example, we do not know a deeper reasoning of the quantum correlations (i.e., of Bose and Fermi statistics), which are very difficult to understand physically. It seems that the problem would be closely related to the fundamental EPR question. On the other hand, the Weinberg–Salam and electroweak theories would indicate that only a special choice of fundamental particles could give us the renormalizability. If so, we can hardly know whether we can discuss the fundamental questions, such as the quantum fluctuation and quantum correlation, irrespective of the renormalizability. In this context, we should understand the recent development as if the renormalizability should become the first principle to cover the whole physics in terms of fundamental particles. We are therefore interested in the following questions: "How deep a level of matter or what kind of particle model is enough to discuss a deeper origin of the quantum fluctuation and the quantum correlation? Besides these kinds of problems, we are also interested in another question, as to whether we can find a deeper origin of the quantum fluctuation in a classic chaotic behavior of the surrounding atmosphere. These questions must be related to a rather philosophical question, *reductionism*. We have to say that quantum mechanics is still developing.

References

1. W. Pauli, *Writings on Physics and Philosophy* (English version), translated by R. Schlapp, Springer-Verlag, Berlin, 1994. Its Japanese version translated by K. Okano was published by Springer-Verlag, Tokyo in 1998.
2. S. Hayakawa, *Prog. Theor. Phys.* **21**, 324 (1959); S. Hayakawa and H. Tanaka, *ibid.* 858 (1961).
3. D. Bohm, *Phys. Rev.* **85**, 166 (1952); E. Nelson, *Phys. Rev.* **150**, 1079 (1966).
4. G. Parisi and Y. Wu, *Sci. Sin.*, 483 (1981); M. Namiki, *Stocastic Quantization*, Springer-Verlag, Heidelberg, 1992; M. Namiki and K. Okano, eds. *Stocastic Quantization, Suppl. Prog. Theor. Phys.*, No. 111 (1993).
5. D. J. E. Callaway and A. Rahman, *Phys. Rev. Lett.* **49**, 613 (1982); *Phys. Rev.* **D28**, 1506 (1983).
6. M. Suzuki, *Prog. Theor. Phys.* **56**, 1454 (1976).
7. M. Namiki and M. Kanenaga, *Phys. Lett.* **A249**, 13 (1998).
8. H. Nakazato, M. Namiki, S. Pascazio, and Y. Yamanaska, *Phys. Lett.* **A222**, 130 (1996).

AUTHOR INDEX

Numbers in parentheses are reference numbers and indicate that the author's work is referred to although his name is not mentioned in the text. Numbers in *italic* show the pages on which the complete references are listed.

Abarbanel, H. D. J., 4(5), 14(5), *19*
Accardi, L., 215(1-2), 216(1-3), 217(1), 218(3-5), 221(6), 222(3), 224(8), 230(3), 237(3), *238*
Afriat, A., 252(20), 255(20), *257*
Agam, O., 118(44-46), *126*
Aguilar, J., 78(17), *106*
Aizawa, Y., 162(1-6), 163(5-6), 164(3), *164*
Alberts, B., 66(42), *73*
Alder, B., 129(2), 158(2), *159*
Alonso Ramirez, D., 118(42), *126*
Altshuler, B. L., 118(44), *126*
An, K., 168(6), 190(6), *195*
Andreev, A. V., 118(44), *126*
Antoine, J. P., 303(7), *319*
Antoniou, I., 34(6-22), 35(10-11), 36(6-22,25-32), 38(8-9), 39(19,21), 41(22), 42-43(7), 44(25), 45(29), 46(30), *46–47*, 78(8-9,12), 97(8-9,12), *105*, 119(52), *126*, 162(4), *164*, 288(6-7,9), 291(9), 298(19), 299, 315(21), *319*
Arecchi, F. T., 203-206(11), *213*
Ariizumi, T., 72(48), *73*
Arimondo, E., 189(60), *197*
Artuso, R., 120(60), 121(61), *127*
Asashima, M., 72(48), *73*
Ashwin, P., 57-58(15), *72*

Baldauf, H. W., 189(53-54), *196–197*
Balescu, R., 4-7(7), 8(7,10), 9(7), 10(7,11), 14(7), 18(7,10), *19*, 22(5), *32*, 110(2), *125*
Balslev, E., 78(17), *106*
Bandtlow, O. F., 6(8), 19(8), *19*, 34-35(11), *46*
Baranyai, A., 110(20), *125*
Baras, F., 112-113(31), *126*
Bardoff, P. J., 182(25), *196*
Barr, G. D., 252(13), 254(13), 257(13), *257*

Barr, G. D., 316(24), *320*
Barucha, C. F., 311(16), *319*
Beck, Ch., 49(1), *51*
Benson, O., 169(9,11), 170(9), 172(9,11), 173-174(9), 175(9,11), 176(11,20), 181(20), 185(34), *195–196*
Berdichevsky, V., 22(4), *32*
Bhamathi, G., *286*
Bigi, I., 252(18), 255(18), *257*
Blum, G., 118(45), *126*
Bohm, A., 37(38), *47*, 78(16), *106*, 281, 303(8), 306(10), 308(10-12), 312(11), 315(21,23), *319–320*
Bohm, D., 323(3), *329*
Boiteux, A., 62(41), *73*
Boon, J.-P., 110(4), *125*
Boya, L. J., *286*
Brattke, S., 182(26), *196*
Bray, D., 66(42), *73*
Brecha, R. J., 169(7), *195*
Breitenbach, G., 211(16), *213*
Breymann, W., 79(24,32,34), 80(32,34), *106*
Briegel, H. J., 176(21), 189(57,59), 190(59,63), 191(59), *196–197*
Briggs, M. E., 112(28), *126*
Brune, M., 170(15), 182(27), *195–196*, 203(8), *213*
Buescu, J., 57-58(15), *72*
Bunimovich, L. A., 110(13), 114(13), *125*
Burgoyne, N., *286*

Calabrese, R. V., 112(28), *126*
Caldeira, A. O., 202(2), *213*, 216(9), *238*
Callaway, D. J. E., 323-324(5), *329*
Campbell, K. H. S., 71(47), *73*
Carnal, O., 168(5), *195*
Caves, C. M., 203(10), *213*

331

Chawanya, T., 60(28), 61(32), *73*
Chernov, N. I., 79(23), *106*, 110(10,14,21), 114(21), 120(57), *125–126*
Childs, J. J., 168(6), 190(6), *195*
Chirikov, B. V., 3(1), 4(1), 14(1), *19*
Chiu, C., 78(15), *105*, *286*
Christenson, J., 252-255(10), *257*, 316(24), *320*
Claus, I., 79(29-30), *106*, 119(56), 123(73), *126–127*
Cohen, E. C. G., 79(24-28), *106*, 110(20), *125*, 129(5), *159*
Cohen-Tannoudji, C., 174(17), *195*, 315(22), *319*
Combes, J. M., 78(17), *106*
Cornfeld, I. P., 84(45), *107*
Courbage, M., 96(51), *107*, 288-289(5), 297(5), *299*
Coveney, P. V., 6(8), 19(8), *19*
Cresser, J. D., 186(37), 188(37), *196*
Crisanti, A., 57(13), *72*
Cronin, J., 252-255(10), *257*, 316(24), *320*
Csordás, A., 49(2), 50(2,7-8), *51*
Cvitanovic, P., 120(59-60), 121(62-63), *126–127*

Daems, D., 79(24), *106*
Dagenais, M., 188(45,47), 191(45), *196*
Dasari, R. R., 168(6), 190(6), *195*
Davidovich, L., 170(15), *195*
De Leener, M., 110(3), *125*, 129(8,13), 132-134(13), 138(13), 140-144(13), 148-149(13), *159*
Dellago, Ch., 110(9), *125*
DeSchepper, I. M., 129(12), 145(12), *159*
Dettmann, C. P., 120(57-58), *126*
Diedrich, F., 188-189(49), 191(49), *196*
Dirac, P. A. M., 37(37), *47*, 264(8), 272(13), *275*, 277-279(1), 281(1,4), *286*
Dmitrieva, L., 34-35(10), *46*
Doebner, H. D., 306(10), 308(10), *319*
Dorfman, J. R., 21(3), *32*, 79(18,22,30,36,41), 80(36), 88(46), 90(36), *106–107*, 110(9,12), 112(28,32), 119(55-56), 122-123(69), *125–127*, 129(3-5,7,9-11, 14-16), 132(14-15), 134(7), *159*
Dreyer, J., 182(27), *196*, 203(8), *213*
Dribe, D., 21(2), *32*
Driebe, D., 34(23), 36(23), 42(23), *47*, 78(10,14), 97(10,14), *105*, 119(53-54), *126*, 289(12), *299*

Duck, I., *286*
Dupont-Roc, J., 174(17), *195*

Eckmann, J.-P., 50(4), *51*, 79(43), 102(43), 104(43), *107*, 112-113(27), 121(63), *126–127*
Eichler, T., 189(53), *196*
Eigen, M., 62(39), *73*
Englert, B.-G., 169(8), 176(8,20-21), 181(20), 183(30), 189(57,59), 190-191(59), *195–197*
Ernst, M. H., 129(3-4,7,12), 134(7), 145(12), *159*
Ershov, S. V., 60(27), *73*
Evans, D. J., 79(25), *106*, 110(20), *125*
Eyink, G. L., 79(23), *106*, 110(21), *125*
Ezekiel, S., 188(43), *196*

Falcioni, M., 57(13), *72*
Feld, M. S., 168(6), 190(6), *195*
Fermi, E., 303(1), 312(17), *318–319*
Ferreira, L. S., 310(15), 312(15), *319*
Filipowicz, P., 170-171(12), *195*
Fischer, M. C., 311(16), *319*
Fishman, S., 118(46), *126*
Fitch, V. L., 252-255(10), *257*, 310(14), 316(24), *319–320*
Fomin, S. V., 84(45), *107*
Fonda, L., 303(6), 310(6), *319*
Fox, R. F., 88(47), *107*
Francis, M. K., 112(28), *126*
Friedrichs, K., 278(7), *286*
Froggatt, C., 252(19), 255(19), *257*
Furusawa, C., 62(37-38), 66(37,44), 67(37), 69(44), *73*

Gadella, M., 37(38), *47*, 78(12,16), 97(12), *105–106*, *281*, 305(9), 308(11), 312(11), *319*
Gallavotti, G., 79(26), *106*
Gammon, R. W., 112(28), *126*
Gamow, G., 281(6), *286*
Garcia-Calderon, G., 313(18), *319*
Gardiner, C. W., 189(61), *197*
Garg, A. K., 203(6), *213*
Gaspard, P., 21(1), *32*, 79(19,22,29-31,33,35,37,41), 80(19,31,33,37), 81(19,37), 84(33), 85(31,37), 87(31,37), 88(33,47), 90(33,37), 101(33), *106–107*, 110(6,11,16-19,23-24),

111(6,25), 112(11,28-32),
113(11,19,25,30-31,33-34), 115(25),
117(25), 118(25,43-43),
119(17-18,25,55-56), 121(19,25,33,62-63),
122(19,25,34,69), 123(34,69,71-73),
124(11), *125–127*
Gelfand, I. M., *281*
George, C., 78(3-5), 99(3-5), *105*, 136(26), *159*,
    262(2), 266(9), 270(2), 271(9), 274(9), *275*
Georgiades, N., 168(5), *195*
Gerz, C., 187(38), *196*
Gheri, K. M., 189(58), 190(62), *197*
Ghirardi, G. C., 303(6), 310(6), *319*
Gibbons, K. L., 310(14), 316(24), *319–320*
Gibbons, L., 252(14), 254(14), 257(14), *257*
Gibbs, H. M., 188(44), *196*
Gilbert, T., 79(30,36,41), 88(46), *106–107*,
    119(55-56), 122-123(69), *126–127*
Ginzel, C., 189(57), *197*
Glauber, R. J., 235(11), *238*
Goldberger, M. L., 310(13), *319*
Goodwin, B., 62(40), *73*
Gordov, E. P., 189(61), *197*
Gorini, V., 78(15), *105, 286*
Goy, P., 170(15), *195*
Grassberger, P., 50(5), *51*
Greenberger, D. M., 182(28-29), *196*
Grove, R. E., 188(43), *196*
Grynberg, G., 174(17), *195*
Gurdon, J. B., 71(47), *73*
Gustafson, K., 36(26), *47*, 239(1-2), 240(1-6),
    242(3,7), 243(5), 244-245(6), 246(1-2,5-6),
    247(1-2), 248(1-4), 250-251(1-2),
    254(3-4), *257*
Györgyi, G., 50(6), *51*

Haake, F., 118(48-49), *126*
Häger, J., 186(37), 188(37), *196*
Hagley, E., 182(27), *196*, 203(8), *213*
Haroche, S., 170(15), 182(27), *195–196*, 203(8),
    213
Harshmann, N. L., 306(10), 308(10), *319*
Hartig, W., 186(36), *196*
Hasegawa, H. H., 4(6), 10(6), 14(6), *19*, 29(8),
    31(8), *32*, 78(7,10), 97(7,10), *105*, 119(53),
    126, 136(27), *159*, 289(12), *299*
Hayakawa, S., 322(2), *329*
Heerlein, R., 176(23), 180(23), *196*
Hegerfeldt, G. C., 303(5), *319*
Heisenberg, W., 277, 280, *286*

Heitler, W., 187(39), *196*, 272(12), *275*
Helmfrid, S. R., 189(53), *196*
Henin, F., 78(3-4), 99(3-4), *105*, 136(26), *159*,
    266(9), 271(9), 274(9), *275*
Hercher, M., 188(41), *196*
Hernandez, E., 313(18), *319*
Hess, B., 62(41), *73*
Hoegy, W. R., 129(7), 134(7), *159*
Höffges, J. T., 189(53-54), *196–197*
Hoover, W. G., 110(8), *125*
Horacek, J., 310(15), 312(15), *319*
Horak, P., 189(58), *197*
Horne, M., 182(28-29), *196*
Horsthemke, W., 61(34), *73*

Ikeda, K., 59(20), *73*
Ito, K., 162(1), *164*

Jakeman, E., 188(46), *196*
Jammer, M., 287(1), *299*
Jarlskog, C., 252(16), 255(16), *257*
Javanainen, J., 170-171(12), *195*
Jessen, P. S., 187(38), *196*
Jex, I., 207-208(14-15), *213*
Jordan, P., 277-280, *286*

Kan, Y., 129(11,16), *159*
Kaneko, K., 55(4-7), 56(5-6,9-11), 57(12,14),
    58(4), 59(21,23), 60(24,26,30), 61(32),
    62(2-3,35-38), 66(3,37,44), 67(37),
    69(44,46), *72–73*
Kanenaga, M., 324(7), 326(7), 328(7), *329*
Kantz, H., 50(5), *51*
Karpov, E., 275(15), *275*, 289(10), 291(10),
    298(10), *299*
Kauffman, S., 67(45), *73*
Kaufmann, Z., 50(9,12), 51(12), *51*
Kawasaki, K., 235(12), *238*
Khalfin, L. A., 297(18), *299*, 303(6), 310(6), *319*
Khazanov, A. M., 189(61), *197*
Khodas, M., 118(46), *126*
Kielanowski, P., 306(10), 308(10), 315(21), *319*
Kimble, H. J., 168(5), 188(45,47), 191(45),
    195–196
King, B. E., 203(7), *213*
Kirkpatrick, T., 129(10,15), 132(15), *159*
Kiss, T., 207-208(14-15), *213*
Kitaev, A. Yu., 121(64), *127*
Klages, R., 79(35), *106*, 123(71), *127*
Klein, N., 168(2), *195*

Klein, O., 279–280
Kleinknecht, K., 316(24), *320*
Knauf, A., 110(15), *125*
Kobayashi, M., 252(12), 255(12), *257*
Koganov, G. A., 189(61), *197*
Koguro, N., 162(4), *164*
Komatsu, T. S., 60(29), *73*
Kondepudi, D., 22(6), *32*
Konishi, T., 59(21), *73*
Koopman, B., 34(1), 43(1), *46*
Kottos, T., 118(47), *126*
Kozyrev, S. V., 216(3), 218(3-5), 222(3), 224(8), 230(3), 237(4), *238*
Krasnopolsky, V. M., 310(15), 312(15), *319*
Krause, J., 180-182(24), *196*
Krylov, N. N., 97(52), 104(52), *107*, 110-111(5), *125*
Kukulin, V. I., 310(15), 312(15), *319*
Kuperin, Yu., 34-35(10), *46*
Kurths, J., 60(25), *73*

Lai, Y.-C., 57(18), *72*
Lange, W., 189(54), *197*
La Porta, A., 204(12), 207(12), *213*
Laskey, R. A., 71(47), *73*
Lasota, A., 34(2), *46*
Lax, P. D., 113(35), *126*
Lebowitz, J. L., 79(23), *106*, 110(21), 122(65), *125, 127*
Lee, T. D., 303(3), 310(3), 312(3), *319*
Lefever, R., 61(34), *73*
Leggett, A. J., 202(2), 203(4-6), *213*, 216(9), *238*
Lett, P. D., 187(38), *196*
Leuchs, G., 186(37), 188(37,48), *196*
Levy, M., 303(4), *319*
Lewis, J., 66(42), *73*
Lichtenberg, A. J., 3(2), 4(2), 14(2), *19*
Lie, C.-B., 29(8), 31(8), *32*
Lieberman, M. A., 3(2), 4(2), 14(2), *19*
Likhoded, A. K., 293(17), *299*
Linblad, G., 224(7), *238*
Livi, R., 110(7), *125*
Lockhart, C., 288(3), 290(3), *299*
Loewe, M., 308(11), 312(11), *319*
Löffler, M., 169(8), 176(8,20), 181(20), 191(65-66), 193(66), 195(67), *195–197*
Loudon, R., 189(55), *197*
Lu, W., 118(50-51), *126*
Lu, Y. G., 215(1-2), 216-217(1), 221(6), *238*
Luders, G., 252(15), *257, 286*

Ludwig, G., 314(19), *319*
Lugiato, L. A., 169-171(13), *195*
Lustfeld, H., 50-51(11-12), *51*

Maali, A., 182(27), *196*, 203(8), *213*
Mabuchi, H., 168(5), *195*
Mackey, M., 34(2), *46*
MacLennan, J. A., 122(66), *127*
Madison, K. W., 311(16), *319*
Mahaux, C., 310(15), 312(15), *319*
Maineri, R., 120(60), *127*
Mainland, G. Bruce, 78(16), *106*
Maitre, X., 203(8), *213*
Mandel, L., 188(45,47,51), 191(45), *196*
Marte, M. A. M., 190(62), *197*
Martini, U., 189(57), *197*
Maskawa, T., 252(12), 255(12), *257*
Mathelitsch, L., 252(17), 255(17), *257*
Matsumoto, K., 59(20), 61(33), *73*
Matyás, L., 79(38,42), 80(38,42), *106–107*
Maxon, S., 308(11), 312(11), *319*
Mayne, F., 78(5), 99(5), *105*
Mayr, E., 182(25), *196*
McWhir, J., 71(47), *73*
Meekhof, D. M., 203(7), *213*
Melnikov, Yu., 34(10,19), 35(10), 36(10,19), 39(19), *46*
Melsheimer, O., 303(7), *319*
Mermin, N. D., 183(31), *196*
Meschede, D., 168(1), *195*
Meyer, G. M., 183(32), 185(33-34), 189(59), 190(59,63), 191(59,64-66), 193(66), *196–197*
Meystre, P., 169(1), 170-171(12), 176(22), 180(22), *195–196*
Mezard, M., 57(8), *72*
Milburn, G. J., 231(10), *238*
Milnor, J., 57(15), *72*
Misra, B., 35(24), *47*, 96(51), *107*, 272(14), *275*, 288(4-5,8), 289(5,11), 297(5,11), *299*
Mlynek, J., 211(16), *213*
Mollow, B. R., 188(40), *196*
Mondragon, A., 313(18), *319*
Monroe, C., 203(7), *213*
Montina, A., 203-206(11), *213*
Morita, S., 60(28), *73*
Morriss, G. P., 79(25), *106*
Morrow, P. R., 311(16), *319*
Mu, Y., 189(56), *197*
Müller, G., 168(1), *195*

Nakagawa, N., 60(29), *73*
Nakamura, H., 310(15), 312(15), *319*
Nakato, M., 162(2), *164*
Nakazato, H., 324(8), *329*
Namiki, M., 323(4,7), 324 (7-8), 325(4), 326(7), 328(7), *329*
Narducci, L. M., 169(10), 180(10), *195*
Nekhoroshev, N. N., 163(7), *164*
Németh, A., 50(10,12), 51(12), *51*
Newton, R. G., 315(20), *319*
Nicolis, G., 79(22,24), *106*, 110(6), 111(6,26), 112(30), 113(30), 116(26), 118(43), *125–126*
Nielson, H., 252(19), 255(19), *257*
Nikolaev, N. N., 310(14), *319*
Niu, Q., 311(16), *319*
Norman, E. B., 310(14), *319*

Oehme, R., 303(3), 310(3), 312(3), *319*
Ogawa, M., 66(43), *73*
Ohtaki, Y., 29(8), 31(9), 32(9), *32*
Ohtsuka, K., 59(20), *73*
Ordonez, G., 262(5-6), 268(5-6), 269(5), 2 71(5), 273-274(5), 275(15), *275*, 289(10), 291(10), 298(10,20), *299*
Ostrovsky, V. N., 310(15), 312(15), *319*
Ott, E., 45(39), *47*, 57(16), *72*

Pance, K., 118(50-51), *126*
Parisi, G., 57(8), *72*, 323(4), 325(4), *329*
Parravicini, G., 78(15), *105*
Pascazio, S., 324(8), *329*
Paul, H., 207-208(14-15), *213*
Pauli, W., *285–286*, 321(1), *329*
Peierls, R., 313(18), *319*
Peik, E., 188-189(52), *196*
Pellizzari, T., 189(58), *197*
Perez, G., 60(31), *73*
Petrosky, T., 34(4-5), *46*, 78(6,11,13), 97(11,13), *105*, 130(20-21), 135(20), 136(20-21,27), *159*, 262(3-6), 264(7), 268(5-7), 269(5,7), 270(7), 271(4-5,7,11), 272(3,11), 273(3,5), 274(4-5), 275(15), *275*, 289(10,13-14), 291(10), 292-293(16), 298(10,14,16,20), *299*
Philipps, W. D., 187(38), *196*
Phillips, R. S., 113(35), *126*
Pianigiani, G., 89(50), *107*
Piasecki, J., 136(28), *159*
Pike, E. R., 188(46), *196*

Pikovsky, A. S., 60(25), *73*
Plessas, W., 252(17), 255(17), *257*
Politi, A., 110(7), *125*
Pollicott, M., 37(33,35), *47*, 89(48), 99(48), *107*, 117-118(36-37), *126*
Polzik, E. S., 168(5), *195*
Pomeau, Y., 129(6), 131(6), *159*
Popper, K., 280
Posch, H. A., 110(8-9), *125*
Potapov, A. B., 60(27), *73*
Prigogine, I., 7(9), *19*, 22(6), *32*, 34(3-6), 35(3), 36(6,28), *46–47*, 78(1-4,6,8,11-13), 96(1,51), 97(2,6,8,11-13), 99(3-4), 104(2), *105*, 110(1), 116(1), 122(70), *125*, *127*, 129(18-19), 130(20-21), 135(19-20), 136(20-21,26), 146(19), *159*, 262(1-6), 264(7), 266(9), 267(1,10), 268(5-7), 269(5,7), 270(2,7), 271(4-5,7,9,11), 272(3,11), 273(3,5), 274(4-5,9), 275(15), *275*, 287(2,5), 288(2,5,9), 289(10-11,13-14), 291(9-10), 292-293(16), 297(5,11), 298(2,10,14,16,20), *299*
Pronko, G. P., 78(12), 97(12), *105*, 293(17), *299*
Provata, A., 118(43), *126*
Pusey, P. N., 188(46), *196*

Qiao, Bi, 34(20-22), 36(20-22), 39(21), 41(22), *47*

Rae, J., 262(2), 270(2), *275*
Raff, M., 66(42), *73*
Rahman, A., 323-324(5), *329*
Raimond, J. M., 170(15), 182(27), *195–196*, 203(8), *213*
Raithel, G., 169(9-11), 170(9), 172(9,11), 173-174(9), 175(9,11), 176(11), 180(10), *195*
Raizen, M., 311(16), *319*
Ralph, T. C., 189(61), *197*
Ramsey, N. F., 173(16), *195*
Rao, D., 240(4), 248(4), 254(4), *257*
Rasmussen, W., 186(36), *196*
Rateike, F. M., 186(37), 188(37,48), *196*
Rechester, A. B., 4(4), 14(4), *19*
Reeves, O. R., 71(47), *73*
Reichl, L. E., 3(3), 4(3), 14(3), *19*
Rempe, G., 168(2-4), 169(3-4), 170(4), 176(22), 180(22), *195–196*
Resenfeld, L., 78(3), 99(3), *105*

Résibois, P., 7(9), *19*, 110(3), *125*,
  129(6,8,13,18), 131(6), 132(13),
  133(13,23-25), 134(13), 138(13), 140-144(13),
  148-149(13), 158(29), *159*
Rice, S. A., 112(29), *126*
Rimini, A., 303(6), 310(6), *319*
Ritchie, W. A., 71(47), *73*
Ritsch, H. J., 189(58,61), 190(62), *197*
Roberts, E., 303(7), *319*
Roberts, K., 66(42), *73*
Rolston, S. L., 187(38), *196*
Rondoni, L., 79(24,28,40), 80(40), *106*
Rose, M., 118(51), *126*
Rosenfeld, L., 266(9), 271(9), 274(9), *275*
Ruelle, D., 37(34,36), *47*, 50(4), *51*,
  79(24,43-44), 89(49), 99(49), 102(43),
  104(43-44), *106–107*, 112-113(27),
  118(38-41), *126*
Ruffo, S., 110(7), *125*

Sadovnichii, V. A., 36(27-28), *47*, 288(7,9), *299*
Saito, A., 59(23), *73*
Sanda, A., 252(18), 255(18), *257*
Saphir, W. C., 4(6), 10(6), 14(6), *19*, 78(7),
  97(7), *105*
Sato, K., 162(1), *164*
Savage, C. M., 189(56,61), *197*
Scheider, R., 186(36), *196*
Schenzle, A., 169(7), 176(18), 189(57), *195, 197*
Schiller, S., 211(16), *213*
Schleich, W. P., 182(25), 185(35), *196*,
  204-206(13), *213*
Schlögl, F., 49(1), *51*
Schmidt-Kaler, F., 168-169(3), 182(27),
  195–196
Schrama, C. A., 188-189(52), *196*
Schreiber, T., 121(62), *127*
Schröder, M., 185(34-35), *196*
Schrödinger, E., 203(3), *213*
Schuda, F., 188(41), *196*
Schuster, P., 62(39), *73*
Schwarzschild, B., 252(8), 254(8), *257*
Schweber, S., 278(9), *286*
Scully, M. O., 169(10,13), 170-171(13),
  180(10,24), 181-182(24), 183(32),
  185(33-35), *195–196*
Seba, P., 118(48), *126*
Sekimoto, K., 22(7), *32*
Selleri, F., 252(20), 255(20), *257*
Sengers, J. V., 112(28), *126*

Shawhan, P., 252(9), 254(9), 257(9), *257*
Shibata, T., 60(26,30), 61(32), *73*
Shimony, M., 182(29), *196*
Shinjo, K., 59(22), *73*
Shkarin, S., 34(19), 36(19,27-28,30), 39(39),
  46(30), *46–47*, 288(7,9), *299*
Short, R., 188(51), *196*
Siegert, A. F. J., 313(18), *319*
Simons, B. D., 118(44), *126*
Sinai, Ya. G., 79(23,44), 84(45), 99(45),
  104(44), *106–107*, 110(13,21), 114(13),
  125
Slusher, R. E., 204(12), 207(12), *213*
Smilansky, U., 118(47), *126*
Smith, W. W., 188-189(52), *196*
Sommerer, J. C., 57(16), *72*
Song, S., 203(10), *213*
Spreuuw, R. J. C., 187(38), *196*
Sridhar, S., 118(50-51), *126*
Sterpi, N., 176(21), 183(30), *196*
Stoler, D., 203(9), 207(9), *213*
Streater, R., *286*
Stroud, C. Jr., 188(41), *196*
Stuart, I., 57-58(15), *72*
Suchanecki, Z., 34(11-20), 35(11),
  36(11-20,25,31-32), 44(25), *46–47*,
  119(52), *126*
Sudarshan, E. C. G., 78(15), *105*, 272(14), *275,
  286*, 290(15), *299*
Sundaram, B., 311(16), *319*
Suzuki, M., 324(6), 327(6), *329*
Szépfalusy, P., 50(6-13), 51(11-12), *51*

Tasaki, S., 34(4,7-9,12-14,18), 36(7-9,12-14,18),
  38(8-9), 42-43(7), *46*, 78(9), 79(31,37),
  80(31,37), 81(37), 85(31,37), 87(31,37),
  88(46-47), 90(37), 97(9), *105–107*,
  110(19,23-24), 113(19), 118(43), 119(52),
  121-122(19), *125–126*, 271-272(11), *275*,
  292-293(16), 298(16,19), *299*
Tél, T., 49(3), 50(8), *51*, 79(24,32,34,38-40,42),
  80(32,34,38-40,42), 81(39), 89(50),
  106–107, 110(22), *126*
Theodosopulu, M., 131(22), 133(25), *159*
Thompson, R. J., 168(5), *195*
Tolstikhin, O. I., 310(15), 312(15), *319*
Torma, P., 207-208(14-15), *213*
Tsuda, I., 55(4), 58(4,19), 59(19), 61(33),
  72–73
Turchette, Q. A., 168(5), *195*

Turing, A. M., 54(1), *72*
Turley, R., 252-255(10), *257*, 316(24), *320*

van Beijeren, H., 110(9), *125*, 129(9,14), 132(14), *159*
van der Hoer, M. A., 129(17), *159*
Van Leeuwen, J., 129(1,7), 134(7), 157(1), *159*
Varcoe, B., 176(20,23), 180(23), 181(20), 182(26), *196*
Vattay, G., 120(60), *127*
Vaugham, J. M., 188(46), *196*
Venkatesan, T. N. C., 188(44), *196*
Viola, L., 118(51), *126*
Virasoro, M. A., 57(8), *72*
Vogel, K., 185(35), *196*
Vollmer, J., 79(24,32,34,38-40,42), 80(32,34,38-40,42), 81(39), *106-107*, 110(22), *126*
Volovich, I. V., 215(1-2), 216(1,3), 217(1), 218(3), 221(6), 222(3), 230(3), *238*
von Neumann, J., 278(3), 281(3), *286*
Vulpiani, A., 57(13), *72*

Wagner, C., 169(7,10), 276(18), 180(10), *195*
Wainwright, T., 129(2), 158(2), *159*
Walls, D. F., 188(50), 189(61), 190(62), *196-197*, 204-206(13), *213*, 231(10), *238*
Walther, H., 168(1-4), 169(3-4,7-11,13), 170(4,9,13), 171(13), 172(9,11), 173-174(9), 175(9,11), 176(8,11,18-23), 180(10,22-24), 181(20,24), 182(24,26), 183(30,32), 185(33-35), 186(36-37), 188(37,42,48-49,52), 189(49,52-54), 190(63), 191(49,64-66), 193(66), 195(67), *195-197*

Watson, J. D., 66(42), *73*
Watson, K. M., 310(13), *319*
Weber, J., 118(48), *126*
Weidinger, M., 176(20,23), 181(20,23), 182(26), *196*
Weisskopf, V., 303(2), 310(2), *319*
Westbrook, C. I., 187(38), *196*
Weyl, H., 280
Weyland, A., 129(1), 157(1), *159*
White, R. B., 4(4), 14(4), *19*
Wightman, A. S., 278(9), *286*
Wigner, E. P., 212(17), *213*, *279*, 303(2), 310(2), *319*
Wilkinson, S. R., 311(16), *319*
Wilmut, I., 71(47), *73*
Wineland, D. J., 203(7), *213*
Winslow, R. L., 57(18), *72*
Wolfenstein, L., 252-253(11), *257*
Wu, F. Y., 188(43), *196*
Wu, Y., 323(4), 325(4), *329*
Wunderlich, C., 203(8), *213*

Yamanaska, Y., 324(8), *329*
Yang, C. N., 303(3), 310(3), 312(3), *319*
Yip, S., 110(4), *125*
Yomo, T., 62(2-3,35), 66(3), *72-73*
Yorke, J. A., 89(50), *107*
Yurke, B., 203(9-10), 204(12-13), 205-206(13), 207(9,12), *213*

Zeilinger, A., 182(28-29), *196*
Zoller, P., 189(61), *197*
Zubarev, D. N., 122(67-68), *127*
Zumino, B., *286*
Zurek, D., 201(1), *213*

SUBJECT INDEX

Accardi-Fedullo probability, operator trigonometry, 251–252
Accessibility, biological irreversibility, cell differentiation, 69
Accretive bounded operator, operator trigonometry, 242–243
Acoustic-optic modulator (AOM), single-atom ion trapping, resonance fluorescence, 189
Age eigenstates, unstable systems
  excited state, average age, 292–295
  time superoperator construction, 291–292
Age fluctuations, unstable systems, 295–297
Antibunching, single-atom ion trapping, resonance fluorescence, 188–189
Anticommutation relations, quantum field theory, 280
Antieigenvectors
  CP symmetry violation
    breaking mechanisms, 255–256
    strangeness total antieigenvectors, 254–255
  operator trigonometry, symmetric positive definite (SPD) matrix, 243–244
Antilinear functionals, irreversibility in multibaker maps, broken time reversal symmetry, spectral theory, 97–99
A priori probability, micromaser experiments, trapping states, 182
Arnold diffusion, log-Weibull distribution, 163–164
Asymmetric conditions, quantum mechanics, probability and time asymmetri, 312–318
Atom-cavity interaction time
  micromasers, quantum jumps and atomic interference, 169–176
  single-atom ion trapping, ion-trap lasers, 190–194
Atomic interference, micromasers, 169, 172–176

Attractor ruins, biological irreversibility, chaotic itinerancy, 58–59
"Auschliessprinzip" (Pauli), quantum field spin-statistics, 285

Backward time evolution, irreversibility in multibaker maps, time reversal symmetry, 91–95
Baker's transformation, unstable system harmonic analysis, spectral decomposition and probabilistic extension, 42–43
Banach space
  operator trigonometry, 240–243
  unstable system harmonic analysis, spectral decomposition and probabilistic extension, Renyi maps, 39
Bandtlow-Coveney equation, iterative maps, 5
  non-Markovian/Markovian evolution equations, 6–9
Bayes theorem, mesoscopic quantum interference, 206–212
Bell's inequality, operator trigonometry, quantum probability, 247–248
  classical probability and, 251–252
Bernoulli polynomial, unstable system harmonic analysis, spectral decomposition and probabilistic extension, 38–39
  tent maps, 40–41
Bessel function, iterative maps, master equation, standard map, 10–11
Bilateral shift, unstable system harmonic analysis, time operator, 43–46
Binary correlation components, moderately dense gas transport, spectral representation, 135–138
Biological irreversibility
  cell differentiation
    dynamic system development, 62–64
    dynamic system representation, 67–69
    macroscopic stability, 67

339

Biological irreversibility (*Continued*)
  microscopic stability, 66–67
  scenario, 64–67
  chaotic itineracy
    characteristics, 58–59
    Milnor attractors, 57–58
  collective dynamics, 60–62
  high-dimensional chaos, 55–57
  phenomenology theory of development, 70–72
  physical properties, 53–55
  thermodynamics, irreducibility to, 69–70
Blackbody radiation, micromaser experiments, trapping states, 176–182
Bohm's theory, canonical quantization, 322–323
Bohr frequencies, stochastic limit, bosonic reservoir model, 219–223
Boltzmann approximation, moderately dense gas transport, theoretical background, 132–133
Boltzmann entropy, cat map thermodynamics, work estimation, isothermal operations, 28–29
Boltzmann-Lorentz operator
  moderately dense gas transport, 141–146
  velocity autocorrelation function, 148–150
Bona fide operators, quantum field theory, 281–282
Born approximation, time asymmetric quantum mechanics, Dirac Golden rule-Gamow ket transition, 307–312
Bosonic reservoirs, stochastic limit
  basic equations, 215–218
  coherence control, 231–232
  Glauber dynamics, 232–237
  model, 218–223
Breit-Wigner energy distribution, time asymmetric quantum mechanics, Dirac Golden rule-Gamow ket transition, 304–312
Broken time reversal symmetry, irreversibility in multibaker maps, 96–102
Burnett coefficients, chaotic systems, Poincaré-Birkhoff mapping, 120

Cabibbo-Kobayashi-Maskawa (CKM) matrices, strangeness total antieigenvectors, 254–255

Canonical quantization, quantum fluctuation, 322–323
Cat map, hamiltonian chaotic system
  thermodynamics, 24–31
  external operations, 25–26
  Hamiltonian equation, 24–25
  large system recovery, 31
  probability density, time evolution, 26–27
  Second Law thermodynamics, 29–31
  work estimation, 27–29
Cauchy's theorem, chaotic systems, Poincaré-Birkhoff mapping, spatially periodic systems, 114–119
Cell differentiation, biological irreversibility dynamic systems
  development, 62–64
  representation, 67–69
  macroscopic stability, 67
  microscopic stability, 66–67
  scenario, 64–67
Chaotic systems
  biological irreversibility
    cell differentiation
      dynamic system development, 62–64
      dynamic system representation, 67–69
      macroscopic stability, 67
      microscopic stability, 66–67
      scenario, 64–67
    chaotic itineracy
      characteristics, 58–59
      Milnor attractors, 57–58
    collective dynamics, 60–62
    high-dimensional chaos, 55–57
    phenomenology theory of development, 70–72
    physical properties, 53–55
    thermodynamics, irreducibility to, 69–70
  critical states, permanent/transient chaos, 49–51
  diffusion, Poincaré-Birkhoff mapping
    eigenvalue consequences, 119–123
    entropy production, 122–123
    higher-order coefficients, 119–120
    nonequilibrium steady states, 121–122
    periodic-orbit theory, 120–121
  microscopic chaos, 111–113
  spatially periodic systems, 114–119
  theoretical background, 109–111
Cheshire Cat effect, stochastic limit, bosonic reservoir model, 222–223

Chirikov-Taylor standard map, properties, 3–5, 19
Classical mechanics
 iterative maps, 4–5
 operator trigonometry, quantum probability, 250–252
 stochastic limit, generic system dynamics, 229–230
 superposition principles, 200–203
Closed equations, iterative maps, 5
Clustered states
 biological irreversibility, high-dimensional chaos, 56–57
 kinetic law formation, 161–164
Coarse-grained entropy, irreversibility research, forward time evolution, decay modes, 90–91
Coherence
 biological irreversibility, high-dimensional chaos, 56–57
 stochastic limit, control of, 230–232
Collective dynamics
 biological irreversibility, 60–62
 moderately dense gas transport
  applications, 138–146
  spectral representation, 133–138
  ternary correlation subspace, 150–152
  theoretical background, 129–133, 131–133
  velocity autocorrelation function, 146–150
  vertices' renormalization, 152–157
Collision operator
 integrable/nonintegrable correlation dynamics, 265–268
 moderately dense gas transport applications, 139–146
 spectral representation, 134–138
 velocity autocorrelation function, 149–150
Configuration space, irreversibility in multibaker maps, measures and time evolution, 84–85
Continuous-time Liouville equation, iterative maps, 4
 non-Markovian/Markovian evolution equations, 5–9
Continuum spectrum, quantum field theory, 280–282
Control theory, operator trigonometry, 241
Cooling mechanisms, single-atom ion trapping, ion-trap lasers, 193–194

Coplanar quantum probability violation, operator trigonometry, quantum probability, 250
Correlation dynamics, integrable/nonintegrable dynamic systems
 basic principles, 261–262
 dissipative systems, 268–274
 Poincaré's theorem, 262–264
 unitary transformation, 264–268
Coulomb potential, operator trigonometry, 241
Counting rate, mesoscopic quantum interference, 207–212
Coupled map lattices (CML), biological irreversibility, high-dimensional chaos, 55–57
CP symmetry, violation, 252–256
 antieigenvector-breaking mechanics, 255–256
 elementary particle physics, 252–254
 strangeness total antieigenvectors, 254–255
CPT theorem
 CP symmetry violation, 252–254
 quantum field theory, 284–285
Critical states, permanent/transient chaos, physical properties, 49–51
Cusp map, unstable system harmonic analysis, time operator, 45–46

Darboux isospectral method, quantum field resonances and poles, 282–284
De Broglie wavelength, micromasers, ultracold atoms, 185–186
Decay states
 irreversibility in multibaker maps, forward time evolution, 88–91
 stochastic limit, generic system dynamics, 228–230
 time asymmetric quantum mechanics
  Dirac Golden rule-Gamow ket transition, 302–312
  transition to microphysical irreversibility, 316–318
Decoherence
 coherence control, 230–232
 mesoscopic quantum interference, 203–212
 principles, 199–203
 stochastic limit
  basic equations, 215–218
  generic systems, 226–230
  Glauber dynamics, 232–237

Decoherence (*Continued*)
  Langevin equation, 223–226
  local operators, subalgebra, 237–238
  model for, 218–223
  "which path" experiment, 212
Degeneracy index, unstable systems
  excited state, average age, 292–295
  time superoperator construction, 289–292
Degrees of freedom, biological irreversibility, 54–55
Density matrix
  integrable/nonintegrable correlation dynamics
    dissipative systems, 270–274
    unitary operator, 264–268
  stochastic limit, 216–218
    generic system dynamics, 227–230
    Glauber dynamics, 234–237
Density profile, iterative maps, 4
  non-Markovian/Markovian evolution equations, 6–9
Destruction operators
  integrable/nonintegrable correlation dynamics, dissipative systems, 268–274
  iterative maps, non-Markovian/Markovian evolution equations, 7–9
Development, biological irreversibility, phenomenology theory, 70–72
Differential dynamic phase, micromaser experiments, atomic interference, 174–176
Diffusion
  chaotic systems, Poincaré-Birkhoff mapping
    eigenvalue consequences, 119–123
      entropy production, 122–123
      higher-order coefficients, 119–120
      nonequilibrium steady states, 121–122
      periodic-orbit theory, 120–121
    microscopic chaos, 111–113
    spatially periodic systems, 114–119
    theoretical background, 109–111
  iterative maps, 4
    master equation, standard map, 10–11
    moderately dense gas transport, 144–146
Dirac delta function
  quantum field theory, 278–279
  unstable system harmonic analysis, spectral decomposition and probabilistic extension, Renyi maps, 38–39
Dirac Golden Rule, time asymmetric quantum mechanics, 302–312

Discrete time master equation, iterative maps, non-Markovian/Markovian evolution equations, 7–9
Dissipative systems, integrable/nonintegrable correlation dynamics, 268–274
  collision operator, 266–268
Distribution theory, quantum field theory, 281–282
Double Darboux isospectral analysis, quantum field resonances and poles, 282–284
Dressed unstable states, integrable/nonintegrable dynamic systems, dissipative systems, 272–274
Drift parameter, stochastic limit, bosonic reservoir model, 222–223
Dynamical systems theory
  biological irreversibility, 55
  cell differentiation
    development of, 62–64
    representation of, 67–69
  collective dynamics, 60–62
  integrable/nonintegrable correlation dynamics
    basic principles, 261–262
    dissipative systems, 268–274
    Poincaré's theorem, 262–264
    unitary transformation, 264–268
  multibaker maps, reversible maps, irreversibility in, 78–80
  stochastic limit
    generic systems, 226–230
    Glauber dynamics, 232–237

Eigenvalues
  chaotic systems, Poincaré-Birkhoff mapping, 119–123
    entropy production, 122–123
    higher-order coefficients, 119–120
    nonequilibrium steady states, 121–122
    periodic-orbit theory, 120–121
  moderately dense gas transport
    collision operators, 142–146
    spectral representation, 136–138
  operator trigonometry, symmetric positive definite (SPD) matrix, 243–244
Eigenvectors
  operator trigonometry, 242–243
  time asymmetric quantum mechanics, Dirac Golden rule-Gamow ket transition, 303–312

SUBJECT INDEX 343

Einstein-Podolsky-Rosen quantum correlations
  operator trigonometry, quantum probability, 247–248
  superposition and decoherence, 200–203
Einstein relation, stochastic limit, coherence control, 231–232
Electroweak theory, quantum fluctuation, 328–329
Elementary particle physics, CP symmetry violation, 252–254
Entanglement, micromaser experiments, 176
Entropy production
  chaotic systems, Poincaré-Birkhoff mapping, 122–123
  unstable systems
    superoperators, 297–298
    time superoperators, 288–289
Escape rate, chaotic systems, Poincaré-Birkhoff mapping, microscopic chaos, 112–113
Euclid measurements
  canonical quantization, 323
  quantum fluctuation, $(D+1)$-dimensional space-time world, 326
Euler equation, operator trigonometry, 242–243
  extended operators, 245–246
Euler polynomial, unstable system harmonic analysis, spectral decomposition and probabilistic extension, tent maps, 40–41
Evolution equations
  iterative maps, 5–9
  stochastic limit, Langevin equation, 223–226
Excited states, unstable systems, average age, 292–295
Exponential law, time asymmetric quantum mechanics, Dirac Golden rule-Gamow ket transition, 310–312
Extended operators, operator trigonometry, 244–246
External operations, cat map thermodynamics, 25–26

Fabry-Perot resonator
  micromasers, ultracold atoms, 185–186
  single-atom ion trapping, resonance fluorescence, 186–189
Fast damping predictions, stochastic limit, generic system dynamics, 227–230
Fermi Golden rule, stochastic limit, 215–218
  bosonic reservoir model, 221–223

Fermi-Pasta-Ulam models, log-Weibull distribution, Arnold diffusion, 163–164
Feynman quantum mechanics, quantum fluctuation, $(D+1)$-dimensional space-time world, 325–326
Fick's law of diffusion
  chaotic systems, Poincaré-Birkhoff mapping, nonequilibrium conditions, 122
  irreversibility in multibaker maps, forward time evolution, 86–88
Field evolution time constant, micromasers, quantum jumps and atomic interference, 171–176
First-order phase transitions, micromasers, quantum jumps and atomic interference, 170–176
Fixed-volume operations, cat map thermodynamics, 25–26
Fluorescence radiation, single-atom ion trapping, 187–189
Fock state, micromaser experiments, trapping mechanisms, 179–182
Fokker-Planck equation
  cat map thermodynamics, 29–31
  integrable/nonintegrable correlation dynamics, unitary operators, 266–268
  integrable/nonintegrable dynamic systems, 274–275
  micromasers, quantum jumps and atomic interference, 170–176
Forward time evolution, irreversibility in multibaker maps
  decay states, 88–91
  steady states, 85–88
Fourier transform
  iterative maps
    master equation, standard map, 9–11
    non-Markovian/Markovian evolution equations, 6–9
    standard map master equation solution, 14–18
  moderately dense gas transport, spectral representation, 135–138
  quantum fluctuation, $(D+1)$-dimensional space-time world, 325–326
  single-atom ion trapping, resonance fluorescence, 187–189
  unstable systems, time superoperator construction, 290–292

Fractal repeller, chaotic systems, Poincaré-Birkhoff mapping, microscopic chaos, 112–113
Fredholm determinant, chaotic systems, Poincaré-Birkhoff mapping, periodic-orbit theory, 120–121
Free chains, multibaker maps, 81–83
Friedrichs model
  integrable/nonintegrable dynamic systems
    dissipative systems, 271–274
    unitary operator, 267–268
  unstable systems, excited state, average age, 292–295
Fringe visibility, mesoscopic quantum interference, 207–212
Frobenius-Perron operator
  biological irreversibility, collective dynamics, 61–62
  cat map thermodynamics
    probability density time evolution, 26–27
    Second Law thermodynamics, 30–31
  chaotic systems, Poincaré-Birkhoff mapping, 109–111
    periodic-orbit theory, 120–121
    spatially periodic systems, 116–119
  critical states, transient and permanent chaos, 50–51
  iterative maps
    master equation, standard map, 9–11
    non-Markovian/Markovian evolution equations, 5–9
  unstable system harmonic analysis, 34–36
    spectral decomposition and probabilistic extension, 36–43
    Baker's transformation, 42–43
    Renyi maps, 38–39
    tent maps, 39–41
Full Markovian approximation, iterative maps
  general solution, 12
  non-Markovian/Markovian evolution equations, 8–9
  standard master equation solution, 16–18

Gallavotti-Cohen hypothesis, irreversibility research, 79–80
Gamow ket, time asymmetric quantum mechanics, 302–312
  transition to microphysical irreversibility, 313–318

Gaussian diffusive profile, iterative maps, standard master equation solution, 16–18
Gaussian distribution
  cat map thermodynamics, 31
  stochastic limit, bosonic reservoir, 218–223
"Gaussian-like" distribution, iterative maps, standard map solution, 13–18
Gauss-Reimann curvature, cluster formation kinetics, 162–164
Generating basis, unstable system harmonic analysis, time operator, shift representation, 43–46
Generic system dynamics, stochastic limit, 226–230
Gibbs distribution, stochastic limit
  generic system dynamics, 230
  Langevin equation, 225–226
Gibbs' entropy, chaotic systems, Poincaré-Birkhoff mapping, 122–123
Glashow-Weinberg-Salam-Yang-Mills model, CP symmetry violation, 252–254
Glauber dynamics, stochastic limit, 232–237
Globally Coupled Map (GCM), biological irreversibility, 55
  chaotic itinerancy, Milnor attractors, 57–59
  collective dynamics, 60–62
  high-dimensional chaos, 55–57
Gram matrix, operator trigonometry
  classical and quantum probability, 252
  triangle inequality, 249–250
Greenberger-Horne-Zeilinger (GHZ) states, micromaser experiments, 182–183
Green-Kubo relation
  chaotic systems, Poincaré-Birkhoff mapping, 109–111
  diffusion coefficients, 120
  moderately dense gas transport
    applications, 138–146
    spectral representation, 133–138
    ternary correlation subspace, 150–152
    theoretical background, 129–133
    velocity autocorrelation function, 146–150
    vertices' renormalization, 152–157
Gustafson theorem, operator trigonometry, 240

Hamiltonian chaotic system, thermodynamics
  basic properties, 21–23
  cat map, 24–31
    external operations, 25–26

Hamiltonian equation, 24–25
  large system recovery, 31
  probability density, time evolution, 26–27
  Second Law thermodynamics, 29–31
  work estimation, 27–29
Hamiltonian equations
  bosonic reservoirs, 215–218
  stochastic limit
    bosonic reservoir model, 218–223
    coherence control, 231–232
    Glauber dynamics, 232–237
Hamiltonian maps, properties, 3–5
Hanbury-Brown protocols, single-atom ion trapping, resonance fluorescence, 189
Hardy space, time asymmetric quantum mechanics
  Dirac Golden rule-Gamow ket transition, 304–312
  transition to microphysical irreversibility, 315–318
Harmonic analysis, unstable systems
  operator theory, 34–36
  spectral decomposition and probabilistic expansion, 36–43
    Baker's transformations, 42–43
    logistic map, 41–442
    Renyi maps, 38–39
    tent maps, 39–41
  time operator and shift-represented evolution, 43–46
    cusp map, 45–46
    Renyi map, 44–45
Hausdorff dimension, chaotic systems, Poincaré-Birkhoff mapping, spatially periodic systems, 119
Heat bath, cat map thermodynamics, 24–25
Heisenberg equation
  quantum field theory, 278–279
  stochastic limit
    basic equation, 216–218
    Langevin equation, 223–226
Hermitian operator
  integrable/nonintegrable correlation dynamics
    dissipative systems, 269–274
    Poincaré's theorem, 264
  operator trigonometry, 242–243
  unstable systems, time superoperator construction, 291–292
High-dimensional chaos, biological irreversibility, 55–57

collective dynamics, 60–62
operator theory, 55–57
Higher-order coefficients
  chaotic systems, Poincaré-Birkhoff mapping, diffusion, eigenvalue problem, 119–120
  collective transport, 157–158
Hilbert space
  canonical quantization, 322–323
  integrable/nonintegrable dynamic systems
    correlation dynamics, 262
    dissipative systems, 271–274
    Poincaré's theorem, 263–264
    unitary operator, 266–268
  irreversibility research, 78–80
    broken time reversal symmetry, spectral theory, 98–99
  moderately dense gas transport, 140–146
  operator trigonometry, 241–243
    classic and quantum probability, 250–252
    extended operators, 245–246
    symmetric positive definite (SPD) matrix, 243–244
    triangle inequality, 249–250
  stochastic limit, Glauber dynamics, 232–237
  superposition principles, 200–203
  time asymmetric quantum mechanics, 301–312
    Dirac Golden rule-Gamow ket transition, 303–312
    transition to microphysical irreversibility, 314–318
  unstable system harmonic analysis
    operator theory, 34–36
    shift representation, 43–46
    spectral decomposition and probabilistic extension, 37–43
    Baker's transformation, 43
Hille-Yosida-Phillips-Lumer theorem, operator trigonometry, 240
Hopf bifurcation, biological irreversibility, collective dynamics, 61–62
Hydrodynamic modes
  chaotic systems, Poincaré-Birkhoff mapping, nonequilibrium conditions, 121–122
  moderately dense gas transport, 138–146
  ternary correlation subspace, 151–152
Hyperbolic systems, irreversibility research, 79–80

Improper states, quantum field theory, 280–282
Inequality derivation, cat map thermodynamics, 30–31
Inhomogeneity components, moderately dense gas transport, 144–146
spectral representation, 135–138
Integrable dynamical systems, correlations
basic principles, 261–262
dissipative systems, 268–274
Poincaré's theorem, 262–264
unitary transformation, 264–268
Integrated distribution function, irreversibility research
measures and time evolution, 84–85
multibaker maps, backward time evolution and time reversal symmetry, 91–95
Interaction stability, biological irreversibility, cell differentiation, 68–69
Interaction time, micromasers, atomic interference, 174–176
Interference decay, decoherence, 201–203
Internal stability, biological irreversibility, cell differentiation, 68–69
Intrinsic diffusive evolution, iterative maps, 4
Inverse Laplace transformation, moderately dense gas transport, 143–146
Ion trap experiments, single atoms
laser techniques, 189–194
resonance fluorescence, 186–189
Irreversibility. *See also* Biological irreversibility
microphysical, time asymmetric quantum mechanics
probability, 312–318
resonances and decay, 301–312
reversible multibaker maps
backward time evolution and reversal symmetry, 91–95
broken time reversal symmetry, 96–102
spectral theory, 97–99
energy coordinates, 80–83
forward time evolution, 85–91
decay modes, 88–91
steady states, 85–88
measure selection and macroscopic properties, 102–104
physical principles, 78–80
time evolution measures, 83–85
unidirectional measures evolution and time reversal symmetry, 95–96

Isothermal operations, cat map thermodynamics, 25–26
work estimation, 27–29
Iterative maps, non-Markovian effects
general master equation solution, 11–12
non-Markovian and Markovian evolution equations, 5–9
properties, 3–5
standard equations, 9–11
solutions, 12–18

Jaynes-Cummings model, micromaser experiments, 168–169
atomic interference, 174–176
trapping states, 182
Jordan blocks, unstable system harmonic analysis, spectral decomposition and probabilistic extension, Baker's transformation, 42–43

Kaon eigenstates
CP symmetry violation, breaking mechanisms, 255–256
strangeness total antieigenvectors, 254–255
time asymmetric quantum mechanics, transition to microphysical irreversibility, 315–318
Kapur-Peierls formalism, quantum field theory, 281–282
Kinetic energy
integrable/nonintegrable correlation dynamics, Poincaré's theorem, 263–264
multibaker maps, 81–83
Kinetic equations
cluster formation, 161–164
iterative maps, 4–5
standard map solution, 14–18
moderately dense gas transport, 143–146
Kinetic operators, integrable/nonintegrable correlation dynamics, dissipative systems, 269–274
Kolmogorov measure
irreversibility research, 79–80
macroscopic properties, 104
operator trigonometry
classic and quantum probability, 250–252
quantum probability, 248
unstable system harmonic analysis
shift representation, 43–46

spectral decomposition and probabilistic extension, Baker's transformation, 42–43
Kolmogorov-Sinai entropy
  biological irreversibility, 70
  chaotic systems, Poincaré-Birkhoff mapping, 110–111
    entropy production, 123
    microscopic chaos, 112–113
  critical states, transient and permanent chaos, 50–51
  irreversibility research, 79–80
Koopman evolution operators, unstable system harmonic analysis, 34–36
  shift representation, 43–46
  spectral decomposition and probabilistic extension, 36–43
    Baker's transformation, 42–43
    logistic map, 41–42
    Renyi maps, 38–39

Lamb-Dicke parameter, single-atom ion trapping, ion-trap lasers, 192–194
Landau theory, micromasers, quantum jumps and atomic interference, 170–176
Langevin equation
  cat map thermodynamics, 29–31
  stochastic limit, 223–226
Laplace transform
  chaotic systems, Poincaré-Birkhoff mapping, spatially periodic systems, 117–119
  iterative maps, non-Markovian/Markovian evolution equations, 6–9
  operator trigonometry, 241
Lasers
  single-atom ion trapping, 189–194
  "which path" experiments, 212
Lebesgue measure
  chaotic systems, Poincaré-Birkhoff mapping
    entropy production, 122–123
    nonequilibrium conditions, 122
    spatially periodic systems, 118–119
  irreversibility research, 79–80
  forward time evolution
    decay modes, 88–91
    steady states, 87–88
    macroscopic properties, 103–104
    measures and time evolution, 84–85
  stochastic limit, bosonic reservoir model, 220–223

unstable system harmonic analysis
  spectral decomposition and probabilistic extension, Baker's transformation, 42–43
  time operator, Renyi map, 44–45
Lebowitz-McLennan Zubarev nonequilibrium, chaotic systems, Poincaré-Birkhoff mapping, 122
Le Chatelier-Braun principle, biological irreversibility, phenomenology theory, 71–72
Linearized Boltzmann collision operator
  moderately dense gas transport, 140–146
  ternary correlation subspace, 150–152
  vertices renormalization, 154–157
Linearized regimes, moderately dense gas transport, 138–146
Liouville equation
  chaotic systems, Poincaré-Birkhoff mapping, 110–111
  spatially periodic systems, 116–119
  integrable/nonintegrable correlation dynamics, unitary operator, 264–268
  integrable/nonintegrable dynamic systems, dissipative systems, 272–274
  irreversibility research, 78–80
  iterative maps, 4
    non-Markovian/Markovian evolution equations, 5–9
  moderately dense gas transport
    spectral representation, 133–138
    theoretical background, 13o–133
  unstable systems
    age fluctuations, 295–297
    time operators, 287–289
    time superoperator construction, 289–292
Liouville operator, unstable system harmonic analysis, 34–36
Liouville-von Neumann space, integrable/nonintegrable correlation dynamics
  dissipative systems, 268–274
  Poincaré's theorem, 263–264
  unitary operator, 264–268
Lippmann-Schwinger equation, time asymmetric quantum mechanics, Dirac Golden rule-Gamow ket transition, 303–312
Local operators, stochastic limit, subalgebra evolution, 237–238

SUBJECT INDEX

Logistic map, unstable system harmonic analysis, spectral decomposition and probabilistic extension, 41–42
Log-Weibull distribution, Arnold diffusion, 163–164
Lorentz gas
  chaotic systems, Poincaré-Birkhoff mapping, 110–111
  diffusion coefficients, 120
  spatially periodic systems, 114–119
  irreversibility in multibaker maps, measures and time evolution, 84–85
  multibaker map energy coordinate, 80–83
Lorenz attractor, unstable system harmonic analysis, time operator, cusp map, 45–46
Lyapunov operators
  biological irreversibility
    collective dynamics, 60–62
    thermodynamic irreducibility, 70
  cat map thermodynamics, Second Law thermodynamics, 31
  chaotic systems, Poincaré-Birkhoff mapping, 110–111
    microscopic chaos, 111–113
  irreversibility research, 79–80
  unstable system harmonic analysis, operator theory, 34–36
  unstable systems, time superoperators, 288–289

Mach-Zender interferometry
  mesoscopic quantum interference, 204–212
  "which path" experiment, 212
Macroscopic stability
  biological irreversibility, cell differentiation, 67
  irreversibility research, measurement selection, 102–104
  stochastic limit
    coherence control, 231–232
    generic system dynamics, 228–2320
Many-body problems, quantum field theory, 279–280
Markovian equations
  collective transport, 157–158
  integrable/nonintegrable correlation dynamics, unitary operator, 266–268
  irreversibility research, 78–80
  iterative maps
    evolution equations, 5–9
    kinetic equations, 5
    standard map solution, 13–18
  moderately dense gas transport
    applications, 139–146
    spectral representation, 138
    theoretical background, 129–133
Markovian generator, stochastic limit
  generic system dynamics, 226–230
  Langevin equation, 224–226
Master equation
  iterative maps
    general solution, 11–12
    non-Markovian/Markovian evolution equations, discrete time, 7–9
    standard map, 9–11
    solution for, 12–18
  moderately dense gas transport, 143–146
  stochastic limit, 216–218
    coherence control, 231–232
    Glauber dynamics, 232–237
    Langevin equation, 223–226
Maxwell demon, cat map thermodynamics, 30–31
Measurement techniques, irreversibility research, 79–80
  macroscopic properties, 102–104
  time evolution and, 83–85
  unidirectional evolution, 95–96
Memory kernel, iterative maps
  master equation, standard map, 10–11
  non-Markovian/Markovian evolution equations, 7–9
Memoryless evolution, iterative maps, standard master equation solution, 16–18
Memory time, iterative maps
  master equation, standard map, 11
  non-Markovian/Markovian evolution equations, 8–9, 19
Mesoscopic quantum interference, optical implementation, 203–212
Microcanonical quantization
  (D+1)-dimensional space-time world, 327–328
  quantum fluctuation, 321–322
Micromasers, 168–186
  entanglement, 176
  GHZ generation, 182–183
  quantum jumps and atomic interferences, 169–176

SUBJECT INDEX

trapping states, 176–182
ultracold atoms, 183–186
Microphysical irreversibility, time asymmetric quantum mechanics
  Gamow kets, 313–318
  Golden rule and Gamow kets, 302–312
  probability, 312–318
  resonances and decay, 301–312
Microscopic chaos, Poincaré-Birkhoff mapping, 111–113
Microscopic stability, biological irreversibility, cell differentiation, 66–67
Milnor attractors
  biological irreversibility, cell differentiation, 68–69
  biological irreversibility, chaotic itinerancy, 57–59
Min-max theorem, operator trigonometry, 242–243
  extended operators, 245–246
Mixmaster universe model, cluster formation kinetics, 162–164
Moderately dense gases, transport theory
  applications, 138–146
  spectral representation, 133–138
  ternary correlation subspace, 150–152
  theoretical background, 129–133
  velocity autocorrelation function, 146–150
  vertices' renormalization, 152–157
Monte Carlo simulations, micromaser experimenets, trapping states, 180–182
Multibaker maps
  irreversibility
    broken time reversal symmetry, 99–102
    energy coordinates, 80–83
    theoretical background, 80
  reversible maps
    backward time evolution and reversal symmetry, 91–95
    broken time reversal symmetry, 96–102
    spectral theory, 97–99
    energy coordinates, 80–83
    forward time evolution, 85–91
      decay modes, 88–91
      steady states, 85–88
    measure selection and macroscopic properties, 102–104
    physical principles, 78–80
    time evolution measures, 83–85

unidirectional measures evolution and time reversal symmetry, 95–96
Multipotency, biological irreversibility, 69–70

N-body Hamiltonian systems, cluster formation, kinetic laws, 161–164
Negative selection, superposition and decoherence, 202–203
Nekhoroshev theorem, log-Weibull distribution, Arnold diffusion, 163–164
Neutrino, quantum fluctuation, 322, 328
Noise-induced transitions, biological irreversibility, collective dynamics, 60–62
Nonequilibrium conditions, chaotic systems, Poincaré-Birkhoff mapping, 112–113
  entropy production, 122–123
  steady state eigenvalue problem, 121–122
Nonhydrodynamic modes, moderately dense gas transport, 138–146
Non-integrable dynamical systems, correlations
  basic principles, 261–262
  dissipative systems, 268–274
  Poincaré's theorem, 262–264
  unitary transformation, 264–268
Nonlocal long-distance correlation, canonical quantization, 323
Non-Markovian equations
  iterative maps
    derivation, 5
    evolution equations, 5–9
    general solution, 11–12
    standard map solution, 13–18
  moderately dense gas transport, theoretical background, 131–133
Nonrelativistic quantum mechanics, quantum field theory, 280–282
Nonunitary operators, integrable/nonintegrable correlation dynamics, 265–268
Nonzero systems, stochastic limit, generic system dynamics, 229–230
Nuclear Spectral Theorem, time asymmetric quantum mechanics, Dirac Golden rule-Gamow ket transition, 308–312
Number operator, quantum field theory, 280

Off-diagonal elements
  integrable/nonintegrable correlation dynamics, unitary operator, 264–268
  stochastic limit, 216–218

SUBJECT INDEX

Off-diagonal elements (*Continued*)
  generic system dynamics, 227–230
  Glauber dynamics, 234–237
One-atom maser. *See* Micromaser
One-dimensional systems, stochastic limit, Glauber dynamics, 235–237
Open quantum systems, superposition principles, 199–203
Operator theory
  quantum field theory, 281–282
  unstable system harmonic analysis, 34–36
Operator trigonometry
  extended operators, 244–246
  historical background, 240–243
  principles of, 243–244
  quantum probability, 246–252
    Bell's inequality, 247–248
    classical probabilities and, 250–252
    general triangle inequality, 248–250
Optical experiments, mesoscopic quantum interference, 203–212
Optical parametric amplifier (OPA)
  mesoscopic quantum interference, 203–212
  "which path" experiment, 212
Ordered phase, biological irreversibility, high-dimensional chaos, 56–57

Partial attractors, biological irreversibility, cell differentiation, 68–69
Partially ordered phase, biological irreversibility
  chaotic itinerancy, 57–58
  high-dimensional chaos, 56–57
Particle distribution, irreversibility in multibaker maps, forward time evolution, 86–88
Pauli matrices
  integrable/nonintegrable dynamic systems, 274–275
    collision operator, 266–268
  stochastic limit, Glauber dynamics, 232–237
Pauli principles, quantum field theory, 285
Paul traps
  mesoscopic quantum interference, 208–212
  single-atom ion trapping, resonance fluorescence, 188–189
Periodic boundary condition, cat map thermodynamics, 24–25
Periodic-orbit theory, chaotic systems, Poincaré-Birkhoff mapping, 120–121
Permanent chaos, critical states, 49–51

Perron-Frobenius operator. *See* Frobenius-Perron operator
Pesin's formula, chaotic systems, Poincaré-Birkhoff mapping, microscopic chaos, 112–113
Phenomenology theory, biological irreversibility, 70–72
Photodetector apparatus
  mesoscopic quantum interference, 206–212
  "which path" experiment, 212
Photon distribution, micromaser experiments, trapping states, 176–182
PIN diodes, mesoscopic quantum interference, 211–212
Poincaré-Birkhoff mapping, chaotic systems
  eigenvalue consequences, 119–123
    entropy production, 122–123
    higher-order coefficients, 119–120
    nonequilibrium steady states, 121–122
    periodic-orbit theory, 120–121
  microscopic chaos, 111–113
  spatially periodic systems, 114–119
  theoretical background, 109–111
Poincaré map
  cat map thermodynamics, 24–25
  unstable system harmonic analysis, time operator, cusp map, 45–46
Poincaré's theorem, integrable/nonintegrable
  correlation dynamics, 262–264
  dissipative systems, 270–274
  unitary operator, 264–268
Pointer states, superposition principles, 201–203
Poisson equation, cluster formation kinetics, 161–164
Poisson suspension, irreversibility in multibaker maps
  broken time reversal symmetry, 99–102
  measures and time evolution, 83–85
Pole characteristics, quantum field theory, 282–284
Pollicott-Ruelle resonances
  chaotic systems, Poincaré-Birkhoff mapping
    diffusion coefficients, 119–120
    periodic-orbit theory, 120–121
    spatially periodic systems, 118–119
  irreversibility research
    broken time reversal symmetry, spectral theory, 99

forward time evolution, decay modes, 89–91
P-projector, iterative maps, master equation, standard map, 10–11
Prigogine-Balescu diagram, velocity autocorrelation function, 146–150
Prigogine-Résibois master equation, iterative maps, non-Markovian/Markovian evolution equations, 7–9
Probabilistic extension, unstable system harmonic analysis, spectral decomposition, 36–43
  Baker's transformations, 42–43
  logistic map, 41–42
  Renyi maps, 38–39
  tent maps, 39–41
Probability density
  cat map thermodynamics, time evolution, 26–27
    Second Law thermodynamics, 29–31
  chaotic systems, Poincaré-Birkhoff mapping, spatially periodic systems, 116–119
  mesoscopic quantum interference, 205–212
  time asymmetric quantum mechanics, 312–318
"Pseudo-Liouvillian" formalism, moderately dense gas transport
  applications, 138–146
  spectral representation, 134–138
Pump parameters
  micromasers
    quantum jumps and atomic interference, 169–176
    ultracold atoms, 183–186
    single-atom ion trapping, ion-trap lasers, 190–194

Quantization rules, quantum fluctuation
  canonical quantization, Bohm's theory, 322–323
  principles of, 321–322
  stochastic and microcanonical quantizations, 324–328
Quantum field theory
  historical background, 277–279
  continuum spectrum, 280–282
  resonances and poles, 282–284
  second quantization, 279–280
  spin-statistics and CPT theorem, 284-285

Quantum jumps, micromasers, 169–176
Quantum probability, operator trigonometry, 246–252
  Bell's inequality, 247–248
  classical probabilities and, 250–252
  general triangle inequality, 248–250
Quantum superposition
  mesoscopic quantum interference, 203–212
  principles, 199–203
  "which path" experiment, 212
Quark mixing matrix
  CP symmetry violation, antieigenvector breaking, 255–256
  strangeness total antieigenvectors, 254–255
Quasilinear diffusion coefficient, iterative maps, standard master equation solution, 14–18
Quasi-structure, cluster formation, 164

Rabi floppy frequency
  micromaser experiments
    Greenberger-Horne-Zeilinger (GHZ) states, 182–183
    trapping states, 176–182
  micromasers
    quantum jumps and atomic interference, 169–176
    ultracold atoms, 184–186
    single-atom ion trapping, 188–189
    ion-trap lasers, 190–194
Ramsey experimental protocol, micromasers, atomic interference, 173–176
Rayleigh-Ritz characterization, operator trigonometry, 242–243
Rayleigh-Schrödinger equation, integrable/nonintegrable dynamic systems, dissipative systems, 273–274
Relaxation time, iterative maps, non-Markovian/Markovian evolution equations, 8–9, 19
Rellich-Kato-Sz.Nagy-Gustafson theorem, operator trigonometry, 240
Renyi maps, unstable system harmonic analysis
  spectral decomposition and probabilistic extension, 38–39
  time operator, 44–45
Resolved-sideband cooling, single-atom ion trapping, ion-trap lasers, 193–194
Resonance fluorescence, single-atom ion trapping, 186–189

Resonances
  quantum field theory, 281–284
  time asymmetric quantum mechanics, Dirac Golden rule-Gamow ket transition, 311–312
  unstable systems, excited state, average age, 293–295
Resonator frequency, micromasers, atomic interference, 173–176
Reversibility properties, quantum field theory, 282–285
Reversible multibaker maps, irreversibility
  backward time evolution and reversal symmetry, 91–95
  broken time reversal symmetry, 96–102
  spectral theory, 97–99
  energy coordinates, 80–83
  forward time evolution, 85–91
    decay modes, 88–91
    steady states, 85–88
  measure selection and macroscopic properties, 102–104
  physical principles, 78–80
  time evolution measures, 83–85
  unidirectional measures evolution and time reversal symmetry, 95–96
Riemannian geometrization
  cluster formation kinetics, 162–164
  quantum field resonances and poles, 282–284
Rigged Hilbert space
  irreversibility research, 78–80
  quantum field theory, 281–282
  time asymmetric quantum mechanics
    Dirac Golden rule-Gamow ket transition, 303–312
    probability and time asymmetry, 312–318
    transition to microphysical irreversibility, 313–318
  unstable system harmonic analysis, spectral decomposition and probabilistic extension, 37–43
    Renyi maps, 38–39
Ring operator
  moderately dense gas transport, 142–146
  velocity autocorrelation function, 146–150
  vertices renormalization, 152–157
Rydberg states, micromaser experimenets, 168–169
  trapping states, 180–182

Sawteeth map, cat map thermodynamics, 31
Scattering experiments, time asymmetric quantum mechanics, Dirac Golden rule-Gamow ket transition, 310–312
Schrödinger equation
  canonical quantization, 323
  mesoscopic quantum interference, 203–212
  quantum field theory, 278–280
    eigenvalues, 282–285
  quantum superposition, 199–203
  stochastic limit, 215–218
  time asymmetric quantum mechanics, probability and time asymmetry, 312–318
Schwartz space property, time asymmetric quantum mechanics, Dirac Golden rule-Gamow ket transition, 305–312
Second harmonic generation, mesoscopic quantum interference, 204–212
Second Law of Thermodynamics, cat map thermodynamics, 29–31
Second quantization, quantum field theory, 279–280
Selfadjoint operator
  operator trigonometry, 242–243
  quantum field theory, 281–282
Self-consistency, biological irreversibility, cell differentiation, 68–69
Self-quenching, single-atom ion trapping, ion-trap lasers, 190–194
Shift representation, unstable system harmonic analysis
  operator theory, 36
  time operator evolution, 43–46
Sinai-Ruelle-Bowen (SRB) measure
  critical states, transient and permanent chaos, 50–51
  irreversibility research, 79–80
  macroscopic properties, 103–104
Single atoms
  ion trap experiments
    laser techniques, 189–194
    resonance fluorescence, 186–189
  micromasers, 168–186
    entanglement, 176
    GHZ generation, 182–183
    quantum jumps and atomic interferences, 169–176
    trapping states, 176–182

SUBJECT INDEX

ultracold atoms, 183–186
physical properties, 167–168
Slow clock technique, stochastic limit, 217–218
Song-Caves-Yurke (SCY) protocol, mesoscopic quantum interference, 203–212
Space-time continuum
  canonical quantization, 323
  quantum fluctuation, $(D+1)$-dimensional space-time world, 323–326
Spatially periodic systems, Poincaré-Birkhoff mapping, 114–119
  Frobenius-Perron operator, 116–119
Special relativity, quantum field theory and, 279–280
Spectral decomposition
  irreversibility research, 78–80
  unstable system harmonic analysis
    operator theory, 35–36
    probabilistic extension, 36–43
      Baker's transformations, 42–43
      logistic map, 41–42
      Renyi maps, 38–39
      tent maps, 39–41
Spectral representation
  irreversibility in multibaker maps, broken time reversal symmetry, 97–99
  moderately dense gas transport, 133–138
Spin operators, stochastic limit, Glauber dynamics, 233–237
Spin-statistics theorem, quantum field theory, 284–285
State reduction techniques, micromaser experimenets, trapping states, 181–182
Steady states
  chaotic systems, Poincaré-Birkhoff mapping, nonequilibrium conditions, 121–122
  irreversibility in multibaker maps
    broken time reversal symmetry, 100–102
    forward time evolution, 85–88
  micromasers
    quantum jumps and atomic interference, 169–176
    trapping states, 176–182
Stem cells, biological irreversibility, cell differentiation, 66–69
Stern-Gerlach apparatus, superposition principles, 200–203
Stieltjes integral, irreversibility in multibaker maps, broken time reversal symmetry, 99–102

Stochasticity parameter
  biological irreversibility, cell differentiation, 66–69
  iterative maps, 4
  master equation, standard map, 10–11
Stochastic limit
  basic equations, 215–218
  bosonic model, 218–223
Stochastic quantization, quantum fluctuation $(D+1)$-dimensional space-time world, 323–326
  principles, 321–322
Strangeness total antieigenvectors, CP symmetry violation, 254–255
Subalgebraic solutions, stochastic limit, 237–238
Subdynamics
  moderately dense gas transport, spectral representation, 137–138
  velocity autocorrelation function, 146–150
  vertices renormalization, 152–157
Sub-Poissonian calculations
  micromaser experiments
    atomic interference, 176
    entanglement phenomena, 176
    single-atom ion trapping, ion-trap lasers, 190–194
Super-Burnett coefficients, chaotic systems, Poincaré-Birkhoff mapping, 120
Superoperators
  integrable/nonintegrable correlation dynamics
    dissipative systems, 268–274
    Poincaré's theorem, 264
  unstable systems
    entropy superoperator, 297–298
    time superoperators, 287–289
Symmetric positive definite (SPD) matrix, operator trigonometry, 242–243
  extended operators, 244–246
  properties of, 243–244

Tent maps, unstable system harmonic analysis, spectral decomposition and probabilistic extension, 39–41
Ternary correlation subspace, collective dynamics, 150–152
Thermodynamics
  biological irreversibility
    irreducibility of, 69–70
    phenomenology theory, 70–72

354 SUBJECT INDEX

Thermodynamics (*Continued*)
  Hamiltonian chaotic system
    basic properties, 21–23
    cat map, 24–31
      external operations, 25–26
      Hamiltonian equation, 24–25
      large system recovery, 31
      probability density, time evolution, 26–27
      Second Law thermodynamics, 29–31
      work estimation, 27–29
Time asymmetric quantum mechanics, microphysical irreversibility
  Gamow kets, 313–318
  Golden rule and Gamow kets, 302–312
  probability, 312–318
  resonances and decay, 301–312
Time-dependent Hamiltonian
  cat map, 24–25
  quantum field theory, 282–285
Time-dependent Schrödinger equation, quantum field theory, 282–285
Time evolution
  cat map thermodynamics, probability density, 26–27
    Second Law thermodynamics, 29–31
  irreversibility in multibaker maps
    backward time evolution and reversal symmetry, 91–95
    decay states, 88–91
    forward time evolution, 85–91
    measures, 83–85
    steady states, 85–88
  single-atom ion trapping, ion-trap lasers, 193–194
  time asymmetric quantum mechanics, Dirac Golden rule-Gamow ket transition, 305–312
  velocity autocorrelation function, 148–150
Time operators, unstable system harmonic analysis
  operator theory, 35–36
  shift-represented evolution, 43–46
Time reversal symmetry, irreversibility in multibaker maps
  backward time evolution, 91–95
  broken time, 96–102
  unidirectional measures evolution, 95–96
Time superoperator, unstable systems, construction of, 289–292

Total operators, operator trigonometry, symmetric positive definite (SPD) matrix, 244
Transfer coefficients, irreversibility research, 79–80
Transient chaos, critical states, 49–51
Transition probability, time asymmetric quantum mechanics, Dirac Golden rule-Gamow ket transition, 302–312
Trapping states
  micromaser experiments, 176–182
  single-atom ion trapping
    ion-trap lasers, 189–194
    resonance fluorescence, 186–189
Triangle inequality, operator trigonometry, quantum probability, 248–250
Turbulent phase, biological irreversibility, high-dimensional chaos, 56–57
Twiss protocols, single-atom ion trapping, resonance fluorescence, 189

Ultracold atoms, micromaser experiments, 183–186
Unidirectional measures evolution, irreversibility in multibaker maps, time reversal symmetry, 95–96
Uniformly hyperfinite (UHF) algebra, stochastic limit, 237–238
Unilateral shift, unstable system harmonic analysis, time operator, 43–46
Unitary operator
  integrable/nonintegrable correlation dynamics, 264–268
  dissipative systems, 271–274
  operator trigonometry
    extended operators, 244–246
    symmetric positive definite (SPD) matrix, 244
Unstable systems
  age fluctuations, 295–297
  aging in excited states, 292–295
  entropy superoperator, 297–298
  harmonic analysis
    operator theory, 34–36
    spectral decomposition and probabilistic expansion, 36–43
      Baker's transformations, 42–43
      logistic map, 41–442
      Renyi maps, 38–39
      tent maps, 39–41

time operator and shift-represented
 evolution, 43–46
 cusp map, 45–46
 Renyi map, 44–45
physical properties, 287–289
quantum field theory, 284–285
time superoperator construction, 289–292

Vacuum components
 moderately dense gas transport, spectral representation, 135–138
 quantum field theory, 278–279
 trapping states, micromaser experiments, 182–183
Vanishing denominators, integrable/nonintegrable correlation dynamics, Poincaré's theorem, 263–264
Velocity autocorrelation function
 collective dynamics, 146–150
 moderately dense gas transport, theoretical background, 132–133
Vertices renormalization, collective dynamics, 152–157
Virial theorem, microcanonical quantization, $(D+1)$-dimensional space-time world, 327–328
Vlasov equation, iterative maps, non-Markovian/Markovian evolution equations, 8–9
Von Neumann's theorem, quantum field theory, 278–279

Wall properties, cat map thermodynamics, 24–25
 probability density, time evolution, 26–27
Weak coupling, unstable systems, excited state, average age, 294–295
Weibull distribution, cluster formation kinetics, 162–164

Weinberg-Salam theory, quantum fluctuation, 328–329
Weisskopf predictions, single-atom ion trapping, resonance fluorescence, 189
Weisskopf-Wigner approximations
 quantum field theory continuum spectrum, 281-282
 time asymmetric quantum mechanics, Dirac Golden rule-Gamow ket transition, 306–312
"Which path" experiment, quantum superposition, 212
White noise equations, stochastic limit, 215–218
Wigner function, operator trigonometry, quantum probability, 247–248
 triangle inequality, 249–250
Work estimation, cat map thermodynamics, isothermal process, 27–29

Yukawa potentials, chaotic systems, Poincaré-Birkhoff mapping, spatially periodic systems, 114–119

Zeno time, integrable/nonintegrable dynamic systems, dissipative systems, 272–274
Zero-Markovian equation, iterative maps
 general solution, 12, 19
 non-Markovian/Markovian evolution equations, 8–9
Zero operation time, cat map thermodynamics, 31
Zeta function, chaotic systems, Poincaré-Birkhoff mapping, periodic-orbit theory, 120–121
Z-transformation, iterative maps, non-Markovian/Markovian evolution equations, 6–9